Probability and Its Applications

Published in association with the Applied Probability Trust

Editors: J. Gani, C.C. Heyde, P. Jagers, T.G. Kurtz

Probability and Its Applications

Anderson: Continuous-Time Markov Chains.
Azencott/Dacunha-Castelle: Series of Irregular Observations.
Bass: Diffusions and Elliptic Operators.
Bass: Probabilistic Techniques in Analysis.
Berglund/Gentz: Noise-Induced Phenomena in Slow-Fast Dynamical Systems.
Biagini/Hu/Øksendal/Zhang: Stochastic Calculus for Fractional Brownian Motion and Applications.
Chen: Eigenvalues, Inequalities and Ergodic Theory.
Choi: ARMA Model Identification.
Costa/Fragoso/Marques: Discrete-Time Markov Jump Linear Systems.
Daley/Vere-Jones: An Introduction to the Theory of Point Processes Volume I: Elementary Theory and Methods, Second Edition.
de la Peña/Giné: Decoupling: From Dependence to Independence.
Durrett: Probability Models for DNA Sequence Evolution.
Galambos/Simonelli: Bonferroni-type Inequalities with Applications.
Gani (Editor): The Craft of Probabilistic Modelling.
Grandell: Aspects of Risk Theory.
Gut: Stopped Random Walks.
Guyon: Random Fields on a Network.
Kallenberg: Foundations of Modern Probability, Second Edition.
Last/Brandt: Marked Point Processes on the Real Line.
Leadbetter/Lindgren/Rootzén: Extremes and Related Properties of Random Sequences and Processes.
Molchanov: Theory of Random Sets.
Nualart: The Malliavin Calculus and Related Topics.
Rachev/Rüschendorf: Mass Transportation Problems. Volume I: Theory.
Rachev/Rüschendorf: Mass Transportation Problems. Volume II: Applications.
Resnick: Extreme Values, Regular Variation and Point Processes.
Schmidli: Stochastic Control in Insurance.
Shedler: Regeneration and Networks of Queues.
Silvestrov: Limit Theorems for Randomly Stopped Stochastic Processes.
Thorisson: Coupling, Stationarity and Regeneration.
Todorovic: An Introduction to Stochastic Processes and Their Applications.

Francesca Biagini   Yaozhong Hu
Bernt Øksendal   Tusheng Zhang

Stochastic Calculus for Fractional Brownian Motion and Applications

Francesca Biagini, PhD
Mathematisches Institut, LMU München
Theresienstr. 39 D 80333
Munich, Germany

Yaozhong Hu, PhD
Department of Mathematics
University of Kansas, 405 Snow Hall
Lawrence, Kansas 66045-2142 USA
and
Center of Mathematics for Applications (CMA)
Department of Mathematics
University of Oslo, Box 1053 Blindern
N-0316, Oslo, Norway

Bernt Øksendal, PhD
Department of Mathematics
University of Oslo, Box 1053 Blindern N-0316,
Oslo
and
Norwegian School of Economics and Business
Administration (NHH)
Helleveien 30, N-5045, Bergen, Norway

Tusheng Zhang, PhD
Department of Mathematics
University of Manchester, Oxford Road
Manchester M13 9PL
and
Center of Mathematics for Applications (CMA)
Department of Mathematics
University of Oslo, Box 1053 Blindern
N-0316, Oslo, Norway

Series Editors:

J. Gani
Stochastic Analysis Group, CMA
Australian National University
Canberra ACT 0200, Australia

C.C. Heyde
Stochastic Analysis Group, CMA
Australian National University
Canberra ACT 0200, Australia

P. Jagers
Mathematical Statistics
Chalmers University of Technology
SE-412 96 Göteberg, Sweden

T.G. Kurtz
Department of Mathematics
University of Wisconsin
480 Lincoln Drive
Madison, WI 53706, USA

ISBN: 978-1-85233-996-8   e-ISBN: 978-1-84628-797-8
DOI: 10.1007/978-1-84628-797-8

British Library Cataloguing in Publication Data
A catalogue record for this book is available from the British Library

Library of Congress Control Number: 2008920683

Mathematics Subject Classification (2000): 60G05; 60G07; 60G15; 60H05; 60H10; 60H40; 60H07; 93E20

©Springer-Verlag London Limited 2008
Apart from any fair dealing for the purposes of research or private study, or criticism or review, as permitted under the Copyright, Designs and Patents Act 1988, this publication may only be reproduced, stored or transmitted, in any form or by any means, with the prior permission in writing of the publishers, or in the case of reprographic reproduction in accordance with the terms of licenses issued by the Copyright Licensing Agency. Enquiries concerning reproduction outside those terms should be sent to the publishers.
The use of registered names, trademarks, etc., in this publication does not imply, even in the absence of a specific statement, that such names are exempt from the relevant laws and regulations and therefore free for general use.
The publisher makes no representation, express or implied, with regard to the accuracy of the information contained in this book and cannot accept any legal responsibility or liability for any errors or omissions that may be made.

Printed on acid-free paper

9 8 7 6 5 4 3 2 1

Springer Science+Business Media
springer.com

To Thilo and to my family,
F.B.

To Jun and to Ruilong,
Y.H.

To Eva,
B.Ø.

To Qinghua,
T.Z.

Preface

Fractional Brownian motion (fBm) appears naturally in the modeling of many situations, for example, when describing

1. The widths of consecutive annual rings of a tree,
2. The temperature at a specific place as a function of time,
3. The level of water in a river as a function of time;
4. The characters of solar activity as a function of time,
5. The values of the log returns of a stock,
6. Financial turbulence, i.e. the empirical volatility of a stock, and other turbulence phenomena,
7. The prices of electricity in a liberated electricity market.

In cases 1 to 5 the corresponding fBm has Hurst coefficient $H > 1/2$, which means that the process is persistent. In cases 6 and 7 the corresponding fBm has Hurst coefficient $H < 1/2$, which means that the process is anti-persistent. For more information about some of these examples we refer to [209].

In addition to the above, it is a mathematically tractable fact that fBm represents a natural one-parameter extension (represented by the Hurst parameter H) of classical Brownian motion. Therefore, it is natural to ask if a stochastic calculus for fBm can be developed. This is not obvious since fBm is not a semimartingale (except when $H = 1/2$, which corresponds to the classical Brownian motion case). Moreover, it is not a Markov process either; so the most useful and efficient classical mathematical machineries and techniques for stochastic calculus are not available in the fBm case. Therefore, it is necessary to develop these techniques from scratch for the fBm. It turns out that this can be done by exploiting the fact that fBm is a Gaussian process.

The purpose of this book is to explain this in detail and to give applications of the resulting theory. More precisely, we will investigate the main approaches used to develop a stochastic calculus for fBm and their relations. We also give some applications, including discussions of the (sometimes controversial) use of fBm in finance, stochastic partial differential equations, stochastic optimal, control and local time for fBm.

As shown by the reference section, there is a large literature concerning stochastic calculus for fBm and its applications. We have tried to cite rigorously every paper, preprint, or book we were aware of, and we apologize if we accidentally overlooked some works.

We want to thank Birgit Beck, Christian Bender, Catriona M. Byrne, Alessandra Cretarola, Robert Elliott, Nils Christian Framstad, Serena Fuschini, Thilo Meyer-Brandis, Kirsten Minkos, Sebastian Queißer, Donna Mary Salopek, Agnès Sulem, Esko Valkeila, John van der Hoek, and three anonymous referees for many valuable communications and comments. Yaozhong Hu acknowledges the support of the National Science Foundation under Grant No. DMS0204613 and DMS0504783. We are also very grateful to our editors Karen Borthwick, Helen Desmond and Stephanie Harding for their patience and support.

Any remaining errors are ours.

Francesca Biagini, Yaozhong Hu, Bernt Øksendal and Tusheng Zhang,

Munich, Lawrence, Oslo, and Manchester, November 2006.

Contents

Preface VII

Introduction 1

Part I Fractional Brownian motion

1 Intrinsic properties of the fractional Brownian motion 5
   1.1 Fractional Brownian motion 5
   1.2 Stochastic integral representation 6
   1.3 Correlation between two increments 8
   1.4 Long-range dependence 9
   1.5 Self-similarity 10
   1.6 Hölder continuity 11
   1.7 Path differentiability 11
   1.8 The fBm is not a semimartingale for $H \neq 1/2$ 12
   1.9 Invariance principle 14

Part II Stochastic calculus

2 Wiener and divergence-type integrals for fractional Brownian motion 23
   2.1 Wiener integrals 23
      2.1.1 Wiener integrals for $H > 1/2$ 27
      2.1.2 Wiener integrals for $H < 1/2$ 34
   2.2 Divergence-type integrals for fBm 37
      2.2.1 Divergence-type integral for $H > 1/2$ 39
      2.2.2 Divergence-type integral for $H < 1/2$ 41

3 Fractional Wick Itô Skorohod (fWIS) integrals for fBm of Hurst index $H > 1/2$ ... 47
- 3.1 Fractional white noise ... 47
- 3.2 Fractional Girsanov theorem ... 59
- 3.3 Fractional stochastic gradient ... 62
- 3.4 Fractional Wick Itô Skorohod integral ... 64
- 3.5 The ϕ-derivative ... 65
- 3.6 Fractional Wick Itô Skorohod integrals in L^2 ... 68
- 3.7 An Itô formula ... 71
- 3.8 L^p estimate for the fWIS integral ... 75
- 3.9 Iterated integrals and chaos expansion ... 78
- 3.10 Fractional Clark Hausmann Ocone theorem ... 83
- 3.11 Multidimensional fWIS integral ... 87
- 3.12 Relation between the fWIS integral and the divergence-type integral for $H > 1/2$ ... 96

4 Wick Itô Skorohod (WIS) integrals for fractional Brownian motion ... 99
- 4.1 The M operator ... 99
- 4.2 The Wick Itô Skorohod (WIS) integral ... 103
- 4.3 Girsanov theorem ... 109
- 4.4 Differentiation ... 110
- 4.5 Relation with the standard Malliavin calculus ... 115
- 4.6 The multidimensional case ... 118

5 Pathwise integrals for fractional Brownian motion ... 123
- 5.1 Symmetric, forward and backward integrals for fBm ... 123
- 5.2 On the link between fractional and stochastic calculus ... 125
- 5.3 The case $H < 1/2$ ... 126
- 5.4 Relation with the divergence integral ... 130
- 5.5 Relation with the fWIS integral ... 132
- 5.6 Relation with the WIS integral ... 137

6 A useful summary ... 147
- 6.1 Integrals with respect to fBm ... 147
  - 6.1.1 Wiener integrals ... 147
  - 6.1.2 Divergence-type integrals ... 150
  - 6.1.3 fWIS integrals ... 151
  - 6.1.4 WIS integrals ... 153
  - 6.1.5 Pathwise integrals ... 154
- 6.2 Relations among the different definitions of stochastic integral ... 155
  - 6.2.1 Relation between Wiener integrals and the divergence ... 156
  - 6.2.2 Relation between the divergence and the fWIS integral ... 156
  - 6.2.3 Relation between the fWIS and the WIS integrals ... 157
  - 6.2.4 Relations with the pathwise integrals ... 158

6.3 Itô formulas with respect to *fBm* 160

Part III Applications of stochastic calculus

7 **Fractional Brownian motion in finance** 169
   7.1 The pathwise integration model ($1/2 < H < 1$) 170
   7.2 The WIS integration model ($0 < H < 1$) 172
   7.3 A connection between the pathwise and the WIS model 179
   7.4 Concluding remarks 180

8 **Stochastic partial differential equations driven by fractional Brownian fields** 181
   8.1 Fractional Brownian fields 181
   8.2 Multiparameter fractional white noise calculus 185
   8.3 The stochastic Poisson equation 189
   8.4 The linear heat equation 194
   8.5 The quasi-linear stochastic fractional heat equation 198

9 **Stochastic optimal control and applications** 207
   9.1 Fractional backward stochastic differential equations 207
   9.2 A stochastic maximum principle 211
   9.3 Linear quadratic control 216
   9.4 A minimal variance hedging problem 218
   9.5 Optimal consumption and portfolio in a fractional Black and Scholes market 221
   9.6 Optimal consumption and portfolio in presence of stochastic volatility driven by *fBm* 232

10 **Local time for fractional Brownian motion** 239
   10.1 Local time for *fBm* 239
   10.2 The chaos expansion of local time for *fBm* 245
   10.3 Weighted local time for *fBm* 250
   10.4 A Meyer Tanaka formula for *fBm* 253
   10.5 A Meyer Tanaka formula for geometric *fBm* 255
   10.6 Renormalized self-intersection local time for *fBm* 258
   10.7 Application to finance................................... 266

Part IV Appendixes

A **Classical Malliavin calculus** 273
   A.1 Classical white noise theory 273
   A.2 Stochastic integration 278
   A.3 Malliavin derivative..................................... 281

B Notions from fractional calculus 285
- B.1 Fractional calculus on an interval 285
- B.2 Fractional calculus on the whole real line 288

C Estimation of Hurst parameter 289
- C.1 Absolute value method 290
- C.2 Variance Method 290
- C.3 Variance residuals methods 290
- C.4 Hurst's rescaled range (R/S) analysis 291
- C.5 Periodogram method 291
- C.6 Discrete variation method 291
- C.7 Whittle method 292
- C.8 Maximum likelihood estimator 293
- C.9 Quasi maximum likelihood estimator 294

D Stochastic differential equations for fractional Brownian motion 297
- D.1 Stochastic differential equations with Wiener integrals 297
- D.2 Stochastic differential equations with pathwise integrals 300
- D.3 Stochastic differential equations via rough path analysis 305
  - D.3.1 Rough path analysis 305
  - D.3.2 Stochastic calculus with rough path analysis 306

References 309

Index of symbols and notation 321

Index 325

Introduction

This book originates from the need of a comprehensive account of the stochastic integration theory of the fractional Brownian motion (fBm). However there are many important aspects of fBm that are not discussed here. For example, for an analysis of the theory and the applications of long-range dependence from a more statistical point of view, we refer to [81]. Our selection of topics is based mainly on our main interests and background in corresponding research papers. However besides our (fractional and standard) white noise approach, we have tried to provide an overview of some of the most important methods of introducing a stochastic integral for fBm.

After reviewing in Chapter 1 the properties of the fractional Brownian motion, in Chapter 2 we start our tour through several definitions of stochastic integral for fBm by the *Wiener integral* since it deals with the simplest case of deterministic integrands. We proceed then to introduce the *divergence type integral* seen as adjoint operator of the stochastic derivative.

In Chapters 3 and 4 we present a stochastic integration based on the white noise theory. In Chapter 3 the stochastic integral is introduced as an element of *fractional Hida distribution space* for Hurst index $1/2 < H < 1$ and then conditions are clarified that guarantee the existence of this type of integral in L^2. In Chapter 4 the integral is defined as an element in the *classical* Hida distribution space by using the *white noise theory* and *Malliavin calculus for standard Brownian motion* introduced in Appendix A. The main advantage of this method with respect to the one presented in Chapter 3 is that it permits us to define the stochastic integral for any $H \in (0,1)$. In addition it doesn't require the introduction of the fractional white noise theory since it uses the well-established theory for the standard case. However, the approach of Chapter 3 can be seen as more intrinsic.

Finally, in Chapter 5 we investigate the definition and the properties of the pathwise integrals, respectively, *symmetric*, *forward*, and *backward integrals*.

All through this part we underline and investigate the relations between the different approaches and in Chapter 6 we provide what in our eyes is a useful summary. Here we present a synthesis of all the definitions and relations of the several kinds of stochastic integration for fBm together with an overview of the Itô formula relative to each approach, trying to emphasize how they can derive from each other by using the connections among the different stochastic integrals.

In the second part we illustrate some application to finance, stochastic partial differential equations, stochastic optimal control, and local time for fBm. In the appendixes we gather the main results concerning the standard white noise theory and Malliavin calculus for Brownian motion and fractional calculus. Without aiming at completeness, for the reader's convenience we also provide a short summary of the main methods used to estimate the Hurst parameter from sequences of data and some results concerning stochastic differential equations for fBm.

In spite of its high level of technicality, we hope that this book will provide a reference text for further development of the theory and the applications of fBm.

Part I

Fractional Brownian motion

1
Intrinsic properties of the fractional Brownian motion

The aim of this book is to provide a comprehensive overview and systematization of stochastic calculus with respect to fractional Brownian motion. However, for the reader's convenience, in this chapter we review the main properties that make fractional Brownian motion interesting for many applications in different fields.
The main references for this chapter are [76], [156], [177], [195], [209], [215]. For further details concerning the theory and the applications of long-range dependence from a more statistical point of view, we also refer to [81].

1.1 Fractional Brownian motion

The fractional Brownian motion was first introduced within a Hilbert space framework by Kolmogorov in 1940 in [141], where it was called *Wiener Helix*. It was further studied by Yaglom in [230]. The name *fractional Brownian motion* is due to Mandelbrot and Van Ness, who in 1968 provided in [156] a stochastic integral representation of this process in terms of a standard Brownian motion.

Definition 1.1.1. *Let H be a constant belonging to $(0,1)$. A fractional Brownian motion (fBm) $(B^{(H)}(t))_{t\geq 0}$ of Hurst index H is a continuous and centered Gaussian process with covariance function*

$$E\left[B^{(H)}(t)B^{(H)}(s)\right] = 1/2(t^{2H} + s^{2H} - |t-s|^{2H}).$$

For $H = 1/2$, the *fBm* is then a standard Brownian motion. By Definition 1.1.1 we obtain that a standard *fBm* $B^{(H)}$ has the following properties:

1. $B^{(H)}(0) = 0$ and $E\left[B^{(H)}(t)\right] = 0$ for all $t \geq 0$.
2. $B^{(H)}$ has homogeneous increments, i.e., $B^{(H)}(t+s) - B^{(H)}(s)$ has the same law of $B^{(H)}(t)$ for $s, t \geq 0$.

6   1 Intrinsic properties of the fractional Brownian motion

3. $B^{(H)}$ is a Gaussian process and $E\left[B^{(H)}(t)^2\right] = t^{2H}$, $t \geq 0$, for all $H \in (0,1)$.
4. $B^{(H)}$ has continuous trajectories.

The existence of the *fBm* follows from the general existence theorem of centered Gaussian processes with given covariance functions (see [196]). We will also give some constructions of the *fBm* through the white noise theory for our special purposes in later chapters. The *fBm* is divided into three very different families corresponding to $0 < H < 1/2$, $H = 1/2$, and $1/2 < H < 1$, respectively, as we will see in the sequel. It was Mandelbrot that named the parameter H of $B^{(H)}$ after the name of the hydrologist Hurst, who made a statistical study of yearly water run-offs of the Nile river (see [129]). He considered the values $\delta_1, \ldots, \delta_n$ of n successive yearly run-offs and their corresponding cumulative value $\Delta_n = \sum_{i=1}^{n} \delta_i$ over the period from the year 662 until 1469. He discovered that the behavior of the normalized values of the amplitude of the deviation from the empirical mean was approximately cn^H, where $H = 0.7$. Moreover, the distribution of $\Delta_n = \sum_{i=1}^{n} \delta_i$ was approximately the same as $n^H \delta_1$, with $H > 1/2$. Hence, this phenomenon could not be modeled by using a process with independent increments, but rather the δ_i could be thought as the increments of a *fBm*. Because of this study, Mandelbrot introduced the name *Hurst index*.

1.2 Stochastic integral representation

Here we discuss some of the integral representations for the *fBm*. In [156], it is proved that the process

$$\begin{aligned}
Z(t) &= \frac{1}{\Gamma(H+1/2)} \int_{\mathbb{R}} \left((t-s)_+^{H-1/2} - (-s)_+^{H-1/2} \right) dB(s) \\
&= \frac{1}{\Gamma(H+1/2)} \left(\int_{-\infty}^{0} \left((t-s)^{H-1/2} - (-s)^{H-1/2} \right) dB(s) \right. \\
&\quad \left. + \int_{0}^{t} (t-s)^{H-1/2} dB(s) \right)
\end{aligned} \tag{1.1}$$

where $B(t)$ is a standard Brownian motion and Γ represents the gamma function, is a *fBm* with Hurst index $H \in (0,1)$. If $B(t)$ is replaced by a complex-valued Brownian motion, the integral (1.1) gives the complex *fBm*. By following [177] we sketch a proof for the representation (1.1). For further detail we refer also to [207]. First we notice that $Z(t)$ is a continuous centered Gaussian process. Hence, we need only to compute the covariance functions. In the following computations we drop the constant $1/\Gamma(H+1/2)$ for the sake of simplicity. We obtain

$$E\left[Z^2(t)\right] = \int_{\mathbb{R}} \left[(t-s)_+^{H-1/2} - (-s)_+^{H-1/2} \right]^2 ds$$

$$= t^{2H} \int_{\mathbb{R}} \left[(1-u)_+^{H-1/2} - (-u)_+^{H-1/2} \right]^2 du$$
$$= C(H) t^{2H},$$

where we have used the change of variable $s = tu$. Analogously, we have that

$$E\left[|Z(t) - Z(s)|^2\right] = \int_{\mathbb{R}} \left[(t-u)_+^{H-1/2} - (s-u)_+^{H-1/2} \right]^2 ds$$
$$= t^{2H} \int_{\mathbb{R}} \left[(t-s-u)_+^{H-1/2} - (-u)_+^{H-1/2} \right]^2 du$$
$$= C(H) |t-s|^{2H}.$$

Now

$$E[Z(t)Z(s)] = -\frac{1}{2} \left\{ E\left[|Z(t) - Z(s)|^2\right] - E\left[Z(t)^2\right] - E\left[Z(s)^2\right] \right\}$$
$$= \frac{1}{2}(t^{2H} + s^{2H} - |t-s|^{2H}).$$

Hence we can conclude that $Z(t)$ is a *fBm* of Hurst index H.

Several other stochastic integral representations have been developed in the literature. By [207], we get the following spectral representation of *fBm*

$$B^{(H)}(t) := \frac{1}{C_2(H)} \int_{\mathbb{R}} \frac{e^{its} - 1}{is} |s|^{1/2 - H} d\tilde{B}(s),$$

where $\tilde{B}(s) = B^1 + iB^2$ is a complex Brownian measure on $\mathbb{R}$ such that $B^1(A) = B^1(-A)$, $B^2(A) = -B^2(-A)$, and $E\left[B^1(A)^2\right] = E\left[B^2(A)^2\right] = |A|/2$ for every $A \in \mathcal{B}(\mathbb{R})$, and

$$C_2(H) = \left(\frac{\pi}{H\Gamma(2H) \sin H\pi} \right)^{1/2},$$

where $B^i(A) := \int_A dB^i(t)$. Equation (1.1) provides an integral representation for *fBm* over the whole real line. By following the approach of [172], we can also represent the *fBm* over a *finite interval*, i.e.,

$$B^{(H)}(t) := \int_0^t K_H(t,s) \, dB(s), \quad t \geq 0,$$

where

1. For $H > 1/2$,

$$K_H(t,s) = c_H s^{1/2 - H} \int_s^t |u-s|^{H-3/2} u^{H-1/2} \, du, \tag{1.2}$$

where $c_H = [H(2H-1)/\beta(2-2H, H-1/2)]^{1/2}$ and $t > s$.

2. For $H < 1/2$,

$$K_H(t,s) = b_H \left[\left(\frac{t}{s}\right)^{H-1/2} (t-s)^{H-1/2} \right.$$
$$\left. - \left(H - \frac{1}{2}\right) s^{1/2-H} \int_s^t (u-s)^{H-1/2} u^{H-3/2}\, du \right] \quad (1.3)$$

with $b_H = [2H/((1-2H)\beta(1-2H, H+1/2))]^{1/2}$ and $t > s$.

For the proof, we refer to [114], [172], and [177]. Note that this representation is canonical in the sense that the filtrations generated by $B^{(H)}$ and B coincide.

In Chapter 2 a definition of stochastic integral with respect to fBm will be introduced by exploiting the stochastic integral representation of $B^{(H)}$ in terms of (1.2) and of (1.3).

Remark 1.2.1. Integral representations that change fBm of arbitrary Hurst index K into fBm of index H have been studied in [191], [133] and [134]. In Theorem 1.1 of [191] it is shown that for any $K \in (0,1)$, there exists a unique K-fBm $\tilde{B}^{(K)}$ such that for all $t \in \mathbb{R}$ there holds

$$B^{(H)}(t) = \tilde{c}_{H,K} \int_\mathbb{R} \left[(t-s)_+^{H-K} - (-s)_+^{H-K} \right] d\tilde{B}^{(K)}(s), \quad \text{a.s.,} \quad (1.4)$$

with $\tilde{c}_{H,K} = 1/\Gamma(H-K+1) \left(\Gamma(2K+1)\sin(\pi K)/\Gamma(2H+1)\sin(\pi H)\right)^{1/2}$. In Theorem 5.1 of [133] integration is carried out on $[0,t]$ and showed that for given $K \in (0,1)$, there exists a unique K-fBm $B^{(K)}(t)$, $t \geq 0$, such that for all $t \geq 0$ we have a.s. that

$$B^{(H)}(t) = c_{H,K} \int_0^t (t-s)^{H-K}$$
$$\cdot F\left(1-K-H, H-K, 1+H-K, \frac{s-t}{s}\right) dB^{(K)}(s), \quad (1.5)$$

where F is Gauss hypergeometric function and

$$c_{H,K} = \frac{1}{\Gamma(H-K+1)} \left(\frac{2H\Gamma(H+1/2)\Gamma(3/2-H)\Gamma(2-2K)}{2K\Gamma(K+1/2)\Gamma(3/2-K)\Gamma(2-2H)} \right)^{1/2}.$$

In [134] an analytical connection between (1.4) and (1.5) is proved.

1.3 Correlation between two increments

For $H = 1/2$, $B^{(H)}$ is a standard Brownian motion; hence, in this case the increments of the process are independent. On the contrary, for $H \neq 1/2$ the increments are not independent. More precisely, by Definition 1.1.1 we know that the covariance between $B^{(H)}(t+h) - B^{(H)}(t)$ and $B^{(H)}(s+h) - B^{(H)}(s)$ with $s + h \leq t$ and $t - s = nh$ is

$$\rho_H(n) = \frac{1}{2} h^{2H} [(n+1)^{2H} + (n-1)^{2H} - 2n^{2H}].$$

In particular, we obtain that two increments of the form $B^{(H)}(t+h) - B^{(H)}(t)$ and $B^{(H)}(t+2h) - B^{(H)}(t+h)$ are positively correlated for $H > 1/2$, while they are negatively correlated for $H < 1/2$. In the first case the process presents an aggregation behavior and this property can be used in order to describe "cluster" phenomena (systems with *memory* and *persistence*). In the second case it can be used to model sequences with *intermittency* and *anti-persistence*.

1.4 Long-range dependence

Definition 1.4.1. *A stationary sequence* $(X_n)_{n \in \mathbb{N}}$ *exhibits* long-range dependence *if the autocovariance functions* $\rho(n) := \mathrm{cov}(X_k, X_{k+n})$ *satisfy*

$$\lim_{n \to \infty} \frac{\rho(n)}{c n^{-\alpha}} = 1$$

for some constant c and $\alpha \in (0,1)$. In this case, the dependence between X_k and X_{k+n} decays slowly as n tends to infinity and

$$\sum_{n=1}^{\infty} \rho(n) = \infty.$$

Hence, we obtain immediately that the increments $X_k := B^{(H)}(k) - B^{(H)}(k-1)$ of $B^{(H)}$ and $X_{k+n} := B^{(H)}(k+n) - B^{(H)}(k+n-1)$ of $B^{(H)}$ have the long-range dependence property for $H > 1/2$ since

$$\rho_H(n) = \frac{1}{2}[(n+1)^{2H} + (n-1)^{2H} - 2n^{2H}] \sim H(2H-1)n^{2H-2}$$

as n goes to infinity. In particular,

$$\lim_{n \to \infty} \frac{\rho_H(n)}{H(2H-1)n^{2H-2}} = 1.$$

Summarizing, we obtain

1. For $H > 1/2$, $\sum_{n=1}^{\infty} \rho_H(n) = \infty$.
2. For $H < 1/2$, $\sum_{n=1}^{\infty} |\rho_H(n)| < \infty$.

There are alternative definitions of long-range dependence. We recall that a function L is *slowly varying* at zero (respectively, at infinity) if it is bounded on a finite interval and if, for all $a > 0$, $L(ax)/L(x)$ tends to 1 as x tends to zero (respectively, to infinity).

We introduce now the *spectral density* of the autocovariance functions $\rho(k)$

$$f(\lambda) := \frac{1}{2\pi} \sum_{k=-\infty}^{\infty} e^{-i\lambda k} \rho(k)$$

for $\lambda \in [-\pi, \pi]$.

Definition 1.4.2. *For stationary sequences $(X_n)_{n \in \mathbb{N}}$ with finite variance, we say that $(X_n)_{n \in \mathbb{N}}$ exhibits* long-range dependence *if one of the following holds:*

1. $\lim_{n \to \infty} (\sum_{k=-n}^{n} \rho(k))/(cn^\beta L_1(n)) = 1$ *for some constant c and $\beta \in (0, 1)$.*
2. $\lim_{k \to \infty} \rho(k)/ck^{-\gamma} L_2(k) = 1$ *for some constant c and $\gamma \in (0, 1)$.*
3. $\lim_{\lambda \to 0} f(\lambda)/c|\lambda|^{-\delta} L_3(|\lambda|) = 1$ *for some constant c and $\delta \in (0, 1)$.*

Here L_1, L_2 are slowly varying functions at infinity, while L_3 is slowly varying at zero.

Lemma 1.4.3. *For fBm $B^{(H)}$ of Hurst index $H \in (1/2, 1)$, the three definitions of long-range dependence of Definition 1.4.2 are equivalent. They hold with the following choice of parameters and slowly varying functions:*

1. $\beta = 2H - 1$, $L_1(x) = 2H$.
2. $\gamma = 2 - 2H$, $L_2(x) = H(2H - 1)$.
3. $\delta = 2H - 1$, $L_3(x) = \pi^{-1} H \Gamma(2H) \sin \pi H$.

Proof. For the proof, we refer to Section 4 in [221]. □

For a survey on theory and applications of long-range dependence, see also [81].

1.5 Self-similarity

By following [209], we introduce the following:

Definition 1.5.1. *We say that an $\mathbb{R}^d$-valued random process $X = (X_t)_{t \geq 0}$ is* self-similar *or satisfies the* property of self-similarity *if for every $a > 0$ there exists $b > 0$ such that*

$$\text{Law}(X_{at}, t \geq 0) = \text{Law}(bX_t, t \geq 0). \tag{1.6}$$

Note that (1.6) means that the two processes X_{at} and bX_t have the same finite-dimensional distribution functions, i.e., for every choice $t_0, \ldots, t_n$ in $\mathbb{R}$,

$$P(X_{at_0} \leq x_0, \ldots, X_{at_n} \leq x_n) = P(bX_{t_0} \leq x_0, \ldots, bX_{t_n} \leq x_n)$$

for every $x_0, \ldots, x_n$ in $\mathbb{R}$.

Definition 1.5.2. *If $b = a^{-H}$ in Definition 1.5.1, then we say that $X = (X_t)_{t \geq 0}$ is a self-similar process with Hurst index H or that it satisfies the property of (statistical) self-similarity with Hurst index H. The quantity $D = 1/H$ is called the* statistical fractal dimension *of X.*

Since the covariance function of the fBm is homogeneous of order $2H$, we obtain that $B^{(H)}$ is a self-similar process with Hurst index H, i.e., for any constant $a > 0$ the processes $B^{(H)}(at)$ and $a^{-H}B^{(H)}(t)$ have the same distribution law.

1.6 Hölder continuity

We recall that according to the Kolmogorov criterion (see [228]), a process $X = (X_t)_{t \in \mathbb{R}}$ admits a continuous modification if there exist constants $\alpha \geq 1$, $\beta > 0$, and $k > 0$ such that

$$E\left[|X(t) - X(s)|^\alpha\right] \leq k|t-s|^{1+\beta}$$

for all $s, t \in \mathbb{R}$.

Theorem 1.6.1. *Let $H \in (0,1)$. The fBm $B^{(H)}$ admits a version whose sample paths are almost surely Hölder continuous of order strictly less than H.*

Proof. We recall that a function $f : \mathbb{R} \to \mathbb{R}$ is Hölder continuous of order α, $0 < \alpha \leq 1$, and write $f \in C^\alpha(\mathbb{R})$, if there exists $M > 0$ such that

$$|f(t) - f(s)| \leq M|t-s|^\alpha,$$

for every $s, t \in \mathbb{R}$. For any $\alpha > 0$ we have

$$E\left[|B^{(H)}(t) - B^{(H)}(s)|^\alpha\right] = E\left[|B^{(H)}(1)|^\alpha\right] |t-s|^{\alpha H};$$

hence, by the Kolmogorov criterion we get that the sample paths of $B^{(H)}$ are almost everywhere Hölder continuous of order strictly less than H. Moreover, by [9] we have

$$\limsup_{t \to 0^+} \frac{|B^{(H)}(t)|}{t^H \sqrt{\log \log t^{-1}}} = c_H$$

with probability one, where c_H is a suitable constant. Hence $B^{(H)}$ cannot have sample paths with Hölder continuity's order greater than H. □

1.7 Path differentiability

By [156] we also obtain that the process $B^{(H)}$ is not mean square differentiable and it does not have differentiable sample paths.

Proposition 1.7.1. *Let $H \in (0,1)$. The fBm sample path $B^{(H)}(.)$ is not differentiable.*

In fact, for every $t_0 \in [0, \infty)$

$$\limsup_{t \to t_0} \left| \frac{B^{(H)}(t) - B^{(H)}(t_0)}{t - t_0} \right| = \infty$$

with probability one.

Proof. Here we recall the proof of [156]. Note that we assume $B^{(H)}(0) = 0$. The result is proved by exploiting the self-similarity of $B^{(H)}$. Consider the random variable

$$\mathcal{R}_{t,t_0} := \frac{B^{(H)}(t) - B^{(H)}(t_0)}{t - t_0}$$

that represents the incremental ratio of $B^{(H)}$. Since $B^{(H)}$ is self-similar, we have that the law of $\mathcal{R}_{t,t_0}$ is the same of $(t - t_0)^{H-1} B^{(H)}(1)$. If one considers the event

$$A(t, \omega) := \left\{ \sup_{0 \le s \le t} \left| \frac{B^{(H)}(s)}{s} \right| > d \right\},$$

then for any sequence $(t_n)_{n \in \mathbb{N}}$ decreasing to 0, we have

$$A(t_n, \omega) \supseteq A(t_{n+1}, \omega),$$

and

$$A(t_n, \omega) \supseteq (|\frac{B^{(H)}(t_n)}{t_n}| > d) = (|B^{(H)}(1)| > t_n^{1-H} d).$$

The thesis follows since the probability of the last term tends to 1 as $n \to \infty$. □

1.8 The *fBm* is not a semimartingale for $H \ne 1/2$

The fact that the *fBm* is not a semimartingale for $H \ne 1/2$ has been proved by several authors. For example, for $H > 1/2$ we refer to [82], [150], [152]. Here we recall the proof of [195] that is valid for every $H \ne 1/2$. In order to verify that $B^{(H)}$ is not a semimartingale for $H \ne 1/2$, it is sufficient to compute the *p*-variation of $B^{(H)}$.

Definition 1.8.1. *Let $(X(t))_{t \in [0,T]}$ be a stochastic process and consider a partition $\pi = \{0 = t_0 < t_1 < \ldots < t_n = T\}$. Put*

$$\mathcal{S}_p(X, \pi) := \sum_{i=1}^{n} |X(t_k) - X(t_{k-1})|^p.$$

The p-variation of X over the interval $[0, T]$ is defined as

1.8 The fBm is not a semimartingale for $H \neq 1/2$

$$\mathcal{V}_p(X, [0,T]) := \sup_\pi \mathcal{S}_p(X, \pi),$$

where π is a finite partition of $[0,T]$. The index of p-variation of a process is defined as

$$I(X, [0,T]) := \inf\{p > 0; \mathcal{V}_p(X, [0,T]) < \infty\}.$$

We claim that

$$I(B^{(H)}, [0,T]) = \frac{1}{H}.$$

In fact, consider for $p > 0$,

$$Y_{n,p} = n^{pH-1} \sum_{i=1}^n \left| B^{(H)}\left(\frac{i}{n}\right) - B^{(H)}\left(\frac{i-1}{n}\right) \right|^p.$$

Since $B^{(H)}$ has the self-similarity property, the sequence $(Y_{n,p})_{n \in \mathbb{N}}$ has the same distribution as

$$\tilde{Y}_{n,p} = n^{-1} \sum_{i=1}^n |B^{(H)}(i) - B^{(H)}(i-1)|^p.$$

By the Ergodic theorem (see, for example, [69]) the sequence $\tilde{Y}_{n,p}$ converges almost surely and in L^1 to $E\left[|B^{(H)}(1)|^p\right]$ as n tends to infinity; hence, it converges also in probability to $E\left[|B^{(H)}(1)|^p\right]$. It follows that

$$V_{n,p} = \sum_{i=1}^n \left| B^{(H)}\left(\frac{i}{n}\right) - B^{(H)}\left(\frac{i-1}{n}\right) \right|^p$$

converges in probability respectively to 0 if $pH > 1$ and to infinity if $pH < 1$ as n tends to infinity. Thus we can conclude that $I(B^{(H)}, [0,T]) = 1/H$. Since for every semimartingale X, the index $I(X, [0,T])$ must belong to $[0,1] \cup \{2\}$, the fBm $B^{(H)}$ cannot be a semimartingale unless $H = 1/2$.

As a direct consequence of this fact, one cannot use the Itô stochastic calculus developed for semimartingales in order to define the stochastic integral with respect to $B^{(H)}$. In the following chapters we will summarize the different approaches developed in the literature in order to overcome this problem.

In [53] it has been introduced the new notion of *weak semimartingale* and shown that $B^{(H)}$ is not even a weak semimartingale if $H \neq 1/2$. A stochastic process $(X(t))_{t \geq 0}$ is said to be a *weak semimartingale* if for every $T > 0$ the family of random variables

$$\left\{ \sum_{i=1}^n a_i [X(t_i) - X(t_{i-1})], n \geq 1, 0 = t_0 < \ldots < t_n = T, |a_i| < 1, a_i \in \mathcal{F}^X_{t_{i-1}} \right\}$$

is bounded in L^0. Here $\mathcal{F}^X$ represents the natural filtration associated to the process X. Moreover, in [53] it is shown that if $B(t)$ is a standard Brownian motion independent of $B^{(H)}$, then the process

$$M^H(t) := B^{(H)}(t) + B(t)$$

is not a weak semimartingale if $H \in (0, 1/2) \cup (1/2, 3/4)$, while it is a semimartingale equivalent in law to B on any finite time interval $[0, T]$ if $H \in (3/4, 1)$. We refer to [53] for further details.

1.9 Invariance principle

Here we present an invariance principle for fBms due to [36].

Assume that $\{X_n, n = 1, 2, ...\}$ is a stationary Gaussian sequence with $E[X_i] = 0$ and $E[X_i^2] = 1$. Define

$$Z_n(t) = \frac{1}{n^H} \sum_{k=0}^{[nt]-1} X_k, \quad 0 \le t \le 1,$$

where $[\cdot]$ stands for the integer part. We will show that if the covariance of $\sum_{k=0}^n X_k$ is proportional to Cn^{2H} for large n, $Z_n(t), t \ge 0$ converges weakly to $\sqrt{C} B_t^{(H)}$ in a suitable metric space. Let us first introduce the the metric space. Let $I = [0, 1]$ and denote by $L^p(I)$ the space of Lebesgue integrable functions with exponent p. For $f \in L^p(I), t \in I$, define

$$\omega_p(f, t) = \sup_{|h| \le t} \left(\int_{I_h} |f(x+h) - f(x)|^p dx \right)^{1/p},$$

where $I_h = \{x \in I, x + h \in I\}$. For $0 < \alpha < 1$ and $\beta > 0$, consider the real-valued function $\omega_\beta^\alpha(\cdot)$ defined by

$$\omega_\beta^\alpha(t) = t^\alpha \left(1 + \log \frac{1}{t} \right)^\beta, \quad t > 0,$$

and we let

$$\|f\|_p^{\omega_\beta^\alpha} = \|f\|_{L^p(I)} + \sup_{0 < t \le 1} \frac{\omega_p(f, t)}{\omega_\beta^\alpha(t)}.$$

The Besov space $Lip_p(\alpha, \beta)$ is the class of functions f in $L^p(I)$ such that $\|f\|_p^{\omega_\beta^\alpha} < \infty$. $Lip_p(\alpha, \beta)$ endowed with the norm $\|\cdot\|_p^{\omega_\beta^\alpha}$ is a nonseparable Banach space. Let $B_p^{\alpha,\beta}$ denote the separable subspace of $Lip_p(\alpha, \beta)$ formed by functions $f \in Lip_p(\alpha, \beta)$ satisfying $\omega_p(f, t) = o(\omega_\beta^\alpha(t))$ as $t \to 0$. For a continuous function f, denote by $\{C_n(f), n \ge 0\}$ the coefficients of the decomposition of f in the Schauder basis given by

$$C_0(f) = f(0), C_1(f) = f(1) - f(0),$$

and for $n = 2^j + k, j \ge 0$, and $k = 0, ..., 2^j - 1$,

$$C_n(f) = 2 \cdot 2^{j/2} \left\{ f\left(\frac{2k-1}{2^{j+1}}\right) - \frac{1}{2}\left[f\left(\frac{2k}{2^{j+1}}\right) + f\left(\frac{2k-2}{2^{j+1}}\right)\right]\right\}.$$

The following characterization theorem proved in [55] will be used.

Theorem 1.9.1. *1. If $\alpha > 1/p$, then $Lip_p(\alpha, \beta)$ is the space of continuous functions with the following equivalence of norms:*

$$\|f\|_p^{\omega_\beta^\alpha} \sim \max\left\{|C_0(f)|, |C_1(f)|,\right.$$

$$\left.\sup_{j \geq 0} \frac{2^{-j(1/2-\alpha+1/p)}}{(1+j)^\beta}\left[\sum_{n=2^j+1}^{2^{j+1}} |C_n(f)|^p\right]^{1/p}\right\}.$$

2. f belongs to $B_p^{\alpha,\beta}$ if and only if

$$\lim_{j \to \infty} \frac{2^{-j(1/2-\alpha+1/p)}}{(1+j)^\beta}\left[\sum_{n=2^j+1}^{2^{j+1}} |C_n(f)|^p\right]^{1/p} = 0.$$

Lemma 1.9.2. *Let $1 \leq p < \infty$, $1/p < \alpha < 1$, and $\beta > 0$. A set F of measurable functions $f : I \to R$ is relatively compact in $B_p^{\alpha,\beta}$ if*

1. $\sup_{f \in F} \|f\|_p^{\omega_\beta^\alpha} < \infty$,
2. $\limsup_{\delta \to 0} \sup_{f \in F} K_\delta(f, \alpha, \beta, p) = 0$, where

$$K_\delta(f, \alpha, \beta, p) = \sup_{0 < t \leq \delta} \frac{\omega_p(f, t)}{\omega_\beta^\alpha(t)}.$$

Proof. It is a consequence of the Frechet–Kolmogorov theorem: a subset $K \subset L^p(I)$ is relatively compact if and only if

$$\sup_{f \in K}\left(\int_I |f(s)|^p ds\right) < \infty,$$

$$\limsup_{t \to 0}{}_{f \in K} \int_I (|f(s+t) - f(s)|^p)\, ds = 0.$$

Now assume (1) and (2) hold for a set F. To prove that F is relatively compact, we need to show that any sequence $\{f_n, n \geq 1\} \subset F$ admits a convergent subsequence. Pick a sequence $\{f_n, n \geq 1\}$ from F. By the Frechet–Kolmogorov theorem, $\{f_n, n \geq 1\}$ has a convergent subsequence in $L^p(I)$. Without loss of generality, we assume that $f_n \to f$ in $L^p(I)$. First we show that $f \in B_p^{\alpha,\beta}$. By the Fatou lemma,

$$\omega_p(f, t) = \sup_{|h| \leq t}\left(\int_{I_h} |f(x+h) - f(x)|^p dx\right)^{1/p}$$

$$\leq \sup_{|h|\leq t} \liminf_{n\to\infty} \left(\int_{I_h} |f_n(x+h) - f_n(x)|^p dx\right)^{1/p}$$

$$\leq \sup_n \sup_{|h|\leq t} \left(\int_{I_h} |f_n(x+h) - f_n(x)|^p dx\right)^{1/p}.$$

This together with assumption (2) implies $\omega_p(f,t) = o(\omega_\beta^\alpha(t))$ as $t \to 0$. Hence, $f \in B_p^{\alpha,\beta}$. We will finish the proof by showing $f_n \to f$ also in $B_p^{\alpha,\beta}$. From the definition of the norm in $B_p^{\alpha,\beta}$, it is sufficient to show

$$\lim_{n\to\infty} \sup_{0<t\leq 1} \frac{\omega_p(f_n - f, t)}{\omega_\beta^\alpha(t)} = 0.$$

For any $0 < \delta < 1$, we have

$$\sup_{0<t\leq 1} \frac{\omega_p(f_n - f, t)}{\omega_\beta^\alpha(t)} \leq \sup_{0<t\leq \delta} \frac{\omega_p(f_n - f, t)}{\omega_\beta^\alpha(t)} + \sup_{\delta<t\leq 1} \frac{\omega_p(f_n - f, t)}{\omega_\beta^\alpha(t)}.$$

Let $\varepsilon > 0$. By assumption (2) we can find $\delta > 0$ such that

$$\sup_{0<t\leq \delta} \frac{\omega_p(f_n - f, t)}{\omega_\beta^\alpha(t)} \leq \frac{\varepsilon}{2},$$

for all $n \geq 1$. On the other hand,

$$\sup_{\delta<t\leq 1} \frac{\omega_p(f_n - f, t)}{\omega_\beta^\alpha(t)} \leq c_\delta \sup_{\delta<t\leq 1} \omega_p(f_n - f, t) \leq 2c_\delta \|f_n - f\|_{L^p(I)}.$$

Thus, there exists $N > 0$ such that for $n \geq N$,

$$\sup_{\delta<t\leq 1} \frac{\omega_p(f_n - f, t)}{\omega_\beta^\alpha(t)} \leq \frac{\varepsilon}{2}.$$

Combining the above arguments, we arrive at

$$\lim_{n\to\infty} \sup_{0<t\leq 1} \frac{\omega_p(f_n - f, t)}{\omega_\beta^\alpha(t)} = 0.$$

□

Lemma 1.9.3. *Let $\alpha > 1/p$ and $0 < \beta < \beta'$. The space $Lip_p(\alpha, \beta)$ is compactly embedded in $B_p^{\alpha, \beta'}$.*

Proof. Let $B = \{f \in Lip_p(\alpha, \beta); \|f\|_p^{\omega_\beta^\alpha} \leq M\}$ be a bounded subset of $Lip_p(\alpha, \beta)$. It is clear that if $\beta < \beta'$, then $\|f\|_p^{\omega_{\beta'}^\alpha} \leq \|f\|_p^{\omega_\beta^\alpha}$. Hence, $\sup_{f \in B} \|f\|_p^{\omega_{\beta'}^\alpha} < \infty$. On the other hand,

$$K_\delta(f,\alpha,\beta',p) = \sup_{0<t\le\delta} \frac{\omega_p(f,t)}{\omega_{\beta'}^\alpha(t)} \le \sup_{0<t\le\delta} \frac{\omega_p(f,t)}{\omega_\beta^\alpha(t)\omega_{\beta-\beta'}^0(\delta)} \le M\omega_{\beta-\beta'}^0(\delta).$$

Therefore,
$$\limsup_{\delta\to 0} \sup_{f\in B} K_\delta(f,\alpha,\beta',p) = 0.$$

By Lemma 1.9.2 this implies that B is relatively compact in $B_p^{\alpha,\beta'}$. □

Lemma 1.9.4. *Let $(X_t^n, t\in I)_{n\ge 1}$ be a sequence of stochastic processes satisfying*

1. $X_0^n = 0$ for all $n\ge 1$.
2. *There exists a positive constant C and $\alpha\in\,]0,1[$ such that for $p\ge 1$, $E[|X_t^n - X_s^n|^p] \le C|t-s|^{p\alpha}$ for all $s,t\in I$.*

Then $(X_t^n, t\in I)_{n\ge 1}$ is tight in $B_p^{\alpha,\beta}$, $\beta>0$ for $p>\max(1/\alpha,1/\beta)$.

Proof. By the assumptions, we have $C_0(X^n) = 0$ and $C_1(X^n) = X_1^n$. To prove the lemma, by Lemma 1.9.3 it is enough to show that there exists a constant $C_p > 0$ such that, for $\lambda > 0$ and $1/p < \beta' < \beta$, we have $P(\|X^n\|_p^{\omega_{\beta'}^\alpha} > \lambda) \le C_p \lambda^{-p}$ for all $n\ge 1$. Applying the characterization Theorem 1.9.1 above, it suffices to show that

$$P\Big(M(X^n) > \lambda\Big) \le C_p\lambda^{-p},$$

where $M(X^n)$ is the maximum of the set

$$\left\{|C_0(X^n)|, |C_1(X^n)|, \sup_{j\ge 0} \frac{2^{-j(1/2-\alpha+1/p)}}{(1+j)^{\beta'}} \left[\sum_{m=2^j+1}^{2^{j+1}} |C_m(X^n)|^p\right]^{1/p}\right\}.$$

Now, by the Chebyshev inequality, we have

$$I = P\left(\sup_{j\ge 0} \frac{2^{-j(1/2-\alpha+1/p)}}{(1+j)^{\beta'}} \left[\sum_{m=2^j+1}^{2^{j+1}} |C_m(X^n)|^p\right]^{1/p} > \lambda\right)$$

$$\le \sum_{j\ge 0} \frac{2^{-jp(1/2-\alpha+1/p)}}{(1+j)^{p\beta'}} \sum_{m=2^j+1}^{2^{j+1}} E[|C_m(X^n)|^p]\lambda^{-p}.$$

Recall that for $m = 2^j + k$,

$$C_m(X^n) = 2\cdot 2^{j/2}\left[X_{(2k-1)/2^{j+1}}^n - \frac{1}{2}(X_{2k/2^{j+1}}^n + X_{(2k-2)/2^{j+1}}^n)\right].$$

Thus,

$$I \leq C_p \lambda^{-p} \sum_{j\geq 0} \frac{2^{-jp(1/2-\alpha+1/p)}}{(1+j)^{p\beta'}} \cdot \sum_{k=1}^{2^j} \left(E\big[|X^n_{(2k-1)/2^{j+1}} - X^n_{2k/2^{j+1}}|^p\big] \right.$$

$$\left. + E\big[|X^n_{(2k-1)/2^{j+1}} - X^n_{(2k-2)/2^{j+1}}|^p\big] \right)$$

$$\leq \lambda^{-p} \left[C_p \sum_{j\geq 0} 1/(1+j)^{p\beta'} \right] \leq C_p \lambda^{-p}$$

which completes the proof. □

Let X_i be a stationary Gaussian sequence with mean 0 and correlations $E[X_k X_l] = r(k-l)$. Define

$$Z_n(t) = \frac{1}{n^H} \sum_{k=0}^{[nt]-1} X_k.$$

Theorem 1.9.5. *Let $H \in]0,1[$, $\beta > 0$ and $p > \max(1/H, 1/\beta)$. Assume*

$$\sum_{k=1}^{n}\sum_{l=1}^{n} r(k-l) \sim Cn^{2H}, \quad a.s. \quad n \to \infty,$$

for some positive constant C. Then $Z_n(t)$ converges weakly to $\sqrt{C}B_t^{(H)}$ in $B_p^{H,\beta}$.

Proof. First we prove that the finite-dimensional distributions of $(Z_n(t), t \in I)$ converge weakly to those of $(\sqrt{C}B_t^{(H)} t \in I)$. Fix any $0 < t_1 < t_2 < \cdots < t_m \leq 1$, we need to show that the distribution of $(Z_n(t_1), Z_n(t_1), ..., Z_n(t_m))$ converges weakly to that of $(\sqrt{C}B_{t_1}^{(H)}, \sqrt{C}B_{t_2}^{(H)}, ..., \sqrt{C}B_{t_m}^{(H)})$. Since they are jointly Gaussian, by considering characteristic functions, it is sufficient to prove that $C_n(t_i, t_j) := E[Z_n(t_i)Z_n(t_j)]$ converges to

$$C(t_i, t_j) := C\frac{1}{2}(t_i^{2H} + t_j^{2H} - |t_i - t_j|^{2H})$$

as $n \to \infty$. Without loss of generality, let us assume $t_j > t_i$. We have

$$C_n(t_i, t_j) = \frac{1}{n^{2H}} \frac{1}{2} \left(E\left[\Big| \sum_{k=0}^{[nt_i]-1} X_k \Big|^2 \right] \right.$$
$$\left. + E\left[\Big| \sum_{k=0}^{[nt_j]-1} X_k \Big|^2 \right] - E\left[\Big| \sum_{k=[nt_i]}^{[nt_j]-1} X_k \Big|^2 \right] \right) \quad (1.7)$$

has the same behavior as

$$\frac{1}{n^{2H}} \frac{1}{2} \{ C[nt_i]^{2H} + C[nt_j]^{2H} - C([nt_j] - [nt_i])^{2H} \}$$

for $n \to \infty$. Hence the limit of (1.7) is

$$C\frac{1}{2}\{t_i^{2H} + t_j^{2H} - |t_i - t_j|^{2H}\} = C(t_i, t_j),$$

as $n \to \infty$. It remains to show that the sequence $(Z_n(t), t \in I)$ is tight in the Banach space $B_p^{H,\beta}$. Let $s, t \in I$ such that $t \geq s$ and $p > 1/H$. Using the stationarity, we see that

$$E[|Z_n(t) - Z_n(s)|^p] = E\left[\left|\frac{1}{n^H} \sum_{k=[ns]}^{[nt]-1} X_k\right|^p\right]$$

$$= E\left[\left|\frac{1}{n^H} \sum_{k=0}^{[nt]-[ns]-1} X_k\right|^p\right]$$

$$= \left|\frac{[nt] - [ns]}{n}\right|^{pH} E\left[\left|\frac{1}{([nt]-[ns])^H} \sum_{k=0}^{[nt]-[ns]-1} X_k\right|^p\right].$$

Note that for n large enough, we have $|([nt] - [ns])/n|^{pH} \leq |t - s|^{pH}$ for all $t, s \in I$. Hence, by Lemma 1.9.4 it suffices to show that there exists a positive constant C such that $E\left[\left|1/n^H \sum_{k=0}^{n-1} X_k\right|^p\right] \leq C$ for all $n \geq 1$, equivalently to show $E[|Z_n(1)|^p] \leq C$ for all $n \geq 1$.

Since $S_n = \sum_{k=0}^{n-1} X_k$ is Gaussian, $E[(S_n)^{2p}]$ is proportional to $(E[(S_n)^2])^p$ for all $p \geq 1$ and $n \geq 1$. By our assumption, $E[(S_n)^2]$ is asymptotically proportional to n^{2H}. Thus,

$$E[|Z_n(1)|^p] \sim \frac{E[|S_n|^p]}{n^{pH}} = \frac{O((E[(S_n)^2])^{p/2})}{n^{pH}} = O(1),$$

as $n \to \infty$. Hence, $\sup_{n \geq 1} E[|Z_n(1)|^p] < \infty$, which proves the theorem. $\square$

Corollary 1.9.6. *Let $H \in (0, 1)$, $\beta > 0$, and $p > \max(1/H, 1/\beta)$. Assume that $\{X_n, n = 1, 2, ...\}$ is a stationary Gaussian sequence with spectral representation*

$$X_n = \int_{-\pi}^{\pi} \exp(in\lambda) |\lambda|^{1/2 - H} B(d\lambda), \quad n = 1, 2, ...,$$

where $B(d\lambda)$ is a Gaussian random measure with $E[B(d\lambda)|^2] = d\lambda$. Then there exists a positive constant C such that $(Z_n(t), t \in [0,1])$ converges weakly to $(CB_t^{(H)}, t \in [0,1])$ in the space $B_p^{H,\beta}$.

Proof. Let $r(k) = E[X_1 X_{k+1}]$ be the covariance function of $\{X_n, n = 1, 2, ...\}$. It suffices to show that

$$\sum_{k=1}^{n} \sum_{l=1}^{n} r(k-l) \sim C^2 n^{2H}.$$

We have

$$n^{-2H}\sum_{k=1}^{n}\sum_{l=1}^{n}r(k-l) = n^{-2H}\sum_{k=1}^{n}\sum_{l=1}^{n}\int_{-\pi}^{\pi}\exp(i(k-l)\lambda)|\lambda|^{1-2H}d\lambda$$

$$= n^{-2H}\int_{-\pi}^{\pi}\left|\sum_{k=1}^{n}\exp(ik\lambda)\right|^{2}|\lambda|^{1-2H}d\lambda$$

$$= n^{-2H}\int_{-\pi}^{\pi}\left|\frac{\exp(i(n+1)\lambda)-\exp(i\lambda)}{(\exp(i\lambda)-1)}\right|^{2}|\lambda|^{1-2H}d\lambda$$

$$= \int_{-\pi}^{\pi}\left|\frac{\exp[i(n+1)\lambda/n]-\exp(i\lambda/n)}{n(\exp(i\lambda/n)-1)}\right|^{2}|\lambda|^{1-2H}d\lambda.$$

Hence $n^{-2H}\sum_{k=1}^{n}\sum_{l=1}^{n}r(k-l)$ tends to

$$\int_{R}\left|\frac{\exp(i\lambda)-1}{i\lambda}\right|^{2}|\lambda|^{1-2H}d\lambda,$$

as $n \to \infty$. Therefore, the proof is complete. □

Part II

Stochastic calculus

2

Wiener and divergence-type integrals for fractional Brownian motion

We start our tour through the different definitions of stochastic integration for *fBm* of Hurst index $H \in (0,1)$ with the *Wiener integrals* since they deal with the simplest case of deterministic integrands. We show how they can be expressed in terms of an integral with respect to the standard Brownian motion, extend their definition also to the case of stochastic integrands, and then proceed to define the stochastic integral by using the *divergence operator*. In both cases we need to distinguish between $H > 1/2$ and $H < 1/2$.

The main references for this chapter are [6], [7], [8], [54], [68], [71], [72], [75], [76].

2.1 Wiener integrals

Here we introduce stochastic integrals with respect to *fBm* by using its Gaussianity. Stochastic integrals of deterministic functions with respect to a Gaussian process were introduced in [171] and are called *Wiener integrals*. In the case of Brownian motion, they coincide with Itô integrals. For *fBm* they were defined for the first time in [76].

Fix an interval $[0,T]$ and let $B^{(H)}(t)$, $t \in [0,T]$, be a *fBm* of Hurst index $H \in (0,1)$ on the probability space $(\Omega, \mathcal{F}^{(H)}, \mathcal{F}_t^{(H)}, \mathbb{P}^H)$ endowed with the natural filtration $(\mathcal{F}_t^{(H)})_{t \in [0,T]}$ and the law $\mathbb{P}^H$ of $B^{(H)}$ (for a construction of the measure $\mathbb{P}^H$ we refer to Chapter 1 and for the case $H > 1/2$, to Section 3.1). Recall that if we define for $s, t > 0$,

$$R_H(t,s) := \frac{1}{2}(s^{2H} + t^{2H} - |t-s|^{2H}), \quad s, t \geq 0,$$

then the covariance $E\left[B^{(H)}(t)B^{(H)}(s)\right] = R_H(t,s)$. By [177], we obtain the following:

1. For $H > 1/2$, the covariance of the fBm can be written as

$$R_H(t,s) = \alpha_H \int_0^t \int_0^s |r-u|^{2H-2}\, du\, dr,$$

where $\alpha_H = H(2H-1)$. We can rewrite

$$|r-u|^{2H-2} = \frac{(ru)^{H-1/2}}{\beta(2-2H, H-1/2)} \qquad (2.1)$$
$$\cdot \int_0^{r\wedge u} v^{1-2H}(r-v)^{H-3/2}(u-v)^{H-3/2}\, dv,$$

where $\beta(\alpha,\gamma) = \Gamma(\alpha+\gamma)/(\Gamma(\beta)\Gamma(\gamma))$ and $\Gamma(\alpha) = \int_0^\infty x^{\alpha-1}e^{-x}dx$ is the Gamma function, since

$$\int_0^u v^{1-2H}(r-v)^{H-3/2}(u-v)^{H-3/2}\, dv$$
$$= (r-u)^{2H-2} \int_{r/u}^\infty (zu-r)^{1-2H} z^{H-3/2}\, dz$$
$$= (ru)^{1/2-H}(r-u)^{2H-2} \int_0^1 (1-x)^{1-2H} x^{H-3/2}\, dx$$
$$= \beta(2-2H, H-1/2)(ru)^{1/2-H}(r-u)^{2H-2},$$

where we have used the change of variable $z = (r-v)/(u-v)$ and $x = r/(uz)$ and supposed $r > u$. Consider now the deterministic kernel

$$K_H(t,s) = c_H s^{1/2-H} \int_s^t (u-s)^{H-3/2} u^{H-1/2}\, du, \qquad (2.2)$$

where $c_H = [H(2H-1)/(\beta(2-2H, H-1/2))]^{1/2}$ and $t > s$. Then we have that

$$R_H(t,s) = \int_0^{t\wedge s} K_H(t,u) K_H(s,u)\, du$$

since by (2.1) it follows that

$$\int_0^{t\wedge s} K_H(t,u) K_H(s,u)\, du$$
$$= c_H^2 \int_0^{t\wedge s} \left(\int_u^t (y-u)^{H-3/2} y^{H-1/2} dy \right)$$
$$\qquad \cdot \left(\int_u^s (z-u)^{H-3/2} z^{H-1/2} dz \right) u^{1-2H}\, du$$
$$= c_H^2 \int_0^t \int_0^s (yz)^{H-1/2} \left(\int_0^{y\wedge z} u^{1-2H}(y-u)^{H-3/2}(z-u)^{H-3/2} du \right) dz\, dy$$
$$= c_H^2 \beta\left(2-2H, H-\frac{1}{2}\right) \int_0^t \int_0^s (y-z)^{2H-2} dz\, dy = R_H(t,s).$$

Note also that with a change of variable in (2.2), $K_H(t,s)$ can be expressed equivalently as

$$K_H(t,s) = c_H(t-s)_+^{H-1/2} \int_0^1 u^{H-3/2}\left(1 - (1-\frac{t}{s})u\right)^{H-1/2} du.$$

See also [72], [76] for further details.

2. For $H < 1/2$, the kernel

$$K_H(t,s) = b_H \left[\left(\frac{t}{s}\right)^{H-1/2}(t-s)^{H-1/2} \right. \\ \left. - \left(H-\frac{1}{2}\right) s^{1/2-H} \int_s^t (u-s)^{H-1/2} u^{H-3/2} du \right] \quad (2.3)$$

with $b_H = \sqrt{2H/((1-2H)\beta(1-2H, H+1/2))}$, and $t > s$ satisfies

$$R_H(t,s) = \int_0^{t \wedge s} K_H(t,u) K_H(s,u)\, du. \quad (2.4)$$

For a detailed proof of equation (2.4), see [76], [189], where it is proved by using the analyticity of both members as functions of the parameter H. See also [177], where a direct proof is given by using the ideas of [172] and the fact that

$$\frac{\partial K_H}{\partial t}(t,s) = c_H(H-1/2)(\frac{t}{s})^{H-1/2}(t-s)^{H-3/2}.$$

3. For $H = 1/2$, we have $K_{1/2}(t,s) = I_{[0,t]}(s)$.

In order to define the Wiener integrals with respect to $B^{(H)}$, we introduce the so-called *reproducing kernel Hilbert space* denoted by $\mathcal{H}$.

Definition 2.1.1. *The **reproducing kernel Hilbert space** (RKHS), denoted by $\mathcal{H}$, associated to $B^{(H)}$ for every $H \in (0,1)$, is defined as the closure of the vector space spanned by the set of functions $\{R_H(t, \cdot), t \in [0,T]\}$ with respect to the scalar product*

$$\langle R_H(t, \cdot), R_H(s, \cdot) \rangle = R_H(t,s) \quad \forall t, s \in [0,T]. \quad (2.5)$$

In the case of standard Brownian motion, there exists a nice characterization of $\mathcal{H}$, which coincides with the space of absolutely continuous functions, vanishing at 0, with square-integrable derivative.

In the case of fBm, it has been proved first in [12] for $H > 1/2$ and then in [75] for every $H \in (0,1)$ that the following holds:

Proposition 2.1.2. *For any $H \in (0,1)$, $\mathcal{H}$ is the set of functions f which can be written as*

$$f(t) = \int_0^t K_H(t,s)\tilde{f}(s)\, ds \qquad (2.6)$$

for some $\tilde{f} \in L^2([0,T])$. By definition, $\|f\|_{\mathcal{H}} = \|\tilde{f}\|_{L^2([0,T])}$.

If $K_H(t,s)$ is of the form (2.2), by [206, p. 187] we have that the integral representation (2.6) induces an isomorphism from $L^2([0,T])$ onto the space $I_{0+}^{H+1/2}(L^2([0,T]))$ introduced in Definition B.1.3. Hence it follows that $\mathcal{H}$ considered as a vector space of functions (and not taking account of its Hilbert space structure and its norm) coincides with the fractional space $I_{0+}^{H+1/2}(L^2([0,T]))$ of functions ψ of the form

$$\psi(x) := \frac{1}{\Gamma(H+1/2)} \int_0^x (x-y)^{H-1/2} f(y)\, dy,$$

for some $f \in L^2([0,T])$. For further details, see also the proofs of [75] and Theorem 3.2 of [76].

Definition 2.1.3. *For any $H \in (0,1)$, the (abstract) Wiener integral with respect to the fBm is defined as the linear extension from $\mathcal{H}$ in $L^2(\mathbb{P}^H)$ of the isometric map $\mathfrak{I}^H$:*

$$\mathfrak{I}^H : \mathcal{H} \longrightarrow L^2(\mathbb{P}^H),$$
$$R_H(t,\cdot) \longmapsto B^{(H)}(t).$$

By Definition 2.1.3, it follows that the abstract Wiener integral with respect to finite combinations of $R_H(t,\cdot)$ is given by

$$\mathfrak{I}^H\left(\sum_{i=1}^n \alpha_i R_H(t_i,\cdot)\right) = \sum_{i=1}^n \alpha_i B^{(H)}(t_i).$$

We consider now a general $u \in \mathcal{H}$. For any $u \in \mathcal{H}$, there exists a sequence $(u_n)_{n \in \mathbb{N}} \subset \mathcal{H}$ such that every u_n is a finite linear combination of functions of the type $R_H(t_i,\cdot)$ that converges to u in $\mathcal{H}$. Hence we can define the *abstract Wiener integral* of $u \in \mathcal{H}$ with respect to $B^{(H)}$ as

$$\mathfrak{I}^H(u) = \lim_{n \to \infty} \mathfrak{I}^H(u_n),$$

where the limit is taken in $L^2(\mathbb{P}^H)$.

If we apply this construction to the standard Brownian motion, it follows by the definition of the RKHS that the space of admissible integrands is given by the deterministic functions which are continuous and whose first derivative is square integrable on $[0,T]$. This characterization determines uniquely the space of integrands in the standard Brownian motion case. This is a consequence of the fact that the properties of Wiener integrals don't change if $\mathcal{H}$ is replaced by an isometrically isomorphic space. Since I_{0+}^1 is a bijective isometry from $L^2([0,T])$ to $\mathcal{H}$, they are usually identified in the construction of the Wiener integral for the Brownian motion. In order to obtain a similar characterization of the space of integrands even in the case of the fBm, we replace $\mathcal{H}$ by an isometrically isomorphic Hilbert space.

Definition 2.1.4. *By a representation of $\mathcal{H}$ we mean a pair $(\mathfrak{F}, i)$ composed of a functional space $\mathfrak{F}$ and a bijective isometry i between $\mathfrak{F}$ and $\mathcal{H}$.*

By Proposition 2.1.2, we immediately get the following:

Theorem 2.1.5. *There exists a canonical isometric bijection between $L^2([0,T])$ and $\mathcal{H}$ given by*

$$i_1 : L^2([0,T]) \longrightarrow \mathcal{H},$$
$$h \longmapsto f(t) = \int_0^t K_H(t,s) h(s) \, ds,$$

where $\mathcal{H}$ is endowed with the scalar product defined in (2.5) and $L^2([0,T])$ with the usual inner product. Hence $(L^2([0,T]), i_1)$ is a representation of $\mathcal{H}$.

According to this representation, for $H = 1/2$ we have $i_1 = I_{0+}^1$, i.e., $i_1(h) = \int_0^t h(s) \, ds$. In general, for any $H \in (0,1)$, $K_H(t, \cdot)$ is associated by i_1 to $R_H(t, \cdot)$.

2.1.1 Wiener integrals for $H > 1/2$

We now focus on the case $H > 1/2$. From now on we denote by $\mathcal{E}$ the space of step functions on $[0,T]$. Another representation of $\mathcal{H}$ is given by

Theorem 2.1.6. *For any $H > 1/2$, consider $L^2([0,T])$ equipped with the twisted scalar product:*

$$\langle f, g \rangle_H := H(2H-1) \int_0^T \int_0^T f(s) g(t) |s-t|^{2H-2} \, ds \, dt. \qquad (2.7)$$

Define the linear map i_2 on the space $\mathcal{E}$ of step functions on $[0,T]$ by

$$i_2 : (L^2([0,T]), <,>_H) \longrightarrow \mathcal{H},$$
$$I_{[0,t]} \longmapsto R_H(t, \cdot).$$

Then the extension of this map to the closure of $(L^2([0,T]), <,>_H)$ with respect to the scalar product defined in (2.7) is a representation of $\mathcal{H}$.

Clearly, we have $\overline{(L^2([0,T]),<,>_H)} = \overline{(\mathcal{E},<,>_H)}$. In the sequel we still denote with i_2 even the extended map. In [188], it is proved that the space $(L^2([0,T]),<,>_H)$ is not complete; hence one needs to take its closure to obtain a Hilbert space. Moreover, in [189] and [190] it is shown that the elements of $\mathcal{H}_2 := \overline{(L^2([0,T]),<,>_H)}$ may not be functions but distributions of negative order.

To summarize, the Wiener integrals with respect to $B^{(H)}$ with $H > 1/2$ can be seen as the extensions of the following isometries:

1. *Wiener integrals of first type*:

$$\mathcal{I}_1^H : L^2([0,T]) \longrightarrow L^2(\Omega, \mathbb{P}^H), \qquad (2.8)$$
$$K_H(t,\cdot) \mapsto B^{(H)}(t),$$

2. *Wiener integrals of second type*:

$$\mathcal{I}_2^H : \overline{(L^2([0,T]),<,>_H)} \longrightarrow L^2(\mathbb{P}^H), \qquad (2.9)$$
$$I_{[0,t]}(\cdot) \mapsto B^{(H)}(t),$$

induced by the representations i_1 and i_2, respectively. So either one keeps the original scalar product on $L^2([0,T])$ and changes the pre-image of $B^{(H)}(t)$ to $K_H(t,\cdot)$ or one changes the scalar product on $L^2([0,T])$ so that $I_{[0,t]}(\cdot)$ remains the predecessor of $B^{(H)}(t)$. This is the main point where the situation for the *fBm* differs from the standard Brownian motion case. Moreover, the integrals of first type (2.8) are not consistent with the isometry property required by the abstract scheme of Wiener integrals since

$$E\left[|\mathcal{I}_1^H(I_{[0,t]})|^2\right] = \|I_{[0,t]}\|^2_{L^2([0,T])} = t \neq R_H(t,t).$$

The process $\mathcal{I}_1^H(I_{[0,t]})$ is a centered Gaussian process with covariance kernel equal to $\min(t,s)$, hence it coincides with a standard Brownian motion. We provide in the following a connection between the two types of Wiener integrals.

Consider the operator K_H induced by the kernel $K_H(t,s)$ on $L^2([0,T])$ for $H \geq 1/2$ as follows:

$$(K_H h)(t) := \int_0^t K_H(t,s) h(s)\, ds.$$

Let $\mathfrak{K}_H^*$ be the adjoint operator of K_H in $L^2([0,T])$, i.e.,

$$\int_0^T (K_H f)(s) g(s)\, ds = \int_0^T f(s) (\mathfrak{K}_H^* g)(s)\, ds \qquad (2.10)$$

for every $f, g \in L^2([0,T])$. By Fubini theorem we obtain that

$$(\mathfrak{K}_H^* g)(t) = \int_t^T K_H(t,s) g(s)\, ds. \qquad (2.11)$$

Since by exploiting fractional calculus (see Appendix B and also [206]) we can rewrite the action of K_H as

$$(K_H f)(t) = c_H \Gamma(h - \frac{1}{2}) I_{0+}^1 x^{H-1/2} I_{0+}^{H-1/2}(x^{1/2-H} f),$$

we deduce that

$$(\mathfrak{K}_H^* g)(t) = c_H \Gamma(h - \frac{1}{2}) x^{1/2-H} I_{T-}^{H-1/2} x^{H-1/2} I_{T-}^1, \qquad (2.12)$$

for $f \in L^2([0,T])$. From a formal point of view, one can think that the two operator are linked by the relation

$$K_H(s,t) = \mathfrak{K}_H^*(\delta_t)(s),$$

where δ_t is the Dirac measure with mass at t. Since $I_{T-}^1 \delta_t = I_{[0,t]}$, we have

$$K_H(t,s) = c_H \Gamma(h - \frac{1}{2}) s^{1/2-H} (I_{T-}^{H-1/2} x^{1/2-H} I_{[0,t]})(s). \qquad (2.13)$$

By using the characterization (2.13) we obtain another representation for $\mathcal{H}$,

$$i_3 : L^2([0,T]) \longrightarrow \mathcal{H},$$
$$h \longmapsto f(t) = c_H \Gamma(h - \frac{1}{2}) t^{1/2-H} (I_{T-}^{H-1/2} x^{1/2-H} h)(t),$$

that is valid even for $H < 1/2$ and can be considered as the "dual" representation of i_1 (see [190]). The dual space of $\mathcal{H}$ contains the linear combinations of Dirac masses as a dense subspace; hence, i_3 is actually an isometrically representation of the dual space of $\mathcal{H}$. However, since a representation is up to isomorphisms, i_3 can also be considered as a representation of $\mathcal{H}$.

By (2.12) we obtain the following relation between Wiener integrals of first and second type.

Theorem 2.1.7. *Let $H > 1/2$. For any function $u \in L^2([0,T])$, we have*

$$\mathfrak{I}_1^H(\mathfrak{K}_H^*(\mathfrak{K}_{1/2}^*)^{-1} u) = \mathfrak{I}_2^H(u), \qquad (2.14)$$

where $\mathfrak{K}_H^*$ is defined by (2.10).

Proof. Equation (2.14) is immediately verified for indicator functions $I_{[0,t]}$ by the definition of $\mathfrak{K}_H^*$. The result then follows by a limiting procedure. See [72] for further details. □

In the sequel we focus on Wiener integrals of second type induced by the isometry (2.9). For the sake of simplicity, from now on *we identify the* RKHS $\mathcal{H}$ *with* $\mathcal{H}_2 = \overline{(L^2([0,T]), <,>_H)}$ *through the representation map* i_2, i.e., we put $\mathcal{H} = \mathcal{H}_2$. Note that the map (2.9) induced by i_2 that associates $I_{[0,t]}$ to

$B^{(H)}(t)$ extended to $\mathcal{H}$ is an isometry between $\mathcal{H}$ and the chaos of first order associated with $B^{(H)}$, i.e., the closed subspace of $L^2(\mathbb{P}^H)$ generated by $B^{(H)}$. Since from now on we don't need to distinguish anymore between Wiener integrals of first and second type, we adopt the notation

$$B^{(H)}(\psi) := \mathfrak{I}_2^H(\psi), \quad \psi \in \mathcal{H}.$$

In order to characterize Wiener integrals of second type, by following the approach of [7] and [177], we now introduce the linear operator K_H^* defined on $\psi \in \mathcal{E}$ as follows:

$$(K_H^*\psi)(s) := \int_s^T \psi(t)\frac{\partial K_H}{\partial t}(t,s)\,dt. \tag{2.15}$$

Then

$$(K_H^* I_{[0,t]})(s) = K_H(t,s)I_{[0,t]}(s). \tag{2.16}$$

By equation (2.16) it follows that the operator K_H^* is an isometry between the space $\mathcal{E}$ of elementary functions and $L^2([0,T])$ that can be extended to the Hilbert space $\mathcal{H}$. This is because

$$\langle K_H^* I_{[0,t]}, K_H^* I_{[0,s]} \rangle_{L^2([0,T])} = \langle K_H(t,\cdot) I_{[0,t]}, K_H(s,\cdot) I_{[0,s]} \rangle_{L^2([0,T])}$$
$$= \int_0^{t\wedge s} K_H(t,u) K_H(s,u)\, du$$
$$= R_H(t,s) = \langle I_{[0,t]}, I_{[0,s]} \rangle_\mathcal{H}.$$

The operator K_H^* can be rewritten by using the means of fractional calculus (see Appendix B). Note that by the representation (2.2) for the square-integrable kernel $K_H(t,s)$ we get

$$\frac{\partial K_H}{\partial t}(t,s) = c_H \left(\frac{t}{s}\right)^{H-1/2} (t-s)^{H-3/2}. \tag{2.17}$$

Hence by equations (2.15) and (2.17) and by the definition of the fractional integral (B.1) of Appendix B with $\alpha = H - 1/2$ and $b = T$, we obtain immediately the following fractional representation for K_H^*:

$$(K_H^*)(\psi) = c_H \Gamma(H - \frac{1}{2}) s^{1/2-H} (I_{T-}^{H-1/2} u^{H-1/2} \psi(u))(s). \tag{2.18}$$

Moreover, by using the following relation between the fractional integral and the fractional derivative

$$D_{T-}^{H-1/2}(I_{T-}^{H-1/2}(\psi)) = \psi$$

for every $\psi \in L^1(0,T)$ [see (B.3) of Appendix B], we also have that

$$(K_H^*)^{-1}(\psi) = \frac{1}{c_H \Gamma(H-1/2)} s^{1/2-H}(D_{T-}^{H-1/2} u^{H-1/2}\psi(u))(s). \qquad (2.19)$$

In particular, we obtain that the indicator function $I_{[0,a]}$ belongs to the image of K_H^* for $a \in [0,T]$ because putting $\psi = I_{[0,a]}$ in (2.19) and using the characterization (B.2) of fractional derivative, we have

$$(K_H^*)^{-1}(I_{[0,a]}) = \frac{1}{c_H \Gamma(H-1/2)} s^{1/2-H}(D_{a-}^{H-1/2} u^{H-1/2})(s) I_{[0,a]}(s).$$

As a consequence of this result, we obtain that the image of the operator K_H^* coincides with $L^2([0,T])$, i.e.,

$$\mathcal{H} = (K_H^*)^{-1}(L^2([0,T])). \qquad (2.20)$$

Remark 2.1.8. The operator K_H^* defined in (2.15) is the adjoint of K_H in the following sense:

Lemma 2.1.9. *For any function $\psi \in \mathcal{E}$ and $h \in L^2([0,T])$ we have*

$$\int_0^T (K_H^* \psi)(t) h(t)\, dt = \int_0^T \psi(t)(K_H h)(dt).$$

Proof. For the proof, we refer to Lemma 1 in [6]. □

Note that the relation between $\mathfrak{K}_H^*$ introduced in (2.11) and K_H^* is then

$$K_H^* = \mathfrak{K}_H^* \circ \mathfrak{K}_{1/2}^*.$$

This follows directly by comparing (2.12) and (2.18) since

$$\mathfrak{K}_{1/2}^* = (I_{0+}^1)^* = I_{T-}^1.$$

Consider now the process $B(t)$ that is associated by the representation i_2 to $(K_H^*)^{-1}(I_{[0,t]})$, i.e.,

$$B(t) := B^{(H)}((K_H^*)^{-1} I_{[0,t]}). \qquad (2.21)$$

Since $B(t)$ is a continuous Gaussian process with covariance given by

$$\begin{aligned}
E[B(t)B(s)] &= E\left[B^{(H)}((K_H^*)^{-1}(I_{[0,t]})) B^{(H)}((K_H^*)^{-1}(I_{[0,s]}))\right] \\
&= \langle (K_H^*)^{-1} I_{[0,t]}, (K_H^*)^{-1} I_{[0,s]} \rangle_{\mathcal{H}} \\
&= \langle I_{[0,t]}, I_{[0,s]} \rangle_{L^2([0,T])} \\
&= s \wedge t,
\end{aligned}$$

we conclude that $B(t)$ is a standard Brownian motion. Analogously, the stochastic process associated to

$$K_H^* I_{[0,t]} = K_H(t,s) I_{[0,t]}(s)$$

by the isometry induced by $B(t)$ on $L^2([0,T])$ is a fBm $B^{(H)}(t)$ with integral representation

$$B^{(H)}(t) = \int_0^T K_H^* I_{[0,t]} dB(s) = \int_0^t K_H(t,s)\, dB(s). \qquad (2.22)$$

Remark 2.1.10. The representation (2.22) holds in law, and it is shown in [76] that it holds in trajectorial sense with a fixed standard Brownian motion constructed on $(\Omega, \mathcal{F}^{(H)}, \mathbb{P}^H)$. This characterization is also quite useful to find a numerical simulation of the fBm paths. By [76] we obtain the following:

Theorem 2.1.11. *Let π_n be an increasing sequence of partitions of $[0,T]$ such that the mesh size $|\pi_n|$ of π_n tends to zero as n goes to infinity. The sequence of processes $(W^n)_{n \in \mathbb{N}}$ defined by*

$$W^n(t) = \sum_{t_i^{(n)} \in \pi_n} \frac{1}{t_{i+1}^{(n)} - t_i^{(n)}} \int_{t_i^{(n)}}^{t_{i+1}^{(n)}} K_H(t,s)\, ds \left[B(t_{i+1}^{(n)}) - B(t_i^{(n)}) \right]$$

converges to $B^{(H)}$ in $L^2(\mathbb{P} \otimes ds)$, where here $\mathbb{P}$ denotes the probability measure induced by the standard Brownian motion B.

Proof. For the proof, we refer to Proposition 3.1 of [76]. □

For further results about discrete approximation for fBm for $H > 1/2$, see also [30], [140] and [165].

By (2.21) and (2.22), it follows that $B(t)$ and $B^{(H)}(t)$ generate the same filtration. Moreover, we obtain an expression of the Wiener integral of second type with respect to $B^{(H)}$ in terms of an integral with respect to the Brownian motion B.

Proposition 2.1.12. *Let $H > 1/2$. If $\psi \in \mathcal{H}$, then*

$$B^{(H)}(\psi) = \mathcal{I}_2^H(\psi) = \int_0^T (K_H^* \psi)(s) dB(s). \qquad (2.23)$$

Since by (2.18) the Hilbert space $\mathcal{H}$ coincides with the space of distributions ψ such that $s^{1/2-H}(I_{t-}^{H-1/2} u^{H-1/2} \psi)(s)$ is a square-integrable function, the integral representation in (2.23) is correctly defined for $\psi \in \mathcal{H}$.

In order to obtain a space of functions contained in $\mathcal{H}$, we consider the linear space $|\mathcal{H}|$ generated by the measurable functions ψ such that

$$\|\psi\|_{|\mathcal{H}|}^2 := \alpha_H \int_0^T \int_0^T |\psi(s)||\psi(t)||s-t|^{2H-2}\, ds\, dt < \infty, \qquad (2.24)$$

where $\alpha_H = H(2H-1)$. The space $|\mathcal{H}|$ is a Banach space with the norm $\|\cdot\|_{|\mathcal{H}|}^2$, and $\mathcal{E}$ is dense in $|\mathcal{H}|$. In [189] it is proved the space $|\mathcal{H}|$ is not complete equipped with the scalar product $\langle \cdot, \cdot \rangle_H$.

We discuss now the relation between $|\mathcal{H}|$, $\mathcal{H}$ and $L^p([0,T]), p \geq 1$.

Proposition 2.1.13. *Let $\psi \in |\mathcal{H}|$. Then*

$$\|\psi\|_{|\mathcal{H}|} \leq \beta_H \|\psi\|_{L^{1/H}([0,T])}$$

for some constant $\beta_H > 0$.

Proof. Here we recall briefly the proof of [161] by following [8]. If one applies the Hölder inequality to

$$\int_0^T \int_0^T |\psi(u)||\psi(r)||r-u|^{2H-2}\, dr\, du$$

with $q = 1/H$, then

$$\|\psi\|_{|\mathcal{H}|}^2 \leq \alpha_H \left(\int_0^T |\psi(r)|^{1/H} dr\right)^H$$

$$\cdot \left\{\int_0^T \left[\int_0^T |\psi(u)||\psi(r)||r-u|^{2H-2} du\right]^{1/(1-H)} dr\right\}^{1-H}.$$

By exploiting fractional calculus (see Appendix B, and also [206]) we have that

$$\|I_{0+}^{2H-1}(\psi)\|_{L^{1/(1-H)}([0,T])}$$

$$= \left\{\int_0^T \left[\int_0^T |\psi(u)||\psi(r)||r-u|^{2H-2} du\right]^{1/(1-H)} dr\right\}^{1-H}.$$

The thesis follows by the Hardy–Littlewood inequality (see [217])

$$\|I_{0+}^{\alpha}(\psi)\|_{L^q(0,\infty)} \leq c_{H,p}\|\psi\|_{L^p(0,\infty)},$$

applied to the particular case when $\alpha = 2H-1$, $q = 1/(1-H)$ and $p = H$. □

As a consequence the following inclusions hold

$$L^2([0,T]) \subset L^{1/H}([0,T]) \subset |\mathcal{H}| \subset \mathcal{H}.$$

The inclusion $L^2([0,T]) \subset |\mathcal{H}|$ can also be seen directly since

$$\int_0^T \int_0^T |\psi(u)||\psi(r)||r-u|^{2H-2}\, dr\, du$$

$$\leq \int_0^T \int_0^T |\psi(u)|^2 |r-u|^{2H-2}\, dr\, du$$

$$\leq \frac{T^{2H-1}}{H-1/2}\int_0^T |\psi(u)|^2\, du.$$

As we have seen, Wiener integrals are introduced for deterministic integrands. In order to extend the definition of the Wiener integral of second type to the general case of *stochastic integrands*, we follow the approach of [72] and use Theorem 2.1.7 to give the following definition.

Definition 2.1.14. *Consider $H > 1/2$. Let u be a stochastic process $u.(\omega)$: $[0,T] \longrightarrow \mathcal{H}$ such that $K_H^* u$ is Skorohod integrable with respect to the standard Brownian motion $B(t)$. Then we define the extended Wiener integral of u with respect to the fBm $B^{(H)}$ as*

$$B^{(H)}(u) := \int_0^T (K_H^* u)(s) \delta B(s),$$

where the integral on the right-hand side must be interpreted as a Skorohod integral with respect to $B(t)$ (Definition A.2.1).

Definition 2.1.14 is an extension of (2.23) to the a case of a stochastic process u seen as a random variable with values in $\mathcal{H}$ and such that $K_H^* u$ is Skorohod integrable. Note that we have used the same symbol for the standard and the extended Wiener integral.

2.1.2 Wiener integrals for $H < 1/2$

We now consider the case when the Hurst index H belongs to the interval $(0, 1/2)$. Main references for this section are [5], [6], [54] and [177].

As for $H > 1/2$, we focus on the following representation for the RKHS $\mathcal{H}$. Consider the space $\mathcal{E}$ of step functions on $[0,T]$ endowed with the inner product
$$\langle I_{[0,t]}, I_{[0,s]} \rangle_H := R_H(t,s), \quad 0 \leq t, s \leq T, \tag{2.25}$$
and the linear map i_2 on $\mathcal{E}$ given by
$$i_2 : (\mathcal{E}, <,>_H) \longrightarrow \mathcal{H},$$
$$I_{[0,t]} \longmapsto R_H(t, \cdot).$$

Then the extension of this map to the closure of $(\mathcal{E}, <,>_H)$ with respect to the scalar product defined in (2.7) is a representation of $\mathcal{H}$. From now on we identify $\mathcal{H} = \overline{(\mathcal{E}, <,>_H)}$, and for $H < 1/2$ we define the Wiener integral for $\psi \in \mathcal{H}$ as the extension of the isometry

$$B^{(H)} : \overline{(\mathcal{E}, <,>_H)} \longrightarrow L^2(\mathbb{P}^H),$$
$$I_{[0,t]}(\cdot) \longmapsto B^{(H)}(t),$$

induced by the representation i_2.

Remark 2.1.15. Note that the use of the same notation for the inner product (2.25) and the twisted product (2.7) is justified since

$$H(2H-1)\int_0^T \int_0^T I_{[0,t]}(u) I_{[0,s]}(v) |u-u|^{2H-2}\, du\, dv = R_H(t,s)$$

for $H > 1/2$.

We now derive also for the case $H < 1/2$ a representation of $B^{(H)}(\psi), \psi \in \mathcal{H}$, in terms of a stochastic integral with respect to a standard Brownian motion that is analogous to the one of Proposition 2.1.12.

For $H < 1/2$, we recall that by equation (2.3) the covariance of the *fBm* is generated by the kernel

$$K_H(t,s) = b_H \left[\left(\frac{t}{s}\right)^{H-1/2} (t-s)^{H-1/2} \right.$$
$$\left. - \left(H - \frac{1}{2}\right) s^{1/2-H} \int_s^t (u-s)^{H-1/2} u^{H-3/2}\, du \right]$$

that can be written in terms of fractional derivatives as

$$K_H(t,s) = b_H \Gamma\left(H + \frac{1}{2}\right) s^{1/2-H} \left(D_{t-}^{1/2-H} u^{H-1/2}\right)(s).$$

Consider the linear operator K_H^* from the space $\mathcal{E}$ of step functions on $[0,T]$ to $L^2([0,T])$ defined by

$$(K_H^*\psi)(s) := K_H(T,s)\psi(s) + \int_s^T (\psi(t) - \psi(s)) \frac{\partial K_H}{\partial t}(t,s)\, dt. \qquad (2.26)$$

It is immediate to verify that (2.26) evaluated for $\psi = I_{[0,t]}$ gives

$$(K_H^* I_{[0,t]})(s) = K_H(t,s) I_{[0,t]}(s).$$

Consequently, we can rewrite the covariance $R_H(t,s)$ as

$$R_H(t,s) = \int_0^{t \wedge s} K_H(t,u) K_H(s,u)\, du$$
$$= \int_0^T K_H^* I_{[0,t]}(u) K_H^* I_{[0,s]}(u)\, du.$$

Hence, the linear operator K_H^* induces an isometry between $\mathcal{E}$ and $L^2([0,T])$ that can be extended to the Hilbert space $\mathcal{H}$ as in the case $H > 1/2$. Since the following equality holds

$$(D_{t_-}^{1/2-H}u^{H-1/2})I_{[0,t]}(s) = (D_{T_-}^{1/2-H}u^{H-1/2}I_{[0,t]})(s),$$

by equation (2.26) we obtain that

$$(K_H^*\psi)(s) = b_H\Gamma(H+\frac{1}{2})s^{1/2-H}(D_{T_-}^{1/2-H}u^{H-1/2}\psi(u))(s).$$

Using (B.3) from Appendix B, we obtain

$$(K_H^*)^{-1}(f)(s) = \frac{1}{b_H\Gamma(H+1/2)}s^{1/2-H}(I_{T_-}^{1/2-H}u^{H-1/2}f(u))(s).$$

By [76] and by Proposition 8 of [6] we obtain, rewriting the kernel K_H as

$$K_H(s,t) = b_H(t-s)^{H-1/2} + s^{H-1/2}F_1(\frac{t}{s}),$$

where

$$F_1(z) = b_H(\frac{1}{2}-H)\int_0^{z-1}u^{H-3/2}(1-(u+1)^{H-1/2})\,du,$$

that one can prove for $H < 1/2$ (see Proposition 8 of [6] and [76])

$$\mathcal{H} = (K_H^*)^{-1}(L^2([0,T])) = I_{T_-}^{1/2-H}(L^2([0,T])). \quad (2.27)$$

In addition (2.27) guarantees that the inner product space $\mathcal{H}$ is complete endowed with

$$\langle f,g\rangle_H = \int_0^T K_H^*f(s)K_H^*g(s)\,ds,$$

as shown in Lemma 5.6 of [188]. Note also that by Definition B.2.1, the scalar product in $\mathcal{H}$ can be written in the simpler form

$$\langle f,g\rangle_H = e_H^2\langle D_-^{1/2-H}f, D_+^{1/2-H}g\rangle_{L^2(\mathbb{R})}, \quad (2.28)$$

with $e_H = C_1(H)\Gamma(1/2+H)$, $f,g \in \mathcal{H}$. Here we consider $f(s) = g(s) = 0$ if $s \notin [0,T]$.

Remark 2.1.16. We emphasize once more that the inner product space $\mathcal{H}$ is *complete* for $H < 1/2$ and *incomplete* if $H > 1/2$. By [189] we obtain that this difference in completeness is a consequence of the following two facts:

1. For $H < 1/2$ the equation

$$s^{1/2-H}(D_{T_-}^{1/2-H}u^{H-1/2}\psi(u))(s) = f(s)$$

has a solution $\psi(s) = s^{H-1/2}(I_{T_-}^{H-1/2}u^{H-1/2}f(u))(s)$ for every $f \in L^2([0,T])$.

2. For $H > 1/2$ there exist functions $f \in L^2([0,T])$ for which the equation

$$s^{1/2-H}(I_{T-}^{H-1/2}u^{H-1/2}\psi(u))(s) = f(s) \qquad (2.29)$$

cannot be solved. In fact, since $I_{T-}^{H-1/2}$ is an *integral* operator, the left-hand side of (2.29) will satisfy some smoothness conditions that may not hold for a general $f \in L^2([0,T])$.

Note also that for the space of Hölder continuous functions of order γ in the interval $[0,T]$ (for a definition, see also the proof of Theorem 1.6.1), it holds

$$C^\gamma([0,T]) \subset \mathcal{H}$$

if $\gamma > 1/2 - H$. As in the case $H > 1/2$, the process

$$B(t) = B^{(H)}((K_H^*)^{-1}(I_{[0,t]}))$$

is a Wiener process and the fractional Brownian has the integral representation

$$B^{(H)}(t) = \int_0^t K_H(t,s)\,dB(s). \qquad (2.30)$$

Hence we can conclude with the following

Proposition 2.1.17. *For $H < 1/2$ the Wiener-type integral $B^{(H)}(\psi)$ with respect to fBm can be defined for functions $\psi \in \mathcal{H} = I_{T-}^{1/2-H}(L^2([0,T]))$ and the following holds:*

$$B^{(H)}(\psi) = \int_0^T (K_H^*\psi)(t)\,dB(t).$$

2.2 Divergence-type integrals for *fBm*

We analyze now the properties and the main results concerning a stochastic integral for *fBm* introduced as dual operator of the stochastic derivative. We also investigate its relations with Wiener integrals and the ones defined in Chapter 3. Main references for this part are [6], [8], [170] and [177].

Consider $H \in (0,1)$ and $\mathcal{H} = \overline{(\mathcal{E}, <,>_H)}$. Let S_H be the set of smooth cylindrical random variables of the form

$$F = f(B^{(H)}(\psi_1), \ldots, B^{(H)}(\psi_n)),$$

where $n \geq 1$, $f \in C_b^\infty(\mathbb{R}^n)$, and $\psi_i \in \mathcal{H}$. The *derivative operator* $D^{(H)}$ of $F \in S_H$ is defined as the $\mathcal{H}$-valued random variable

$$D^{(H)}F = \sum_{i=1}^n \frac{\partial f}{\partial x_i}(B^{(H)}(\psi_1), \ldots, B^{(H)}(\psi_n))\psi_i. \qquad (2.31)$$

The derivative operator $D^{(H)}$ is then a closable unbounded operator from $L^p(\Omega, \mathbb{P}^H)$ in $L^p(\Omega; \mathcal{H})$ for any $p \geq 1$. We denote by $D^{(H),k}$ the iteration of the derivative operator. The iterate derivative operator $D^{(H),k}$ maps $L^p(\Omega, \mathbb{P}^H)$ into $L^p(\Omega, \mathcal{H}^{\otimes k})$.

Definition 2.2.1. *For any $k \in \mathbb{N}$ and $p \geq 1$ we denote by $\mathbb{D}_H^{k,p}$ the Sobolev space generated by the closure of S_H with respect to the norm*

$$\|F\|_{k,p}^p = E\left[|F|^p\right] + \sum_{i=1}^{k} E\left[\|(D^{(H)})^j F\|_{\mathcal{H}^{\otimes j}}^p\right]$$

and by $\mathbb{D}^{k,p}(\mathcal{H})$ the corresponding Sobolev space of $\mathcal{H}$-valued random variables.

We introduce the adjoint operator of the derivative.

Definition 2.2.2. *We say that a random variable $u \in L^2(\Omega; \mathcal{H})$ belongs to the domain $\operatorname{dom} \delta_H$ of the divergence operator if*

$$|E\left[\langle D^{(H)} F, u \rangle_H\right]| \leq c_u \|F\|_{L^2(\mathbb{P}^H)}$$

for any $F \in S_H$.

Definition 2.2.3. *Let $u \in \operatorname{dom} \delta_H$. Then $\delta_H(u)$ is the element in $L^2(\mathbb{P}^H)$ defined by the duality relationship*

$$E[F \delta_H(u)] = E\left[\langle D^{(H)} F, u \rangle_H\right]$$

for any $F \in \mathbb{D}_H^{1,2}$.

Hence the *divergence operator* δ_H is the adjoint of the derivative operator $D^{(H)}$. Note that by Definition 2.2.3 we obtain immediately that the space $\mathbb{D}^{1,2}(\mathcal{H})$ of $\mathcal{H}$-valued random variables is included in $\operatorname{dom} \delta_H$ and for $u \in \mathbb{D}^{1,2}(\mathcal{H})$ the following holds:

$$E\left[\delta_H(u)^2\right] \leq E\left[\|u\|_{\mathcal{H}}^2\right] + E\left[\|D^{(H)} u\|_{\mathcal{H} \otimes \mathcal{H}}^2\right].$$

By the Meyer inequalities (see, for example, [176]), we also get for all $p > 1$ that

$$\|\delta_H(u)\|_{L^p(\mathbb{P}^H)} \leq c_p \|u\|_{\mathbb{D}^{1,p}(\mathcal{H})}.$$

If u is a simple $\mathcal{H}$-valued random variable of the form

$$u = \sum_{j=1}^{n} F_j X_j,$$

where $F_j \in \mathbb{D}_H^{1,2}$ and $X_j \in \mathcal{H}$, then u belongs to $\operatorname{dom} \delta_H$ and by Definition 2.2.3 we obtain

$$\delta_H(u) = \sum_{j=1}^n F_j \delta_H(X_j) - \langle D^{(H)} F_j, X_j \rangle_H. \qquad (2.32)$$

Moreover, if $F \in \mathbb{D}_H^{1,2}$ and $u \in \operatorname{dom} \delta_H$ are such that Fu, and $F\delta_H(u) + \langle D^{(H)}F, u \rangle_H$ are square integrable, then $Fu \in \operatorname{dom} \delta_H$ and (2.32) extends to

$$\delta_H(Fu) = F\delta_H(u) - \langle D^{(H)}F, u \rangle_H. \qquad (2.33)$$

We now investigate separately the cases $H > 1/2$ and $H < 1/2$. We focus first on the case $H > 1/2$; for $H < 1/2$ the definition of the divergence is more delicate and will be studied in Section 2.2.2.

2.2.1 Divergence-type integral for $H > 1/2$

In the particular case of the Brownian motion, the divergence operator is an extension of the Itô integral in the sense that the set of square-integrable adapted processes is included in $\operatorname{dom} \delta_H$ and the divergence operator restricted to this set coincides with the Itô stochastic integral. More precisely, in the case of the Brownian motion, the divergence operator coincides with the Skorohod integral introduced in [213] (for a survey, see Section A.2). For further details for the Brownian motion case, we refer also to [179].

The same relation holds for the *fBm* for $H > 1/2$, i.e., the divergence operator coincides with the generalized Wiener integral introduced in Definition 2.1.14. To show this, we proceed as follows. First of all, by Definition 2.2.3 we immediately obtain that

$$\delta_H(\psi) = B^{(H)}(\psi)$$

and, in particular,

$$\delta_H\left(\sum_{i=1}^n a_i I_{[t_i, t_{i+1}]}\right) = \sum_{i=1}^n a_i (B^{(H)}(t_{i+1}) - B^{(H)}(t_i)) \qquad (2.34)$$

for $0 \leq t_1 \leq t_2 \leq \cdots \leq t_{n+1} \leq T$. Then we study the relation between the derivatives and the divergence operators defined with respect to $B^{(H)}$ and to B, respectively. By (2.20) we have that

$$\mathcal{H} = (K_H^*)^{-1}(L^2(0, T)).$$

Hence, it also follows that

$$\mathbb{D}_H^{1,2} = (K_H^*)^{-1}(\mathbb{L}^{1,2}), \qquad (2.35)$$

where $\mathbb{L}^{1,2} := \mathbb{D}^{1,2}(L^2([0, T]))$. Moreover, by [6] we obtain the following relation between the derivative $D^{(H)}$ with respect to $B^{(H)}$ and the Malliavin derivative D with respect to the standard Brownian motion B.

Proposition 2.2.4. For any $F \in \mathbb{D}_H^{1,2}$, we have
$$K_H^* D^{(H)} F = DF. \tag{2.36}$$

Proof. By following [6], let
$$F = f(B^{(H)}(t)) \tag{2.37}$$
for $f \in C^1$, and let $u \in \mathcal{H}$. Then
$$\begin{aligned}
E\left[\langle u, D^{(H)} F\rangle_H\right] &= E\left[\langle u, D^{(H)} f(B^{(H)}(t))\rangle_H\right] \\
&= E\left[\langle K_H^* u, f'(B^{(H)}(t)) K_H^* I_{[0,t]}\rangle_{L^2([0,T])}\right] \\
&= E\left[\langle K_H^* u, f'(B^{(H)}(t)) K_H(t,\cdot) I_{[0,t]}\rangle_{L^2([0,T])}\right] \\
&= E\left[\langle K_H^* u, Df(B^{(H)}(t))\rangle_{L^2([0,T])}\right].
\end{aligned}$$

Since
$$E\left[\langle u, D^{(H)} F\rangle_H\right] = E\left[\langle K_H^* u, K_H^* D^{(H)} F\rangle_{L^2([0,T])}\right],$$
then (2.36) holds for random variables of the form (2.37). The result extends to every $F \in \mathbb{D}_H^{1,2}$ by a density argument. □

Hence, if $F \in \mathbb{D}_H^{1,2}$, we have
$$E\left[\langle u, D^{(H)} F\rangle_H\right] = E\left[\langle K_H^* u, DF\rangle_{L^2([0,T])}\right]$$
for any $u \in \mathcal{H}$ and the equality $K_H^* D^{(H)} F = DF$ holds. This implies that
$$\operatorname{dom} \delta_H = (K_H^*)^{-1}(\operatorname{dom} \delta),$$
where $\delta = \delta_{1/2}$ denotes the divergence operator with respect to the standard Brownian motion B. Hence, for any $\mathcal{H}$-valued random variable $u \in \operatorname{dom} \delta_H$ it holds
$$\delta_H(u) = \delta(K_H^* u) = \int_0^T K_H^* u(s) \delta B(s), \tag{2.38}$$
where the integral on the right-hand side must be interpreted as the Skorohod integral with respect to the standard Brownian motion (Definition A.2.1). Hence, we have proved the following:

Proposition 2.2.5. *Let $u \in \operatorname{dom} \delta_H$. Then $\delta_H(u)$ coincides with the extended Wiener integral of u (Definition 2.1.14), i.e.,*
$$\delta_H(u) = B^{(H)}(u).$$

Note that by (2.35) and (2.38) we obtain that $K_H^*(\mathbb{L}^{1,2})$ is included in the domain $\operatorname{dom} \delta_H$.

2.2.2 Divergence-type integral for $H < 1/2$

In this section we describe the approach of [54], [177] in order to introduce the divergence-type integral for $H < 1/2$. The standard divergence integral of $B^{(H)}$ with respect to itself does not exist if $H < 1/4$ because the paths becomes too irregular. In [54] the standard divergence operator is extended by a change of the order of integration in the duality relationship that defines the divergence operator as the adjoint of the Malliavin derivative. For the extended divergence operator, a Fubini theorem and an Itô formula hold with any $H \in (0, 1/2)$ (see Chapter 6). In this approach, the integral of divergence type is always a random variable, while in Chapter 4 the stochastic integral is defined as an Hida distribution. Here we reformulate the results of [54] for processes u_t with t in $[0, T]$ for the sake of homogeneity with the previous sections. In [54] the following results hold for processes $(u_t)_{t \in \mathbb{R}}$ defined on the whole real line. Recall that $\mathcal{H} = \overline{(\mathcal{E}, <,>_H)}$.

Proposition 2.2.6. *Consider $H < 1/2$. Let $0 < a < b < T$ and set*

$$u(t) = B^{(H)}(t) I_{(a,b]}(t), \quad t \in \mathbb{R}.$$

Then

$$\mathbb{P}^H(u \in \mathcal{H}) = 1, \quad H \in \left(\frac{1}{4}, \frac{1}{2}\right)$$

and

$$\mathbb{P}^H(u \in \mathcal{H}) = 0, \quad H \in \left(0, \frac{1}{4}\right].$$

Proof. Here we sketch for $(u_t)_{t \in [0,T]}$ the proof provided in Proposition 3.2 of [54] for processes defined on the whole real line. Let $H \in (1/4, 1/2)$. By Kolmogorov's Continuity Theorem there exists a measurable set $\tilde{\Omega} \subset \Omega$ with $\mathbb{P}^H(\tilde{\Omega}) = 1$ such that for all $\omega \in \tilde{\Omega}$ there exists constant $\tilde{C}(\omega)$ such that

$$\sup_{t \in (a,b]} |B^{(H)}(t, \omega)| \leq \tilde{C}(\omega)$$

and

$$\sup_{t,s \in (a,b], t \neq s} \frac{|B^{(H)}(t, \omega) - B^{(H)}(s, \omega)|}{|t-s|^{1/4}} \leq \tilde{C}(\omega).$$

Fix an $\omega \in \Omega$ and set

$$\psi(t) := u_t(\omega) = B^{(H)}(t) I_{(a,b]}(t), \quad t \in \mathbb{R},$$

and

$$\hat{C} := \frac{\alpha}{\Gamma(1-\alpha)} \tilde{C}(\omega).$$

Let $\epsilon > 0$. For $t \in (0, a]$, consider

$$D^\alpha_{+,\epsilon} \psi(t) = 0.$$

For the definition of $D^\alpha_{+,\epsilon}$ we refer to the Appendix B. For $t \in (a, b]$,

$$|D^\alpha_{+,\epsilon}\psi(t)| \leq \frac{\alpha}{\Gamma(1-\alpha)} \left(I_{\{t-a>\epsilon\}} \int_\epsilon^{t-a} \left|\frac{\psi(t) - \psi(t-s)}{s^{1+\alpha}}\right| ds \right.$$
$$\left. + |\psi(t)| \int_{(t-a)\vee\epsilon}^T s^{-1-\alpha} ds \right)$$
$$\leq \hat{C} \left(I_{\{t-a>\epsilon\}} \int_\epsilon^{t-a} s^{-3/4-\alpha} ds + \int_{(t-a)\vee\epsilon}^T s^{-1-\alpha} ds \right)$$
$$\leq \hat{C} \left[\frac{1}{1/4-\alpha}(t-a)^{1/4-\alpha} + \frac{1}{\alpha}(t-a)^{-\alpha} \right].$$

For $t \in (b, T)$,

$$|D^\alpha_{+,\epsilon}\psi(t)| \leq \frac{\alpha}{\Gamma(1-\alpha)} \int_{t-b}^{t-a} \frac{|\psi(t-s)|}{s^{1+\alpha}} ds$$
$$\leq \hat{C} \int_{t-b}^{t-a} s^{-1-\alpha} ds = \frac{\hat{C}}{\alpha} \left((t-b)^{-\alpha} - (t-a)^{-\alpha} \right).$$

Hence, for all $\epsilon > 0$ and for all $t \in [0, T]$, we have

$$|D^\alpha_{+,\epsilon}\psi(t)| \leq \bar{\psi}(t),$$

where

$$\bar{\psi}(t) = \begin{cases} 0 & \text{if } t \in (0, a), \\ C[(t-a)^{1/4-\alpha} + (t-a)^{-\alpha}] & \text{if } t \in (a, b], \\ C[(t-b)^{-\alpha} - (t-a)^{-\alpha}] & \text{if } t \in (b, T], \end{cases}$$

with $C = \hat{C}\left[1/(1/4-\alpha) \vee 1/\alpha\right]$. Since $\bar{\psi} \in L^2([0, T])$, by Theorem 6.2 of [206] it follows that $\psi \in \mathcal{H}$.

We now consider the case $H \in (0, 1/4]$. The process

$$\tilde{B}^{(H)}(t) := B^{(H)}(t+a) - B^{(H)}(a), \quad t \in \mathbb{R},$$

is also a *fBm* with Hurst parameter H. Since it is self-similar, for all $t \in (0, b-a)$, the random variable

$$t^{-2H} \int_0^{b-a-t} [\tilde{B}^{(H)}(s+t) - \tilde{B}^{(H)}(s)]^2 ds$$

has the same distribution as

$$\int_0^{b-a-t} [\tilde{B}^{(H)}(\tfrac{s}{t}+1) - \tilde{B}^{(H)}(\tfrac{s}{t})]^2 ds$$
$$= t \int_0^{(b-a)/t-1} [\tilde{B}^{(H)}(x+1) - \tilde{B}^{(H)}(x)]^2 dx \qquad (2.39)$$
$$= \frac{b-a-t}{(b-a)/t-1} \int_0^{(b-a)/t-1} [\tilde{B}^{(H)}(x+1) - \tilde{B}^{(H)}(x)]^2 dx$$

The process $(\tilde{B}^{(H)}(x+1) - \tilde{B}^{(H)}(x))_{x\geq 0}$ is stationary and mixing. Therefore, it follows from the ergodic theorem that (2.39) converges to

$$(b-a)E\left[[\tilde{B}^{(H)}(1)]^2\right] > 0$$

in L^1 as $t \to 0$. Hence, it follows that there exists a measurable set $\tilde{\Omega} \subset \Omega$ with $\mathbb{P}^H(\tilde{\Omega}) = 1$ and a sequence of positive numbers $(t_k)_{k\in\mathbb{N}}$ that converges to 0 such that for all $\omega \in \tilde{\Omega}$ and $k \in \mathbb{N}$,

$$\int_0^T [u_{s+t_k}(\omega) - u_s(\omega)]^2 ds \geq \int_a^{b-t_k} [\tilde{B}^{(H)}_{s+t_k}(\omega) - \tilde{B}^{(H)}_s(\omega)]^2 ds$$
$$= \int_0^{b-a-t_k} [\tilde{B}^{(H)}_{s+t_k}(\omega) - \tilde{B}^{(H)}_s(\omega)]^2 ds$$
$$\geq \frac{b-a}{2} E\left[(\tilde{B}^{(H)}_1)^2\right] t_k^{2H}. \qquad (2.40)$$

Assume that there exists an $\omega \in \tilde{\Omega}$ such that $u(\omega) \in \mathcal{H}$. By (6.40) of [206], the function $u(\omega)$ has the property

$$\int_0^T [u_{s+t}(\omega) - u_s(\omega)]^2 ds = o(t^2\alpha) \quad \text{as } t \to 0.$$

But $u(\omega)$ can only satisfy both (2.39) and (2.40) at the same time if $H > \alpha = 1/2 - H$, which contradicts $H \leq 1/4$. Therefore, $u(\omega) \notin \mathcal{H}$ for all $\omega \in \tilde{\Omega}$. This concludes the proof. $\square$

By Proposition 2.2.6 it follows that processes of the form

$$B^{(H)}(t)I_{(a,b]}(t), \quad t \in \mathbb{R},$$

cannot be in $\operatorname{dom} \delta_H$ (Definition 2.2.2) if $H \leq 1/4$. Hence, we consider now the following extension of the divergence δ_H to an operator whose domain also contains processes with paths that are not in $\mathcal{H}$ by using the approach of [54] and [177].

By (2.28) and the fractional integration by parts formula (B.5), for any $f, g \in \mathcal{H}$ we obtain that

$$\langle f, g \rangle_H = \langle K_H^* f, K_H^* g \rangle_{L^2([0,T])}$$
$$= d_H^2 \langle s^{1/2-H} D_{T-}^{1/2-H} s^{H-1/2} f, s^{1/2-H} D_{T-}^{1/2-H} s^{H-1/2} g \rangle_{L^2([0,T])}$$
$$= d_H^2 \langle f, s^{H-1/2} s^{1/2-H} D_{0+}^{1/2-H} (s^{1-2H} D_{T-}^{1/2-H} s^{H-1/2} g) \rangle_{L^2([0,T])},$$

where for $H \leq 1/2$ the operator K_H^* is introduced in (2.26). This implies that the adjoint $K_H^{*,a}$ of the operator K_H^* in $L^2([0,T])$ is

$$(K_H^{*,a} f)(s) = d_H s^{1/2-H} D_{0+}^{1/2-H} (s^{1-2H} D_{T-}^{1/2-H} s^{H-1/2} f).$$

In order to introduce an extended domain for the divergence operator δ_H as in the approach of [54], we introduce the space

$$\mathcal{K} := (K_H^*)^{-1}(K_H^{*,a})^{-1}[L^2([0,T])].$$

Let $S_{\mathcal{K}}$ the space of smooth cylindrical random variables of the form

$$F = f(B^{(H)}(\psi_1), \ldots, B^{(H)}(\psi_n)),$$

where $n \geq 1$, $f \in C_b^\infty(\mathbb{R}^n)$, i.e., f is bounded with smooth bounded partial derivatives, and $\psi_i \in \mathcal{K}$.

Definition 2.2.7. Let $u(t), t \in [0,T]$ be a measurable process such that $E\left[\int_0^T u^2(t)\,dt\right] < \infty$. We say that $u \in \operatorname{dom}^* \delta_H$ if there exists a random variable $\delta_H(u) \in L^2(\mathbb{P}^H)$ such that for all $F \in S_{\mathcal{K}}$ we have

$$\int_0^T E\left[u(t) K_H^{*,a} K_H^* D_t^{(H)} F\right] dt = E\left[\delta_H(u) F\right].$$

Note that if $u \in \operatorname{dom}^* \delta_H$, then $\delta_H(u)$ is unique and the mapping

$$\delta_H : \operatorname{dom}^* \delta_H \to \cup_{p>1} L^p(\mathbb{P}^H)$$

is linear. This extended domain satisfies the following natural requirements:

1. $\operatorname{dom} \delta_H \subset \operatorname{dom}^* \delta_H$, and δ_H restricted to $\operatorname{dom} \delta_H$ coincides with the divergence operator (Proposition 3.5 of [54]).
2. In particular, $\operatorname{dom} \delta_H = \operatorname{dom}^* \delta_H \cap [\cup_{p>1} L^p(\Omega; \mathcal{H})]$ (Proposition 3.5 of [54]).
3. If $u \in \operatorname{dom}^* \delta_H$ such that $E[u] \in L^2([0,T])$, then $E[u]$ belongs to $\mathcal{H}$ (Proposition 3.6 of [54]).
4. If u is a deterministic process, then $u \in \operatorname{dom}^* \delta_H$ if and only if $u \in \mathcal{H}$.
5. The extended divergence operator δ_H is closed in the following sense. Let $p \in (1, \infty]$ and $q \in (2/(1+2H), \infty]$. Let $(u^k)_{k\geq 1}$ be a sequence in $\operatorname{dom}^* \delta_H \cap L^p(\Omega, L^q([0,T]))$ and $u \in L^p(\Omega, L^q([0,T]))$ such that

$$\lim_{k\to\infty} u^k = u \quad \text{in} \quad L^p(\Omega, L^q([0,T])).$$

It follows that for all $n \in \mathbb{N}_0$ and $F \in S_H$, we have

$$\lim_{k\to\infty} u^k(t) K_H^{*,a} K_H^* D^{(H)} F = u(t) K_H^{*,a} K_H^* D^{(H)} F$$

in $L^1(\Omega \times [0,T])$. If there exists a $\hat{p} \in (1, \infty]$ and an $X \in L^{\hat{p}}(\Omega)$ such that

$$\lim_{k\to\infty} \delta_H(u^k) = X \quad \text{in} \quad L^{\hat{p}}(\Omega),$$

then $u \in \operatorname{dom}^* \delta_H$ and $\delta_H(u) = X$.

By Theorem 3.7 of [54] we also get the following theorem:

Theorem 2.2.8 (Fubini Theorem). *Let $(Y, \mathcal{Y}, \mu)$ be a measure space and $u = u(\omega, t, y) \in L^0(\Omega \times [0, T] \times Y)$ such that*

1. *For almost $y \in Y$, $u(\cdot, \cdot, y) \in \text{dom}^* \delta_H$.*
2. *For almost all $(\omega, t) \in \Omega \times [0, T]$, $u(\omega, t, \cdot) \in L^1(Y)$ and $\int_Y |u(\omega, t, y)| \, d\mu(y) \in L^2(\Omega \times [0, T])$.*
3. *For almost all $\omega \in \Omega$, $\delta_H(u)(\omega) \in L^1(Y)$ and $\int_Y \delta_H(u) \, d\mu(y) \in L^2(\Omega)$.*

Then $\int_Y u(\omega, t, y) d\mu(y) \in \text{dom}^* \delta_H$ and

$$\delta_H \left[\int_Y u(\omega, t, y) \, d\mu(y) \right] = \int_Y \delta_H(u) \, d\mu(y).$$

3

Fractional Wick Itô Skorohod (fWIS) integrals for *fBm* of Hurst index $H > 1/2$

In this chapter we introduce the definition of stochastic integral with respect to the *fBm* for Hurst index $1/2 < H < 1$ by using the white noise analysis method. At this purpose we define the *fractional white noise* and stochastic integral as an element in the *fractional Hida distribution space*.

To obtain a classical Itô formula, we need the stochastic integral to be an ordinary random variable. Hence the ϕ-derivative is introduced to handle the existence of the Wick product in L^2. Classical Itô type formulas are obtained and applications are discussed. The main references for this chapter are [32], [83] and [121].

3.1 Fractional white noise

Fix H with $1/2 < H < 1$. We put

$$\phi(s,t) = \phi_H(s,t) = H(2H-1)|s-t|^{2H-2}, \quad s,t \in \mathbb{R}, \tag{3.1}$$

and recall that for $s, t > 0$,

$$\int_0^t \int_0^s \phi(u,v)\,du\,dv = \frac{1}{2}(s^{2H} + t^{2H} - |t-s|^{2H}) = R_H(t,s). \tag{3.2}$$

Let $\mathcal{S}(\mathbb{R})$ be the Schwartz space of rapidly decreasing smooth functions on $\mathbb{R}$, and if $f \in \mathcal{S}(\mathbb{R})$, denote

$$\|f\|_H^2 := \int_\mathbb{R} \int_\mathbb{R} f(s)f(t)\phi(s,t)\,ds\,dt < \infty. \tag{3.3}$$

If we equip $\mathcal{S}(\mathbb{R})$ with the inner product

$$\langle f, g \rangle_H := \int_\mathbb{R} \int_\mathbb{R} f(s)g(t)\phi(s,t)\,ds\,dt, \quad f, g \in \mathcal{S}(\mathbb{R}), \tag{3.4}$$

then the completion of $\mathcal{S}(\mathbb{R})$, denoted by $L^2_\phi(\mathbb{R})$, becomes a separable Hilbert space. Similarly, we can define $L^2_\phi(\mathbb{R}_+)$ or $L^2_\phi([0,T])$ on a finite interval. We remark that elements of $L^2_\phi(\mathbb{R})$ may be distributions (see, for example, [190]). For further comments we refer to Chapter 2. From now on we denote by $L^2_H(\mathbb{R})$ the subspace of deterministic functions contained in $L^2_\phi(\mathbb{R})$.

Remark 3.1.1. We remark that in (3.3) and (3.4) we use the same notation as in Theorem 2.1.6, since the inner product (3.4) extends (2.7) to the case of functions defined on the whole real axis. Hence we start here with an analogous setting to the one for Wiener integrals in Chapter 2, but we provide a different construction of the stochastic integral for *fBm* for $H > 1/2$. Moreover, by (2.7), (3.1), and (3.4) we obtain that $L^2_\phi([0,T]) = \mathcal{H} = \overline{(L^2([0,T]), <,>_H)}$ and $L^2_\phi(\mathbb{R}) \supseteq L^2_\phi([0,T]) = \mathcal{H}$ if we identify $\psi \in L^2_\phi([0,T])$ with $\psi I_{[0,T]}$.

In particular we obtain the following representation of $L^2_\phi(\mathbb{R})$.

Lemma 3.1.2. *Let*

$$I_-^{H-1/2} f(u) = c_H \int_u^\infty (t-u)^{H-3/2} f(t)\, dt,$$

where $c_H = \sqrt{H(2H-1)\Gamma(3/2-H)/(\Gamma(H-1/2)\Gamma(2-2H))}$, *and* Γ *denotes the gamma function. Then* $I_-^{H-1/2}$ *is an isometry from* $L^2_\phi(\mathbb{R})$ *to* $L^2(\mathbb{R})$.

Proof. By a limiting argument, we may assume that f and g are continuous with compact support. By definition,

$$\langle I_-^{H-1/2}(f), I_-^{H-1/2}(g) \rangle_{L^2(\mathbb{R})}$$
$$= c_H^2 \int_\mathbb{R} \left\{ \int_u^\infty (s-u)^{H-3/2} f(s)\, ds \int_u^\infty (t-u)^{H-3/2} g(t)\, dt \right\} du$$
$$= c_H^2 \int_{\mathbb{R}^2} f(s) g(t) \left\{ \int_{-\infty}^{s \wedge t} (s-u)^{H-3/2} (t-u)^{H-3/2}\, du \right\} ds\, dt$$
$$= \int_\mathbb{R} \int_\mathbb{R} f(s) g(t) \phi(s,t)\, ds\, dt = (f,g)_H,$$

where we have used the identity (see [103, p. 404])

$$c_H^2 \int_{-\infty}^{s \wedge t} (s-u)^{H-3/2}(t-u)^{H-3/2} du = \phi(s,t).$$

$\square$

Now let $\Omega = \mathcal{S}'(\mathbb{R})$ be the dual of $\mathcal{S}(\mathbb{R})$ (considered as Schwartz space), i.e., Ω is the space of *tempered distributions* on $\mathbb{R}$. The map $f \mapsto \exp(-1/2\|f\|_H^2)$, with $f \in \mathcal{S}(\mathbb{R})$, is positive definite on $\mathcal{S}(\mathbb{R})$, and by the Bochner–Minlos theorem (see [144] or [109]) there exists a probability measure $\mathbb{P}^H$ on the Borel subsets $\mathcal{B}(\Omega)$ of Ω such that

$$\int_\Omega e^{i\langle\omega,f\rangle}d\mathbb{P}^H(\omega) = e^{-1/2\|f\|_H^2} \quad \forall f \in \mathcal{S}(\mathbb{R}), \tag{3.5}$$

where $\langle\omega,f\rangle$ denotes the usual pairing between $\omega \in \mathcal{S}'(\mathbb{R})$ and $f \in \mathcal{S}(\mathbb{R})$ and $\|f\|_H$ is defined in (3.3). It follows from (3.5) that

$$E\left[\langle\cdot,f\rangle\right] = 0 \quad \text{and} \quad E\left[\langle\cdot,f\rangle^2\right] = \|f\|_H^2, \tag{3.6}$$

where E denotes the expectation under the probability measure $\mathbb{P}^H$. Using this we see that we may define

$$\tilde{B}_t^{(H)} = \tilde{B}^{(H)}(t,\omega) = \langle\omega, I_{[0,t]}(\cdot)\rangle$$

as an element of $L^2(\mathbb{P}^H)$ for each $t \in \mathbb{R}$, where

$$I_{[0,t]}(s) = \begin{cases} 1 & \text{if } 0 \leq s \leq t, \\ -1 & \text{if } t \leq s \leq 0, \text{ except } t = s = 0, \\ 0 & \text{otherwise.} \end{cases}$$

By Kolmogorov's continuity theorem $\tilde{B}_t^{(H)}$ has a t-continuous version, which we will denote by $B_t^{(H)}$, $t \in \mathbb{R}$. From (3.6) we see that $B_t^{(H)}$ is a Gaussian process with

$$E\left[B_t^{(H)}\right] = 0$$

and

$$E\left[B_s^{(H)}B_t^{(H)}\right] = \frac{1}{2}\left(|t|^{2H} + |s|^{2H} - |t-s|^{2H}\right).$$

It follows that $B_t^{(H)}$ is a *fBm*. From now on we endow Ω with the natural filtration $\mathcal{F}_t^{(H)}$ of $B^{(H)}$.

The stochastic integral with respect to *fBm* for deterministic function is easily defined (see also Section 2.1).

Lemma 3.1.3. *If f, g belong to $L_H^2(\mathbb{R})$, then $\int_\mathbb{R} f_s\, dB_s^{(H)}$ and $\int_\mathbb{R} g_s\, dB_s^{(H)}$ are well-defined zero mean, Gaussian random variables with variances $\|f\|_H^2$ and $\|g\|_H^2$, respectively, and*

$$E\left[\int_\mathbb{R} f_s\, dB_s^{(H)} \int_\mathbb{R} g_s\, dB_t^{(H)}\right] = \int_\mathbb{R}\int_\mathbb{R} f(s)g(t)\phi(s,t)\, ds\, dt = \langle f,g\rangle_H.$$

Proof. This lemma is verified in [103]. It can be proved directly by verifying it for simple functions $\sum_{i=1}^n a_i I_{[t_i,t_{i+1}]}(s)$ and then proceeding with a passage to the limit. □

Let $L^p(\mathbb{P}^H) = L^p$ be the space of all random variables $F: \Omega \to \mathbb{R}$ such that

$$\|F\|_{L^p(\mathbb{P}^H)} = E\left[|F|^p\right]^{1/p} < \infty.$$

For any $f \in L_H^2(\mathbb{R})$, define $\varepsilon : L_H^2(\mathbb{R}) \to L^1(\mathbb{P}^H)$ as

$$\varepsilon(f) := \exp\left(\int_{\mathbb{R}} f_t \, dB_t^{(H)} - \frac{1}{2} \int_{\mathbb{R}} \int_{\mathbb{R}} f_s f_t \phi(s,t) \, ds \, dt\right) \\ = \exp\left(\int_{\mathbb{R}} f_t \, dB_t^{(H)} - \frac{1}{2} \|f\|_H^2\right). \quad (3.7)$$

If $f \in L_H^2(\mathbb{R})$, then $\varepsilon(f) \in L^p(\mathbb{P}^H)$ for each $p \geq 1$ and $\varepsilon(f)$ is called an exponential functional (e.g., [160]). Let $\mathcal{E}$ be the linear span of the exponentials, that is,

$$\mathcal{E} = \left\{\sum_{k=1}^n a_k \varepsilon(f_k) : n \in \mathbb{N},\ a_k \in \mathbb{R},\ f_k \in L_\phi^2(\mathbb{R}) \text{ for } k \in \{1,\ldots,n\}\right\}.$$

Theorem 3.1.4. $\mathcal{E}$ *is a dense set of $L^p(\mathbb{P}^H)$ for each $p \geq 1$. In particular, $\mathcal{E}$ is a dense set of $L^2(\mathbb{P}^H)$.*

Proof. A functional $F : \Omega \to \mathbb{R}$ is said to be a polynomial of the fBm if there is a polynomial $p(x_1, x_2, \ldots, x_n)$ such that

$$F = p(B_{t_1}^{(H)}, B_{t_2}^{(H)}, \ldots, B_{t_n}^{(H)})$$

for some $0 \leq t_1 < t_2 < \cdots < t_n$. Since $(B_t^{(H)}, t \geq 0)$ is a Gaussian process, it is well known that the set of all polynomial fractional Brownian functionals is dense in $L^p(\mathbb{P}^H)$ for $p \geq 1$. In this case, the denseness of the polynomials follows from the continuity of the process and the Stone–Weierstrass theorem. To prove the theorem it is only necessary to prove that any polynomial can be approximated by the elements in $\mathcal{E}$. Since the Wick product of exponentials is still an exponential, it is easy to see that it is only necessary to show that for any $t > 0$, $B_t^{(H)}$ can be approximated by elements in $\mathcal{E}$.

Let $f_\delta(s) = I_{[0,t]}(s)\, \delta$, $\delta > 0$. Clearly f_δ is in $L_H^2(\mathbb{R})$. Then $\varepsilon(f_\delta) = c(\delta) e^{\delta B_t^{(H)}}$ for some positive constant $c(\delta)$. It is easy to see that

$$F_\delta = \frac{\varepsilon(f_\delta) - c(\delta)}{c(\delta)\delta} = \frac{e^{\delta B_t^{(H)}} - 1}{\delta}$$

is in $\mathcal{E}$. If $\delta \to 0$, then $F_\delta \to B_t^{(H)}$ in $L^p(\mathbb{P}^H)$ for each $p \geq 1$. This completes the proof. $\square$

The following theorem is also interesting.

Theorem 3.1.5. *If $f_1, f_2, \ldots, f_n$ are elements in $L_H^2(\mathbb{R})$ such that $\|f_i - f_j\|_H \neq 0$ for $i \neq j$, then $\varepsilon(f_1), \varepsilon(f_2), \ldots, \varepsilon(f_n)$ are linearly independent in $L^2(\mathbb{P}^H)$.*

Proof. This theorem is known to be true if the *fBm* is replaced by a standard Brownian motion, (e.g., [160]).

Let $f_1, f_2, \ldots, f_k$ be distinct elements in $L_H^2(\mathbb{R})$. Let $\lambda_1, \lambda_2, \ldots, \lambda_k$ be real numbers such that

$$\|\lambda_1 \varepsilon(f_1) + \lambda_2 \varepsilon(f_2) + \cdots + \lambda_k \varepsilon(f_k)\|_{L^2(\mathbb{P}^H)} = 0.$$

Thus for any $g \in L_H^2(\mathbb{R})$,

$$E\left[(\lambda_1 \varepsilon(f_1) + \lambda_2 \varepsilon(f_2) + \cdots + \lambda_k \varepsilon(f_k))\, \varepsilon(g)\right] = 0.$$

By an elementary computation for Gaussian random variables it follows that

$$\lambda_1 e^{\langle f_1, g \rangle_H} + \lambda_2 e^{\langle f_2, g \rangle_H} + \cdots + \lambda_k e^{\langle f_k, g \rangle_H} = 0.$$

Replace g by δg for $\delta \in \mathbb{R}$ to obtain

$$\lambda_1 e^{\delta \langle f_1, g \rangle_H} + \lambda_2 e^{\delta \langle f_2, g \rangle_H} + \cdots + \lambda_k e^{\delta \langle f_k, g \rangle_H} = 0.$$

Expand the above identity in the powers of δ and compare the coefficients of δ^p, for $p \in \{0, 1, \ldots, k-1\}$ to obtain the family of equations

$$\lambda_1 \langle f_1, g \rangle_H^p + \lambda_2 \langle f_2, g \rangle_H^p + \cdots + \lambda_k \langle f_k, g \rangle_H^p = 0$$

for $p = 0, 1, \ldots, k-1$. This is a linear system of k equations and k unknowns. By the Vandermonde formula, the determinant of this linear system is

$$\det\left(\langle f_i, g \rangle_H^p\right) = \prod_{i<j} \langle f_i - f_j, g \rangle_H^p.$$

For every pair (i,j) with $i \neq j$, the set $\{g \in L_H^2(\mathbb{R}) : \langle f_i - f_j, g \rangle_H \neq 0\}$ is the complement of a hyperplane in $L_H^2(\mathbb{R})$. Since the intersection of finitely many complements of hyperplanes in $L_H^2(\mathbb{R})$ is not empty, there is a $g \in L_H^2(\mathbb{R})$ such that $\langle f_i - f_j, g \rangle_H \neq 0$ for all pairs i and j such that $i \neq j$. Thus $\lambda_1 = \lambda_2 \cdots = \lambda_k = 0$. This proves the theorem. □

In the following we let

$$h_n(x) = (-1)^n e^{x^2/2} \frac{d^n}{dx^n}\left(e^{-x^2/2}\right), \quad n = 0, 1, 2, \ldots, \tag{3.8}$$

be the *Hermite polynomials*. For further details on Hermite polynomials, we refer to Appendix A.

Lemma 3.1.6. *There is an orthonormal basis $\{e_i\}_{i=1}^{\infty}$ of $L_\phi^2(\mathbb{R})$ such that for any $t \in \mathbb{R}$ there exists $C_t < \infty$ such that*

$$\left|\int_{\mathbb{R}} e_n(s)\phi(s,t)\,ds\right| < C_t n^{1/6} \tag{3.9}$$

Proof. Define the Hermite functions as in (A.6),

$$\xi_n(x) = \pi^{-1/4}((n-1)!)^{-1/2} h_{n-1}(\sqrt{2}x) e^{-x^2/2}, \quad n = 1, 2, \ldots.$$

Then from [223], $\{\xi_n(x), n = 1, 2, \ldots\}$ is an orthonormal basis of $L^2(\mathbb{R})$ and

$$|\xi_n(x)| \leq \begin{cases} C n^{-1/12} & \text{when } |x| \leq 2\sqrt{n}, \\ C e^{-\gamma x^2} & \text{when } |x| > 2\sqrt{n} \end{cases}$$

where γ and C are certain positive constants, independent of n. (See, for example, [219] or [223, p. 26, Lemma 1.5.1]). Set

$$e_n(u) = (I_-^{H-1/2})^{-1}(\xi_n)(u). \tag{3.10}$$

Then by Lemma 3.1.2, $\{e_n, n = 1, 2, \ldots\}$ is an orthonormal basis of $L_\phi^2(\mathbb{R})$. We also have

$$\int_{\mathbb{R}} e_n(s) \phi(s,t) \, ds = c_H^2 \int_{\mathbb{R}} e_n(s) \int_{-\infty}^{s \wedge t} (s-u)^{H-3/2} (t-u)^{H-3/2} \, du \, ds$$

$$= c_H^2 \int_{-\infty}^{t} (t-u)^{H-3/2} du \int_{u}^{\infty} (s-u)^{H-3/2} e_n(s) \, ds$$

$$= c_H \int_{-\infty}^{t} (t-u)^{H-3/2} \Gamma_\phi(e_n)(u) \, du$$

$$= c_H \int_{-\infty}^{t} (t-v)^{H-3/2} \xi_n(v) \, dv$$

$$\leq C_t \left[\int_{v \leq 2\sqrt{n}} |v-t|^{H-3/2} n^{-1/12} \, dv + \int_{|v| > 2\sqrt{n}} |v-t|^{H-3/2} e^{-\gamma |v|^2} \, dv \right]$$

$$\leq C_t n^{1/6}.$$

This proves the lemma. □

From now on we let $\{e_n\}_{n=1}^\infty$ be the orthonormal basis of $L_\phi^2(\mathbb{R})$ defined in (3.10). Then the e_i's are smooth. Moreover, we see that

$$t \to \int_{\mathbb{R}} e_i(s) \phi(s,t) \, ds \quad \text{is continuous for each } i. \tag{3.11}$$

Let $\mathcal{J} = \left(\mathbb{N}_0^{\mathbb{N}}\right)_c$ denote the set of all (finite) multi-indices $\alpha = (\alpha_1, \ldots, \alpha_m)$ of nonnegative integers ($\mathbb{N}$ is the set of natural numbers and $\mathbb{N}_0 = \mathbb{N} \cup \{0\}$). Then if $\alpha = (\alpha_1, \ldots, \alpha_m) \in \mathcal{J}$, we put

$$\widetilde{\mathcal{H}}_\alpha(\omega) := h_{\alpha_1}(\langle \omega, e_1 \rangle) \cdots h_{\alpha_m}(\langle \omega, e_m \rangle). \tag{3.12}$$

In particular, if we let $\varepsilon^{(i)} := (0,\ldots,0,1,0,\ldots,0)$ denote the ith unit vector, then we get
$$\widetilde{\mathcal{H}}_{\varepsilon^{(i)}}(\omega) = h_1(\langle \omega, e_i \rangle) = \langle \omega, e_i \rangle.$$

Remark 3.1.7. Note that $\widetilde{\mathcal{H}}_\alpha(\omega)$ and $\mathcal{H}_\alpha(\omega)$ introduced in (A.8) are defined on different probability spaces, but they have the same law.

The following result is a fractional Wiener Itô chaos expansion theorem. It is well-known in a more general context (see, e.g., Theorem 2.6 of [131]).

Theorem 3.1.8. *Let $F \in L^2(\mathbb{P}^H)$. Then there exist constants $c_\alpha \in \mathbb{R}$, $\alpha \in \mathcal{J}$, such that*
$$F(\omega) = \sum_{\alpha \in \mathcal{J}} c_\alpha \widetilde{\mathcal{H}}_\alpha(\omega), \tag{3.13}$$

where the convergence holds in $L^2(\mathbb{P}^H)$. Moreover,
$$\|F\|^2_{L^2(\mathbb{P}^H)} = \sum_{\alpha \in \mathcal{J}} \alpha! c_\alpha^2,$$

where $\alpha! = \alpha_1! \alpha_2! \cdots \alpha_m!$ if $\alpha = (\alpha_1, \ldots, \alpha_m) \in \mathcal{J}$.

Proof. The following proof is standard. For the reader's convenience we repeat it in the *fBm* framework. Consider
$$\mathcal{E}(a_k e_k) = \exp\left(a_k \langle \omega, e_k \rangle - \frac{1}{2} a_k^2\right) = \sum_{n=0}^{\infty} \frac{a_k^n}{n!} h_n(\langle \omega, e_k \rangle), \tag{3.14}$$

where $a_k \in \mathbb{R}$, $k = 1, 2, \ldots$. If $f \in L^2_H(\mathbb{R})$ has the expansion $f = \sum_{k=1}^{\infty} a_k e_k$, then
$$\mathcal{E}(f) = \exp\left(\sum_{k=1}^{\infty} a_k \langle \omega, e_k \rangle - \frac{1}{2} \sum_{k=1}^{\infty} a_k^2\right)$$
$$= \lim_{N \to \infty} \prod_{k=1}^{\infty} \left(\sum_{n=0}^{N} \frac{a_k^n}{n!} h_n(\langle \omega, e_k \rangle)\right)$$
$$= \lim_{N \to \infty} \sum_{\alpha \in \mathcal{J}^{(N)}} \prod_{k=1}^{\infty} \frac{a_k^{\alpha_k}}{\alpha_k!} h_{\alpha_k}(\langle \omega, e_k \rangle)$$
$$= \lim_{N \to \infty} \sum_{\alpha \in \mathcal{J}^{(N)}} c_\alpha \widetilde{\mathcal{H}}_\alpha(\omega) \quad \text{(limit in } L^2(\mathbb{P}^H)\text{)}, \tag{3.15}$$

where $\mathcal{J}^{(N)}$ denotes the set of all multi-indices $\alpha = (\alpha_1, \ldots, \alpha_m)$ of nonnegative integers with $\alpha_i \leq N$ and we have put
$$c_\alpha = \prod_{k=1}^{\infty} \frac{a_k^{\alpha_k}}{\alpha_k!} \quad \text{if} \quad \alpha = (\alpha_1, \ldots, \alpha_m).$$

If we combine Theorem 3.1.4 with (3.15), we obtain that the linear span of $\{\widetilde{\mathcal{H}}_\alpha\}_{\alpha \in \mathcal{J}}$ is dense in $L^2(\mathbb{P}^H)$.

It remains to prove that

$$E\left[\widetilde{\mathcal{H}}_\alpha \widetilde{\mathcal{H}}_\beta\right] = 0 \quad \text{if} \quad \alpha \neq \beta \tag{3.16}$$

and

$$E\left[\widetilde{\mathcal{H}}_\alpha^2\right] = \alpha!\,.$$

To this end note that from (3.5) it follows that

$$E[f(\langle \omega, e_1 \rangle, \ldots, \langle \omega, e_m \rangle)] = \int_{\mathbb{R}^m} f(x)\, d\lambda_m(x)$$

for all $f \in L^1(\lambda_m)$, where λ_m is the normal distribution on $\mathbb{R}^m$, i.e.,

$$d\lambda_m(x) = (2\pi)^{-m/2} e^{-1/2|x|^2}\, dx_1 \cdots dx_m, \quad x = (x_1, \ldots, x_m) \in \mathbb{R}^m.$$

Therefore, if $\alpha = (\alpha_1, \ldots, \alpha_m)$ and $\beta = (\beta_1, \ldots, \beta_m)$ we have

$$
\begin{aligned}
E\left[\widetilde{\mathcal{H}}_\alpha \widetilde{\mathcal{H}}_\beta\right] &= E\left[\prod_{k=1}^m h_{\alpha_k}(\langle \omega, e_k \rangle) h_{\beta_k}(\langle \omega, e_k \rangle)\right] \\
&= \int_{\mathbb{R}^m} \prod_{k=1}^m h_{\alpha_k}(x_k) h_{\beta_k}(x_k)\, d\lambda_m(x_1, \ldots, x_m) \\
&= \prod_{k=1}^m \int_{\mathbb{R}} h_{\alpha_k}(x_k) h_{\beta_k}(x_k)\, d\lambda_m(x_k) \\
&= \prod_{k=1}^m \delta_{\alpha_k, \beta_k} \alpha_k! = \begin{cases} 0 & \text{if } \alpha \neq \beta \\ \alpha! & \text{if } \alpha = \beta, \end{cases}
\end{aligned}
$$

where we have used the following orthogonality relation for Hermite polynomials:

$$\int_{\mathbb{R}} h_i(x) h_j(x) e^{-1/2x^2}\, dx = \delta_{ij} \sqrt{2\pi} j!\,.$$

$\square$

Example 3.1.9. Note that by orthogonality of the family $\{\widetilde{\mathcal{H}}_\alpha\}_{\alpha \in \mathcal{J}}$ in $L^2(\mathbb{P}^H)$ we have that the coefficients c_α in the expansion (3.13) of F are given by

$$c_\alpha = \frac{1}{\alpha!} E\left[F \widetilde{\mathcal{H}}_\alpha\right].$$

Choose $f \in L_H^2(\mathbb{R})$ and put $F(\omega) = \langle \omega, f \rangle = \int_{\mathbb{R}} f(s)\, dB_s^{(H)}$. Then F is Gaussian and by (3.16) we deduce that

$$E\left[F\widetilde{\mathcal{H}}_{\varepsilon^{(i)}}\right] = E\left[\langle\omega,f\rangle\langle\omega,e_i\rangle\right]$$
$$= \langle f,e_i\rangle_H = \int_{-\infty}^{\infty}\int_{-\infty}^{\infty} f(u)e_i(v)\phi(u,v)\,du\,dv.$$

Moreover,
$$E\left[F\widetilde{\mathcal{H}}_{\varepsilon^{(i)}}\right] = 0 \quad \text{if} \quad |\alpha| > 1.$$

We conclude that we have the expansion
$$\int_{\mathbb{R}} f(s)\,dB_s^{(H)} = \sum_{i=1}^{\infty}\langle f,e_i\rangle_H \widetilde{\mathcal{H}}_{\varepsilon^{(i)}}(\omega), \quad f \in L_H^2(\mathbb{R}).$$

In particular, for fBm we get, by choosing $f = I_{[0,t]}$,
$$B_t^{(H)} = \sum_{i=1}^{\infty}\left[\int_0^t\left(\int_{-\infty}^{\infty} e_i(v)\phi(u,v)\,dv\right)du\right]\widetilde{\mathcal{H}}_{\varepsilon^{(i)}}(\omega).$$

We proceed to define the fractional Hida test function and distribution spaces (compare with Definition A.1.4).

Definition 3.1.10. *1. The fractional Hida test function space: Define* $(\mathcal{S})_H$ *to be the set of all* $\psi(\omega) = \sum_{\alpha\in\mathcal{J}} a_\alpha \widetilde{\mathcal{H}}_\alpha(\omega) \in L^2(\mathbb{P}^H)$ *such that*
$$\|\psi\|_{H,k}^2 := \sum_{\alpha\in\mathcal{J}} \alpha! a_\alpha^2 (2\mathbb{N})^{k\alpha} < \infty \quad \text{for all} \quad k \in \mathbb{N},$$

where
$$(2\mathbb{N})^\gamma = \prod_j (2j)^{\gamma_j} \quad \text{if} \quad \gamma = (\gamma_1,\ldots,\gamma_m) \in \mathcal{J}.$$

2. The fractional Hida distribution space: Define $(\mathcal{S})_H^*$ *to be the set of all formal expansions*
$$G(\omega) = \sum_{\beta\in\mathcal{J}} b_\beta \widetilde{\mathcal{H}}_\beta(\omega)$$

such that
$$\|G\|_{H,-q}^2 := \sum_{\beta\in\mathcal{J}} \beta! b_\beta^2 (2\mathbb{N})^{-q\beta} < \infty \quad \text{for some} \quad q \in \mathbb{N}.$$

We equip $(\mathcal{S})_H$ with the projective topology and $(\mathcal{S})_H^*$ with the inductive topology. Then $(\mathcal{S})_H^*$ can be identified with the dual of $(\mathcal{S})_H$ and the action of $G \in (\mathcal{S})_H^*$ on $\psi \in (\mathcal{S})_H$ is given by
$$\langle\!\langle G,\psi\rangle\!\rangle := \langle G,\psi\rangle_{(\mathcal{S})_H,(\mathcal{S})_H^*} := \sum_{\alpha\in\mathcal{J}} \alpha! a_\alpha b_\alpha.$$

In particular, if G belongs to $L^2(\mathbb{P}^H) \subset (\mathcal{S})_H^*$ and $\psi \in (\mathcal{S})_H \subset L^2(\mathbb{P}^H)$, then

$$\langle\!\langle G,\psi\rangle\!\rangle = E\left[G\psi\right] = (G,\psi)_{L^2(\mathbb{P}^H)}.$$

The construction of the fractional Hida space is analogous to the one for the standard Brownian motion case. If we compare Definition 3.1.10 and Definition A.1.4), we note that here the fractional Hida distribution space $(\mathcal{S})_H^*$ is an extension of $L^2(\mathbb{P}^H)$, while in the standard Brownian motion case $(\mathcal{S})^*$ extends $L^2(\mathbb{P})$, where $\mathbb{P}$ is the Wiener measure associated to the Brownian motion B.

We can in a natural way define $(\mathcal{S})_H^*$-valued integrals as follows:

Definition 3.1.11. *Suppose $Z : \mathbb{R} \to (\mathcal{S})_H^*$ is a given function with property that*

$$\langle\!\langle Z(t),\psi\rangle\!\rangle \in L^1(\mathbb{R},dt) \quad \text{for all } \psi \in (\mathcal{S})_H. \tag{3.17}$$

Then $\int_\mathbb{R} Z(t)\,dt$ is defined to be the unique element of $(\mathcal{S})_H^*$ such that

$$\left\langle\!\!\!\left\langle \int_\mathbb{R} Z(t)\,dt, \psi \right\rangle\!\!\!\right\rangle = \int_\mathbb{R} \langle\!\langle Z(t),\psi\rangle\!\rangle\,dt \quad \text{for all} \quad \psi \in (\mathcal{S})_H. \tag{3.18}$$

Just as in Proposition 8.1 of [109], one can show that (3.18) defines $\int_\mathbb{R} Z(t)\,dt$ as an element of $(\mathcal{S})_H^*$.

If (3.17) holds, then we say that $Z(t)$ is dt-*integrable* in $(\mathcal{S})_H^*$.

Example 3.1.12. The *fractional white noise* $W^{(H)}(t)$ at time t is defined by

$$W^{(H)}(t) = \sum_{i=1}^\infty \left[\int_\mathbb{R} e_i(v)\phi(t,v)\,dv\right]\widetilde{\mathcal{H}}_{\varepsilon^{(i)}}(\omega). \tag{3.19}$$

We see that for $q > 4/3$ we have

$$\|W^{(H)}(t)\|_{H,-q}^2 = \sum_{i=1}^\infty \varepsilon^{(i)}!\left[\int_\mathbb{R} e_i(v)\phi(t,v)\,dv\right]^2 (2\mathbb{N})^{-q\varepsilon^{(i)}}$$

$$= \sum_{i=1}^\infty \left[\int_\mathbb{R} e_i(v)\phi(t,v)\,dv\right]^2 (2i)^{-q} < \infty$$

by (3.9). Hence, $W^{(H)}(t) \in (\mathcal{S})_H^*$ for all t. Moreover, by (3.11) it follows that $t \to W^{(H)}(t)$ is a continuous function from $\mathbb{R}$ into $(\mathcal{S})_H^*$. Hence, $W^{(H)}(t)$ is integrable in $(\mathcal{S})_H^*$ for $0 \le s \le t$ and

$$\int_0^t W^{(H)}(s)\,ds = \sum_{i=1}^\infty \left\{\int_0^t \left[\int_\mathbb{R} e_i(v)\phi(u,v)\,dv\right]du\right\}\widetilde{\mathcal{H}}_{\varepsilon^{(i)}}(\omega) = B_t^{(H)} \tag{3.20}$$

by Example 3.1.9. Therefore, $t \to B_t^{(H)}$ is differentiable in $(\mathcal{S})_H^*$ and

$$\frac{d}{dt}B_t^{(H)} = W^{(H)}(t) \quad \text{in} \quad (\mathcal{S})_H^*. \tag{3.21}$$

This justifies the name *fractional white noise* for $W^{(H)}(t)$.

We remark that in (4.16) of Chapter 4 we have introduced the fractional white noise $W^{(H)}$ as an element in the Hida space $(\mathcal{S})^*$. Here the fractional white noise $W^{(H)}$ is considered as an element of the fractional Hida space $(\mathcal{S})_H^*$.

Definition 3.1.13. *Let*
$$F(\omega) = \sum_{\alpha \in \mathcal{J}} a_\alpha \widetilde{\mathcal{H}}_\alpha(\omega) \quad \text{and} \quad G(\omega) = \sum_{\beta \in \mathcal{J}} b_\beta \widetilde{\mathcal{H}}_\beta(\omega)$$
be two members of $(\mathcal{S})_H^*$. *Then we define the Wick product* $F \diamond G$ *of* F *and* G *by*
$$(F \diamond G)(\omega) = \sum_{\alpha,\beta \in \mathcal{J}} a_\alpha b_\beta \widetilde{\mathcal{H}}_{\alpha+\beta}(\omega) = \sum_{\gamma \in \mathcal{J}} \left(\sum_{\alpha+\beta=\gamma} a_\alpha b_\beta \right) \widetilde{\mathcal{H}}_\gamma(\omega). \quad (3.22)$$

Just as for the usual white noise theory one can now prove Lemma 2.4.4 of [109]

Lemma 3.1.14. *1.* $F, G \in (\mathcal{S})_H^* \implies F \diamond G \in (\mathcal{S})_H^*$.
2. $\psi, \eta \in (\mathcal{S})_H \implies \psi \diamond \eta \in (\mathcal{S})_H$.

Example 3.1.15. Let $f, g \in L_H^2(\mathbb{R})$. Then by (3.12)

$$\left(\int_\mathbb{R} f \, dB^{(H)} \right) \diamond \left(\int_\mathbb{R} g \, dB^{(H)} \right) = \left(\sum_{i=1}^\infty \langle f, e_i \rangle_H \widetilde{\mathcal{H}}_{\varepsilon(i)} \right) \diamond \left(\sum_{j=1}^\infty \langle g, e_j \rangle_H \widetilde{\mathcal{H}}_{\varepsilon(j)} \right)$$
$$= \sum_{i,j=1}^\infty \langle f, e_i \rangle_H \langle g, e_j \rangle_H \widetilde{\mathcal{H}}_{\varepsilon(i)+\varepsilon(j)}$$
$$= \sum_{\substack{i,j=1 \\ i \neq j}}^\infty \langle f, e_i \rangle_H \langle g, e_j \rangle_H \langle \omega, e_i \rangle \langle \omega, e_j \rangle$$
$$+ \sum_{i=1}^\infty \langle f, e_i \rangle_H \langle g, e_i \rangle_H \left(\langle \omega, e_i \rangle^2 - 1 \right)$$
$$= \left(\sum_{i=1}^\infty \langle f, e_i \rangle_H \langle \omega, e_i \rangle \right) \left(\sum_{j=1}^\infty \langle g, e_j \rangle_H \langle \omega, e_j \rangle \right)$$
$$- \sum_{i=1}^\infty \langle f, e_i \rangle_H \langle g, e_i \rangle_H.$$

We conclude that
$$\left(\int_\mathbb{R} f \, dB^{(H)} \right) \diamond \left(\int_\mathbb{R} g \, dB^{(H)} \right)$$
$$= \left(\int_\mathbb{R} f \, dB^{(H)} \right) \cdot \left(\int_\mathbb{R} g \, dB^{(H)} \right) - \langle f, g \rangle_H. \quad (3.23)$$

Example 3.1.16. If $X \in (\mathcal{S})_H^*$, then we define its Wick powers $X^{\diamond n}$ by

$$X^{\diamond n} = X \diamond X \diamond \cdots \diamond X \quad (n \text{ factors})$$

and we define its Wick exponential $\exp^\diamond(X)$ by

$$\exp^\diamond(X) = \sum_{n=0}^{\infty} \frac{1}{n!} X^{\diamond n}$$

provided that the series converges in $(\mathcal{S})_H^*$. Note that by definition of the Wick product we have

$$\langle \omega, e_k \rangle^{\diamond n} = \left(\widetilde{\mathcal{H}}_{\varepsilon(k)} \right)^{\diamond n} = \widetilde{\mathcal{H}}_{n\varepsilon(k)} = h_n(\langle \omega, e_k \rangle).$$

Therefore, if $c_k \in \mathbb{R}$, we get

$$\exp^\diamond(c_k \langle \omega, e_k \rangle) = \sum_{n=0}^{\infty} \frac{c_k^n}{n!} \langle \omega, e_k \rangle^{\diamond n}$$

$$= \sum_{n=0}^{\infty} \frac{c_k^n}{n!} h_n(\langle \omega, e_k \rangle)$$

$$= \exp\left(c_k \langle \omega, e_k \rangle - \frac{1}{2} c_k^2 \right)$$

by the generating property of Hermite polynomials. More generally, if $f \in L_H^2(\mathbb{R})$, we get

$$\exp^\diamond(\langle \omega, f \rangle) = \exp^\diamond \left(\sum_k \langle f, e_k \rangle_H \langle \omega, e_k \rangle \right)$$

$$= \prod_k {}^\diamond \exp^\diamond \left(\langle f, e_k \rangle_H \langle \omega, e_k \rangle \right)$$

$$= \prod_k \exp^\diamond \left(\langle f, e_k \rangle_H \langle \omega, e_k \rangle \right)$$

$$= \prod_k \exp \left(\langle f, e_k \rangle_H \langle \omega, e_k \rangle - \frac{1}{2} \langle f, e_k \rangle_H^2 \right)$$

$$= \exp \left(\sum_k \langle f, e_k \rangle_H \langle \omega, e_k \rangle - \frac{1}{2} \sum_k \langle f, e_k \rangle_H^2 \right)$$

$$= \exp \left(\langle \omega, f \rangle - \frac{1}{2} \|f\|_H^2 \right). \tag{3.24}$$

Thus

$$\exp^\diamond(\langle \omega, f \rangle) = \mathcal{E}(f) \quad \text{for all} \quad f \in L_H^2(\mathbb{R}). \tag{3.25}$$

More generally, if $g : \mathbb{C} \to \mathbb{C}$ is an entire function ($\mathbb{C}$ is the set of complex numbers) with the power series expansion

$$g(z_1, \ldots, z_n) = \sum_\alpha c_\alpha z_1^{\alpha_1} \cdots z_n^{\alpha_n} =: \sum_\alpha c_\alpha z^\alpha,$$

where we have put $z^\alpha = z_1^{\alpha_1} \cdots z_n^{\alpha_n}$ if $\alpha = (\alpha_1, \ldots, \alpha_n) \in \mathcal{J}$, then we define, for $X = (X_1, \ldots, X_n) \in ((\mathcal{S})_H^*)^n$,

$$g^\diamond(X_1, \ldots, X_n) = \sum_\alpha c_\alpha X^{\diamond \alpha}.$$

It is useful to note that with this notation we in fact have

$$\widetilde{\mathcal{H}}_\alpha(\omega) = \langle \omega, e_1 \rangle^{\diamond \alpha_1} \diamond \cdots \diamond \langle \omega, e_n \rangle^{\diamond \alpha_n}$$

if $\alpha = (\alpha_1, \ldots, \alpha_n) \in \mathcal{J}$. Or, if we define

$$\xi_k(\omega) = \langle \omega, e_k \rangle, \quad \xi = (\xi_1, \xi_2, \ldots),$$

then

$$\widetilde{\mathcal{H}}_\alpha(\omega) = \xi^{\diamond \alpha}.$$

3.2 Fractional Girsanov theorem

With the white noise machinery just established in Section 3.1 one can now verify that the proof of the Benth–Gjessing version of the Girsanov formula, as presented in Corollary 2.10.5 in [109], applies to the fractional case.

At this purpose we now introduce the translation operator by following [109]. Let $\omega_0 \in \mathcal{S}'(\mathbb{R})$. For $F \in (\mathcal{S})_H$, we define

$$T_{\omega_0} F(\omega) = F(\omega + \omega_0), \quad \omega \in \mathcal{S}'(\mathbb{R}).$$

It is easy to verify, as in the proof of Theorem 2.10.1 of [109], that $f \to T_{\omega_0} f$ is a continuous linear map from $(\mathcal{S})_H$ to $(\mathcal{S})_H$. We then define the adjoint translation operator $T_{\omega_0}^*$ from $(\mathcal{S})_H^*$ to $(\mathcal{S})_H^*$ by

$$\langle\!\langle T_{\omega_0}^* X, F \rangle\!\rangle = \langle\!\langle X, T_{\omega_0} F \rangle\!\rangle, \quad X \in (\mathcal{S})_H^*, \ F \in (\mathcal{S})_H.$$

Lemma 3.2.1. *Let $\omega_0 \in L_H^2(\mathbb{R})$ and define $\tilde{\omega}_0(t) = \int_\mathbb{R} \omega_0(u) \phi(t, u) \, du$. Then*

$$T_{\tilde{\omega}_0}^* X = X \diamond \exp^\diamond\left(\langle \omega, \omega_0 \rangle\right).$$

Proof. By a density argument it suffices to show that

$$\langle\!\langle T_{\tilde{\omega}_0}^* X, F \rangle\!\rangle = \langle\!\langle X \diamond \exp^\diamond\left(\langle \omega, \omega_0 \rangle\right), F \rangle\!\rangle$$

for

$$X = \exp\left(\langle \omega, g\rangle - \frac{1}{2}\|g\|_H^2\right), \quad F = \exp\left(\langle \omega, f\rangle - \frac{1}{2}\|f\|_H^2\right),$$

where $f, g \in L_H^2(\mathbb{R})$ and $\omega_0 \in \mathcal{S}(\mathbb{R})$. In this case, we have by definition

$$\langle\langle T_{\tilde{\omega}_0}^* X, F\rangle\rangle = \langle\langle X, F(\omega + \tilde{\omega}_0)\rangle\rangle = e^{\langle \tilde{\omega}_0, f\rangle}\langle\langle X, F(\omega)\rangle\rangle = e^{\langle \omega_0, f\rangle + \langle f, g\rangle_H}.$$

On the other hand,

$$\begin{aligned}\langle\langle X \diamond \exp^\diamond(\langle \omega, \omega_0\rangle), F\rangle\rangle &= \langle\langle e^{\langle \omega, g + \omega_0\rangle - \frac{1}{2}\|g + \omega_0\|_H^2}, F\rangle\rangle \\ &= e^{\langle g + \omega_0, f\rangle_H}.\end{aligned}$$

This shows the lemma. □

Theorem 3.2.2 (Fractional Girsanov formula I). *Let $\psi \in L^p(\mathbb{P}^H)$ for some $p > 1$ and let $\gamma \in L_\phi^2(\mathbb{R}) \cap C(\mathbb{R}) \subset \mathcal{S}'(\mathbb{R})$. Let $\tilde{\gamma}$ be defined by $\tilde{\gamma}(t) = \int_\mathbb{R} \phi(t, s)\gamma(s)\, ds$. Then the map $\omega \to \psi(\omega + \tilde{\gamma})$ belongs to $L^\rho(\mathbb{P}^H)$ for all $\rho < p$ and*

$$\int_{\mathcal{S}'(\mathbb{R})} \psi(\omega + \tilde{\gamma})\, d\mathbb{P}^H(\omega) = \int_{\mathcal{S}'(\mathbb{R})} \psi(\omega) \cdot \exp^\diamond(\langle \omega, \gamma\rangle)\, d\mathbb{P}^H(\omega).$$

Proof. By Lemma 3.2.1, we have

$$\langle\langle X, T_{\tilde{\gamma}}\psi\rangle\rangle = \langle\langle X \diamond \exp^\diamond(\langle \omega, \gamma\rangle), \psi\rangle\rangle. \tag{3.26}$$

Let $X = 1$. We see that the left-hand side of (3.26) is $\int_{\mathcal{S}'(\mathbb{R})} \psi(\omega + \tilde{\gamma})\, d\mathbb{P}^H(\omega)$ and the right-hand side of (3.26) is $\int_{\mathcal{S}'(\mathbb{R})} \psi(\omega) \cdot \exp^\diamond(\langle \omega, \gamma\rangle)\, d\mathbb{P}^H(\omega)$. This completes the proof of this theorem. □

Corollary 3.2.3. *Let $g : \mathbb{R} \to \mathbb{R}$ be bounded and let $\gamma \in L_\phi^2(\mathbb{R}) \cap C(\mathbb{R})$. Then, with $\mathcal{E}(\cdot)$ as in (3.7),*

$$E\left[g(B_t^{(H)} + \int_0^t \tilde{\gamma}(s)\, ds)\right] = E\left[g(B_t^{(H)})\mathcal{E}(\gamma)\right]. \tag{3.27}$$

Proof. Define $\psi(\omega) = g(\langle \omega, I_{[0,t]}\rangle) = g(B_t^{(H)})$. Then

$$\psi(\omega + \tilde{\gamma}) = g(\langle \omega + \tilde{\gamma}, I_{[0,t]}\rangle) = g(B_t^{(H)} + \int_0^t \tilde{\gamma}(s)\, ds);$$

so the result follows from (3.25) and Theorem 3.2.2. □

Theorem 3.2.4 (Fractional Girsanov formula II). *Let $T > 0$ and let γ be a continuous function with $\operatorname{supp} \gamma \subset [0, T]$. Let K be a function with $\operatorname{supp} K \subset [0, T]$ and such that*

$$\langle K, f\rangle_H = \langle \gamma, f\rangle_{L^2(\mathbb{R})}, \quad \text{for all } f \in \mathcal{S}(\mathbb{R}),\ \operatorname{supp} f \subset [0, T],$$

i.e.,
$$\int_{\mathbb{R}} K(s)\phi(s,t)\, ds = \gamma(t), \quad 0 \le t \le T.$$

On the σ-algebra $\mathcal{F}_T^{(H)}$ generated by $\{B_s^{(H)} : 0 \le s \le T\}$, define a probability measure $\mathbb{P}^{H,\gamma}$ by
$$\frac{d\mathbb{P}^{H,\gamma}}{d\mathbb{P}^H} = \exp^\diamond(-\langle \omega, K \rangle)$$

Then $\hat{B}^{(H)}(t) = B_t^{(H)} + \int_0^t \gamma_s\, ds$, $0 \le t \le T$, is a fBm under $\mathbb{P}^{H,\gamma}$.

Proof. It suffices to show that for any $G(\omega) = \exp(\langle \omega, f \rangle)$ with $f \in \mathcal{S}(\mathbb{R})$, supp $f \subset [0,T]$, we have
$$E_{H,\gamma}[G(\omega + \gamma)] = E[G(\omega + \gamma)\exp^\diamond[-\langle \omega, K \rangle]] = E[G(\omega)],$$
where $E_{H,\gamma}[\cdot]$ denotes the expectation under $\mathbb{P}^{H,\gamma}$. But in this case
$$E[G(\omega + \gamma)\exp^\diamond[-\langle \omega, K \rangle]] = E\left[\exp\left(\langle \omega + \gamma, f \rangle - \langle \omega, K \rangle - \frac{1}{2}\|K\|_H^2\right)\right]$$
$$= E\left[\exp\left(\langle \omega, f - K \rangle + \langle \gamma, f \rangle_{L^2(\mathbb{R})} - \frac{1}{2}\|K\|_H^2\right)\right]$$
$$= \exp\left(\frac{1}{2}\|f - K\|_H^2 + \langle \gamma, f \rangle_{L^2(\mathbb{R})} - \frac{1}{2}\|K\|_H^2\right)$$
$$= \exp\left(-\langle K, f \rangle_H + \frac{1}{2}\|f\|_H^2 + \langle \gamma, f \rangle_{L^2(\mathbb{R})}\right)$$
$$= e^{1/2\|f\|_H^2} = E\left[e^{\langle \omega, f \rangle}\right] = E[G(\omega)]. \qquad \square$$

Remark 3.2.5. Since $B_t^{(H)}$ is not a martingale, unlike in the standard Brownian motion case, the restriction of $d\mathbb{P}^{H,\gamma}/d\mathbb{P}^H$ to $\mathcal{F}_t^{(H)}$, $0 < t < T$ is in general not given by $\exp^\diamond(-\langle \omega, I_{[0,t]}K \rangle)$.

Remark 3.2.6. In [172], a special case of (3.27) was obtained.

Lemma 3.2.7 (Wick products on different white noise spaces). *Let $P = \mathbb{P}^H$, $Q = \mathbb{P}^{H,\gamma}$ and $\hat{B}^{(H)}(t) = B_t^{(H)} + \int_0^t \gamma_s\, ds$ be as in Theorem 3.18. Let the Wick products corresponding to P and Q be denoted by $\diamond_P$ and $\diamond_Q$, respectively. Then*
$$F \diamond_P G = F \diamond_Q G$$
for all $F, G \in (\mathcal{S})_H^*$.

Proof. First let $F = \exp(\int_{\mathbb{R}} f(s)\, dB_s^{(H)})$ and $G = \exp(\int_{\mathbb{R}} g(s)\, dB_s^{(H)})$, where f and g are in $L_H^2(\mathbb{R})$. Then the Wick product of F and G is

$$F \diamond_P G = \exp\left(\int_{\mathbb{R}} [f(s) + g(s)] \, dB_s^{(H)} - \langle f, g \rangle_H\right).$$

On the other hand

$$F = \exp\left(\int_{\mathbb{R}} f(s) \, d\hat{B}^{(H)}(s) - \langle f, \gamma \rangle_{L^2(\mathbb{R})}\right)$$

and

$$G = \exp\left(\int_{\mathbb{R}} g(s) \, d\hat{B}^{(H)}(s) - \langle g, \gamma \rangle_{L^2(\mathbb{R})}\right).$$

Thus,

$$F \diamond_Q G = \exp\left(\int_{\mathbb{R}} [f(s) + g(s)] \, d\hat{B}^{(H)}(s) - \langle f, g \rangle_H - \langle f + g, \gamma \rangle_{L^2(\mathbb{R})}\right)$$

$$= \exp\left(\int_{\mathbb{R}} [f(s) + g(s)] \, dB_s^{(H)} - \langle f, g \rangle_H\right).$$

Thus $F \diamond_Q G = F \diamond_P G$ for exponential functions. Since the Wick product is linear with respect to F and G, we know that $F \diamond_Q G = F \diamond_P G$ for linear combinations of exponential functions. The identity follows by a density argument. □

3.3 Fractional stochastic gradient

Now that the basic fractional white noise theory is established, we can proceed as in [1] to define stochastic gradient.

Definition 3.3.1. *Let $F : \mathcal{S}'(\mathbb{R}) \to \mathbb{R}$ be a given function and let $\gamma \in \mathcal{S}'(\mathbb{R})$. We say that F has a* directional derivative *in the direction γ if*

$$D_\gamma^{(H)} F(\omega) := \lim_{\varepsilon \to 0} \frac{F(\omega + \varepsilon \gamma) - F(\omega)}{\varepsilon}$$

exists in $(\mathcal{S})_H^*$. If this is the case, we call $D_\gamma^{(H)} F$ the directional derivative of F in the direction γ.

Example 3.3.2. If $F(\omega) = \langle \omega, f \rangle = \int_{\mathbb{R}} f(t) \, dB_t^{(H)}$ for some $f \in \mathcal{S}(\mathbb{R})$ and $\gamma \in L^2(\mathbb{R}) \subset \mathcal{S}'(\mathbb{R})$, then

$$D_\gamma^{(H)} F(\omega) = \lim_{\varepsilon \to 0} \frac{1}{\varepsilon} [\langle \omega + \varepsilon \gamma, f \rangle - \langle \omega, f \rangle]$$

$$= \lim_{\varepsilon \to 0} \frac{1}{\varepsilon} [\langle \varepsilon \gamma, f \rangle] = \langle \gamma, f \rangle = \int_{\mathbb{R}} f(t) \gamma(t) \, dt.$$

Definition 3.3.3. *We say that $F : \mathcal{S}'(\mathbb{R}) \to \mathbb{R}$ is differentiable if there exists a map $\Psi : \mathbb{R} \to (\mathcal{S})_H^*$ such that*

$$\Psi(t)\gamma(t) = \Psi(t,\omega)\gamma(t) \quad \text{is } (\mathcal{S})_H^* \text{-integrable}$$

and

$$D_\gamma^{(H)} F(\omega) = \int_\mathbb{R} \Psi(t,\omega)\gamma(t)\,dt \quad \forall \gamma \in L^2(\mathbb{R}).$$

In this case we put

$$D_t^{(H)} F(\omega) := \frac{dF}{d\omega}(t,\omega) := \Psi(t,\omega),$$

and we call $D_t^{(H)} F(\omega) = dF(t,\omega)/d\omega$ the stochastic gradient (or the Hida Malliavin derivative) of F at t.

Example 3.3.4. Let $F(\omega) = \langle \omega, f \rangle$ with $f \in \mathcal{S}(\mathbb{R})$. Then by Example 3.3.2 F is differentiable and its stochastic gradient is

$$D_t^{(H)} F(\omega) = f(t) \quad \text{for all most all} \quad (t,\omega).$$

Just as in Lemma 3.6 of [1], we now get

Lemma 3.3.5 (The chain rule I). *Let $P(y) = \sum_\alpha c_\alpha y^\alpha$ be a polynomial in n variables $y = (y_1, \ldots, y_n) \in \mathbb{R}^n$. Choose $f_i \in \mathcal{S}(\mathbb{R})$ and put $Y = (Y_1, \ldots, Y_n)$ with*

$$Y_i(\omega) = \langle \omega, f_i \rangle = \int_\mathbb{R} f_i(t)\,dB_t^{(H)}, \quad 1 \le i \le n.$$

Then $P^\diamond(Y)$ and $P(Y)$ are differentiable and

$$D_t^{(H)} P^\diamond(Y) = \sum_{i=1}^n \frac{\partial P^\diamond}{\partial x_i}(Y_1, \ldots, Y_n) f_i(t) = \sum_\alpha c_\alpha \sum_i \alpha_i Y^{\diamond(\alpha - \varepsilon^{(i)})} f_i(t),$$

$$D_t^{(H)} P(Y) = \sum_{i=1}^n \frac{\partial P}{\partial x_i}(Y_1, \ldots, Y_n) f_i(t) = \sum_\alpha c_\alpha \sum_i \alpha_i Y^{\alpha - \varepsilon^{(i)}} f_i(t).$$

Similarly, if we define $Y^{(t)} = (Y_1^{(t)}, \ldots, Y_n^{(t)})$ with

$$Y_i^{(t)}(\omega) = \int_0^t f_i(s)\,dB_s^{(H)} = \int_\mathbb{R} f_i(s) I_{[0,t]}(s)\,dB_s^{(H)}, \quad 1 \le i \le n,$$

then we obtain, as in Lemma 3.7 of [1]

Lemma 3.3.6 (Chain rule II).

$$\frac{d}{dt} P^\diamond(Y^{(t)}) = \sum_{j=1}^n f_j(t) \left(\frac{\partial P}{\partial x_j}\right)^\diamond (Y^{(t)}) \diamond W^{(H)}(t).$$

We remark that Lemma 3.3.5 shows that definition 3.3.3 extends Definition (2.31) of Chapter 2 of the derivative operator as an element of $(\mathcal{S})_H^*$.

3.4 Fractional Wick Itô Skorohod integral

The following definition is an extension of the fractional stochastic integral of Itô type introduced in Theorem 3.6.1.

Definition 3.4.1. *Suppose $Y : \mathbb{R} \to (\mathcal{S})_H^*$ is a given function such that $Y(t) \diamond W^{(H)}(t)$ is dt-integrable in $(\mathcal{S})_H^*$ (Definition 3.1.11). Then we define its fractional Wick Itô Skorohod (fWIS) integral, $\int_{\mathbb{R}} Y(t) \, dB_t^{(H)}$, by*

$$\int_{\mathbb{R}} Y(t) \, dB_t^{(H)} := \int_{\mathbb{R}} Y(t) \diamond W^{(H)}(t) \, dt. \tag{3.28}$$

In particular, the integral on an interval can be defined as

$$\int_0^T Y(t) \, dB_t^{(H)} = \int_{\mathbb{R}} Y(t) I_{[0,T]}(t) \, dB_t^{(H)}.$$

Example 3.4.2. Suppose

$$Y(t) = \sum_{i=1}^n F_i(\omega) I_{[t_i, t_{i+1})}(t), \quad \text{where} \quad F_i \in (\mathcal{S})_H^*.$$

Then by (3.20) we see that

$$\int_{\mathbb{R}} Y(t) \, dB_t^{(H)} = \sum_{i=1}^n F_i(\omega) \diamond \left(B_{t_{i+1}}^{(H)} - B_{t_i}^{(H)} \right).$$

We illustrate how the fractional Wick calculus works in $(\mathcal{S})_H^*$ by means of two simple examples.

Example 3.4.3. We compute the following integral.

$$\begin{aligned}
\int_0^t B_s^{(H)} \, dB_s^{(H)} &= \int_0^t B_s^{(H)} \diamond W^{(H)}(s) \, ds \\
&= \int_0^t B_s^{(H)} \diamond \frac{d}{ds} B_s^{(H)} \, ds = \frac{1}{2} (B_t^{(H)})^{\diamond 2} \\
&= \frac{1}{2} (B_t^{(H)})^2 - \frac{1}{2} t^{2H},
\end{aligned} \tag{3.29}$$

where we have used (3.28), (3.21), standard Wick calculus, (3.23), and finally the fact that

$$\int_0^t \int_0^t \phi(u,v) \, du \, dv = t^{2H}.$$

Example 3.4.4 (Geometric fractional Brownian motion). Consider the fractional stochastic differential equation

$$dX(t) = \mu X(t)\,dt + \sigma X(t)\,dB_t^{(H)}, \quad X(0) = x > 0,$$

where x, μ and σ are constants. We rewrite this as the following equation in $(\mathcal{S})_H^*$:

$$\frac{dX(t)}{dt} = \mu X(t) + \sigma X(t) \diamond W^{(H)}(t)$$

or

$$\frac{dX(t)}{dt} = \left(\mu + \sigma W^{(H)}(t)\right) \diamond X(t).$$

Using Wick calculus, we see that the solution of this equation is

$$X(t) = x \exp^\diamond \left(\mu t + \sigma \int_0^t W^{(H)}(s)\,ds\right) = x \exp^\diamond \left(\mu t + \sigma B_t^{(H)}\right). \quad (3.30)$$

By (3.25) and (3.29) this can be written

$$X(t) = x \exp\left(\sigma B_t^{(H)} + \mu t - \frac{1}{2}\sigma^2 t^{2H}\right).$$

Note that

$$E[X(t)] = xe^{\mu t}.$$

3.5 The ϕ-derivative

The stochastic integral defined (3.28) is a $(\mathcal{S})_H^*$-valued stochastic process. To discuss the Itô formula for usual functions, we need to have stochastic integral with value in L^2 or as an ordinary random variable. For this reason we study the Malliavin derivative in more detail.

Definition 3.5.1. *Let $g \in L_H^2(\mathbb{R})$. The ϕ-derivative of a random variable $F \in L^p(\mathbb{P}^H)$ in the direction of Φg is defined as*

$$D_{\Phi g} F(\omega) = \lim_{\delta \to 0} \frac{1}{\delta} \left\{ F\left(\omega + \delta \int_0^{\cdot} (\Phi g)(u)\,du\right) - F(\omega) \right\}$$

if the limit exists in $L^p(\mathbb{P}^H)$. Furthermore, if there is a process $(D_s^\phi F, s \geq 0)$ such that

$$D_{\Phi g} F = \int_{\mathbb{R}} D_s^\phi F g_s\,ds \quad \text{almost surely}$$

for all $g \in L_H^2(\mathbb{R})$, then F is said to be ϕ-differentiable.

Remark 3.5.2. Note that by comparing Definition 3.3.3 and 3.5.1 we obtain that

$$D_t^\phi F = \int_{\mathbb{R}} \phi(t,v) D_v^{(H)} F\,dv. \quad (3.31)$$

The higher order derivatives can be defined in a similar manner.

Definition 3.5.3. *Let $F : [0,T] \times \Omega \to \mathbb{R}$ be a stochastic process. The process F is said to be ϕ-differentiable if for each $t \in [0,T]$, $F(t,\cdot)$ is ϕ-differentiable and $D_s^\phi F_t$ is jointly measurable.*

It is easy to verify an elementary version of a chain rule, that is, if $f : \mathbb{R} \to \mathbb{R}$ is a smooth function and $F : \Omega \to \mathbb{R}$ is ϕ-differentiable then $f(F)$ is also ϕ-differentiable and
$$D_{\Phi g} f(F) = f'(F) D_{\Phi g} F$$
and
$$D_s^\phi f(F) = f'(F) D_s^\phi F$$
and the iterated directional derivatives are
$$D_{\Phi g_1} D_{\Phi g_2} f(F) = f'(F) D_{\Phi g_1} D_{\Phi g_2} F + f''(F) D_{\Phi g_1} F D_{\Phi g_2} F.$$

The following rules for differentiation, which can be verified as in the proof of Proposition 3.5.4, are useful later:
$$D_{\Phi g} \int_\mathbb{R} f_s \, dB_s^{(H)} = \int_\mathbb{R} \int_\mathbb{R} \phi(u,v) f_u g_v \, du \, dv = \langle f, g \rangle_H;$$
$$D_s^\phi \int_\mathbb{R} f_u \, dB_u^{(H)} = \int_\mathbb{R} \phi(u,s) f_u \, du = (\Phi f)(s);$$
$$D_{\Phi g} \varepsilon(f) = \varepsilon(f) \int_\mathbb{R} \int_\mathbb{R} \phi(u,v) f_u g_v \, du \, dv = \varepsilon(f) \langle f, g \rangle_H,$$
$$D_s^\phi \varepsilon(f) = \varepsilon(f) \int_\mathbb{R} \phi(u,s) f_u \, du = \varepsilon(f)(\Phi f)(s),$$
where $f, g \in L_H^2(\mathbb{R})$.

We note that the Wick product of two exponentials $\varepsilon(f)$ and $\varepsilon(g)$ has the following form
$$\varepsilon(f) \diamond \varepsilon(g) := \varepsilon(f+g). \tag{3.32}$$
Since for distinct $f_1, f_2, \ldots, f_n$ in $L_H^2(\mathbb{R})$, $\varepsilon(f_1), \varepsilon(f_2), \ldots, \varepsilon(f_n)$ are linearly independent, this property uniquely defines the Wick product $F \diamond G$ of two functionals F and G in $\mathcal{E}$.

Proposition 3.5.4. *If $g \in L_H^2(\mathbb{R})$, $F \in L^2(\mathbb{P}^H)$ and $D_{\Phi g} F \in L^2(\mathbb{P}^H)$, then*
$$F \diamond \int_\mathbb{R} g_s \, dB_s^{(H)} = F \int_\mathbb{R} g_s \, dB_s^{(H)} - D_{\Phi g} F \tag{3.33}$$

Proof. By (3.32),
$$\varepsilon(f) \diamond \varepsilon(\delta g) = \varepsilon(f + \delta g), \quad \delta \in \mathbb{R}. \tag{3.34}$$
Differentiate the above identity with respect to δ and evaluate at $\delta = 0$, to obtain
$$\begin{aligned}\varepsilon(f) \diamond \int_{\mathbb{R}} g_s \, dB_s^{(H)} &= \varepsilon(f) \left[\int_{\mathbb{R}} g_s \, dB_s^{(H)} - \langle f, g \rangle_H \right] \\ &= \varepsilon(f) \int_{\mathbb{R}} g_s \, dB_s^{(H)} - \varepsilon(f) \langle f, g \rangle_H. \end{aligned} \tag{3.35}$$

By (3.5) it follows that the last term of the above expression is $D_{\Phi g} \varepsilon(f)$. Thus the following equality is satisfied:
$$\varepsilon(f) \diamond \int_{\mathbb{R}} g_s \, dB_s^{(H)} = \varepsilon(f) \int_{\mathbb{R}} g_s \, dB_s^{(H)} - D_{\Phi g} \varepsilon(f). \tag{3.36}$$

If $F \in \mathcal{E}$ is a finite linear combination of $\varepsilon(f_1), \varepsilon(f_2), \ldots, \varepsilon(f_n)$, then extend (3.36) by linearity
$$\begin{aligned}F \diamond \int_{\mathbb{R}} g_s \, dB_s^{(H)} &= F \int_{\mathbb{R}} g_s \, dB_s^{(H)} - D_{\Phi g} F \\ &= F \int_{\mathbb{R}} g_s \, dB_s^{(H)} - \int_{\mathbb{R}} D_s^{\phi} F g_s \, ds. \end{aligned} \tag{3.37}$$

The proof of the proposition is completed by Theorem 3.1.4. $\square$

Now we compute the second moment of (3.33). Note that by a simple computation for Gaussian random variables, it follows that
$$E\left[\varepsilon(f)\varepsilon(g)\right] = \exp\left(\langle f, g \rangle_H\right).$$
Thus for $\delta, \gamma \in \mathbb{R}$
$$\begin{aligned}E\left[(\varepsilon(f) \diamond \varepsilon(\gamma g))(\varepsilon(h) \diamond \varepsilon(\delta g))\right] &= E\left[\varepsilon(f + \gamma g)\varepsilon(h + \delta g)\right] \\ &= \exp(\langle f + \gamma g, h + \delta g \rangle_H).\end{aligned}$$

Both sides of this equality are functions of γ and δ. Taking the partial derivative $\partial^2/(\partial \gamma \partial \delta)$ evaluated at $\gamma = \delta = 0$, it follows that
$$\begin{aligned}E\left[\left(\varepsilon(f) \diamond \int_{\mathbb{R}} g_s \, dB_s^{(H)}\right)\left(\varepsilon(h) \diamond \int_{\mathbb{R}} g_s \, dB_s^{(H)}\right)\right] \\ = \exp\left(\langle f, h \rangle_H\right)\left\{\langle f, g \rangle_H \langle h, g \rangle_H + \langle g, g \rangle_H\right\} \\ = E\left[D_{\Phi g}\varepsilon(f) D_{\Phi g}\varepsilon(h) + \varepsilon(f)\varepsilon(h)\langle g, g \rangle_H\right].\end{aligned}$$

Thus

$$E\left[\left(\varepsilon(f) \diamond \int_{\mathbb{R}} g_s \, dB_s^{(H)}\right)\left(\varepsilon(h) \diamond \int_{\mathbb{R}} g_s \, dB_s^{(H)}\right)\right]$$
$$= E\left[(D_{\Phi g}\varepsilon(f) D_{\Phi g}\varepsilon(h) + \varepsilon(f)\varepsilon(h)\langle g, g\rangle_H)\right].$$

By bilinearity, for any F and G in $\mathcal{E}$, the following equality is satisfied

$$E\left[\left(F \diamond \int_{\mathbb{R}} g_s \, dB_s^{(H)}\right)\left(G \diamond \int_{\mathbb{R}} g_s \, dB_s^{(H)}\right)\right] = E\left[D_{\Phi g}F D_{\Phi g}G + FG\langle g, g\rangle_H\right].$$

Let F be equal to G. Then

$$E\left[\left(F \diamond \int_{\mathbb{R}} g_s \, dB_s^{(H)}\right)^2\right] = E\left[(D_{\Phi g}F)^2 + F^2\|g\|_H^2\right].$$

This result is stated in the following theorem.

Theorem 3.5.5. *Let $g \in L_H^2(\mathbb{R})$ and let $\mathcal{E}_g$ be the completion of $\mathcal{E}$ under the norm*

$$\|F\|_g^2 = E\left[(D_{\Phi g}F)^2 + F^2\right],$$

where F is a random variable. Then for any element $F \in \mathcal{E}_g$, $F \diamond \int_{\mathbb{R}} g_s \, dB_s^{(H)}$ is well-defined and

$$E\left[\left(F \diamond \int_{\mathbb{R}} g_s \, dB_s^{(H)}\right)^2\right] = E\left[(D_{\Phi g}F)^2 + F^2\|g\|_H^2\right]. \tag{3.38}$$

By the polarization technique [160], there is the following corollary.

Corollary 3.5.6. *Let $g, h \in L_H^2(\mathbb{R})$ and $F, G \in \mathcal{E}$. Then*

$$E\left[\left(F \diamond \int_{\mathbb{R}} g_s \, dB_s^{(H)}\right)\left(G \diamond \int_{\mathbb{R}} h_s \, dB_s^{(H)}\right)\right]$$
$$= E\left[D_{\Phi h}F D_{\Phi g}G + FG\langle g, h\rangle_H\right]. \tag{3.39}$$

3.6 Fractional Wick Itô Skorohod integrals in L^2

We now wish to define the stochastic integral $\int_0^T F_s \, dB_s^{(H)}$, for a suitable integrand F_t, as an element in $L^2(\mathbb{P}^H)$. We recall that we are assuming $H > 1/2$.

Consider an arbitrary partition of $[0, T]$, $\pi : 0 = t_0 < t_1 < t_2 < \ldots < t_n = T$ and the Riemann sum

$$S(F, \pi) = \sum_{i=0}^{n-1} F_{t_i} \diamond (B_{t_{i+1}}^{(H)} - B_{t_i}^{(H)}).$$

Note first that if $Y(t) = F_i I_{\{t_i \leq t < t_{i+1}\}}$, then by (3.28) we have

$$\int_{\mathbb{R}} Y(t)\, dB_t^{(H)} = \int_{\mathbb{R}} Y(t) \diamond W^{(H)}(t)\, dt = F_i \diamond \int_{t_i}^{t_{i+1}} W^{(H)}(t)\, dt$$
$$= F_i \diamond (B_{t_{i+1}}^{(H)} - B_{t_i}^{(H)})$$

Under some assumptions on ϕ-derivative of F_i we see that $\int_{\mathbb{R}} Y(t)\, dB_t^{(H)}$ is square integrable. For the general case we shall use equality (3.39). From (3.34) it easily follows that that for any F and G in $\mathcal{E}$, $E[F \diamond G] = E[F]E[G]$. This identity extends to more general F and G such that $F \diamond G$ is well-defined (e.g. [109, p. 83]). Thus for any partition π,

$$E\left[\sum_{i=0}^{n-1} F_{t_i} \diamond (B_{t_{i+1}}^{(H)} - B_{t_i}^{(H)})\right] = \sum_{i=0}^{n-1} E\left[F_{t_i} \diamond (B_{t_{i+1}}^{(H)} - B_{t_i}^{(H)})\right]$$
$$= \sum_{i=0}^{n-1} E[F_{t_i}] E\left[B_{t_{i+1}}^{(H)} - B_{t_i}^{(H)}\right] = 0.$$

To compute the L^2 norm of $S(F,\pi)$, denote

$$\sigma_{ij} = E\left[\left(F_{t_i} \diamond (B_{t_{i+1}}^{(H)} - B_{t_i}^{(H)})\right)\left(F_{t_j} \diamond (B_{t_{j+1}}^{(H)} - B_{t_j}^{(H)})\right)\right].$$

By Corollary 3.5.6, it follows that

$$\sigma_{ij} = E\left[\int_{t_j}^{t_{j+1}} D_s^\phi F_{t_i}\, ds \int_{t_i}^{t_{i+1}} D_t^\phi F_{t_j}\, dt + F_{t_i} F_{t_j} \int_{t_i}^{t_{i+1}} \int_{t_j}^{t_{j+1}} \phi(u,v)\, du\, dv\right].$$

Thus

$$E[S(F,\pi)^2] = \sum_{i,j=0}^{n-1} E\left[\int_{t_j}^{t_{j+1}} D_s^\phi F_{t_i}\, ds \int_{t_i}^{t_{i+1}} D_t^\phi F_{t_j}\, dt \right.$$
$$\left. + F_{t_i} F_{t_j} \int_{t_i}^{t_{i+1}} \int_{t_j}^{t_{j+1}} \phi(u,v)\, du\, dv\right].$$

Denote $|\pi| := \max_i (t_{i+1} - t_i)$ and $F_t^\pi = F_{t_i}$ if $t_i \leq t < t_{i+1}$. Assume that as $|\pi| \to 0$, $E\left[\|F^\pi - F\|_H^2\right] \to 0$ and

$$E\left[\left|\sum_{i,j=0}^{n-1} \int_{t_i^{(n)}}^{t_{i+1}^{(n)}} \int_{t_j^{(n)}}^{t_{j+1}^{(n)}} D_s^\phi F_{t_i^{(n)}}^\pi D_t^\phi F_{t_j^{(n)}}^\pi\, ds\, dt - \int_0^T \int_0^T D_s^\phi F_t D_t^\phi F_s\, ds\, dt\right|\right]$$

converges to 0. Then from the above it is easy to see that if $(\pi_n, n \in \mathbb{N})$ is a sequence of partitions such that $|\pi_n| \to 0$ as $n \to \infty$, then $(S(F,\pi_n), n \in \mathbb{N})$ is a Cauchy sequence in $L^2(\mathbb{P}^H)$. The limit of this sequence in $L^2(\mathbb{P}^H)$ is defined as $\int_0^T F_s\, dB_s^{(H)}$, that is, define

$$\int_0^T F_s \, dB_s^{(H)} = \lim_{|\pi| \to 0} \sum_{i=0}^{n-1} F_{t_i}^\pi \diamond (B_{t_{i+1}}^{(H)} - B_{t_i}^{(H)}) \qquad (3.40)$$

so that

$$E\left[|\int_0^T F_s \, dB_s^{(H)}|^2\right] = E\left[\int_0^T \int_0^T D_s^\phi F_t D_t^\phi F_s \, ds \, dt + \|f\|_H^2\right].$$

Let $\mathcal{L}_\phi(0,T)$ be the family of stochastic processes F on $[0,T]$ with the following properties: $F \in \mathcal{L}_\phi(0,T)$ if and only if $E\left[\|F\|_H^2\right] < \infty$, F is ϕ-differentiable, the trace of $D_s^\phi F_t, 0 \leq s, t \leq T$, exists, and $E\left[\int_0^T \int_0^T |D_s^\phi F_t|^2 \, ds \, dt\right] < \infty$ and for each sequence of partitions $(\pi_n, n \in \mathbb{N})$ such that $|\pi_n| \to 0$ as $n \to \infty$,

$$\sum_{i,j=0}^{n-1} E\left[\int_{t_i^{(n)}}^{t_{i+1}^{(n)}} \int_{t_j^{(n)}}^{t_{j+1}^{(n)}} \left| D_s^\phi F_{t_i^{(n)}}^\pi D_t^\phi F_{t_j^{(n)}}^\pi - D_s^\phi F_t D_t^\phi F_s \right| ds \, dt\right]$$

and

$$E\left[\|F^\pi - F\|_H^2\right]$$

tend to 0 as $n \to \infty$, where $\pi_n : 0 = t_0^{(n)} < t_1^{(n)} < \ldots < t_{n-1}^{(n)} < t_n^{(n)} = T$.

The following result summaries the above construction of a stochastic integral.

Theorem 3.6.1. *Let $(F_t, t \in [0,T])$ be a stochastic process such that $F \in \mathcal{L}_\phi(0,T)$. The limit (3.40) exists, and this limit is defined as $\int_0^T F_s \, dB_s^{(H)}$. Moreover, this integral satisfies*

$$E\left[\int_0^T F_s \, dB_s^{(H)}\right] = 0$$

and

$$\|\int_0^T F_s \, dB_s^{(H)}\|_{\mathcal{L}_\phi(0,T)} := E\left[|\int_0^T F_s \, dB_s^{(H)}|^2\right] \qquad (3.41)$$

$$= E\left[\int_0^T \int_0^T D_s^\phi F_t D_t^\phi F_s \, ds \, dt + \|1_{[0,T]}F\|_H^2\right].$$

The following properties follow directly from the above theorem.

1. If $F, G \in \mathcal{L}_\phi(0,T)$, then

$$\int_0^t (aF_s + bG_s) \, dB_s^{(H)} = a \int_0^t F_s \, dB_s^{(H)} + b \int_0^t G_s \, dB_s^{(H)} \quad \text{a.s.}$$

for any constants a and b.

2. If $F \in \mathcal{L}_\phi(0,T)$, $E\left[\sup_{0\leq s\leq T}|F_s|^2\right] < \infty$ and $\sup_{0\leq s,t\leq T} E\left[|D_s^\phi F_t|^2\right] < \infty$, then $(\int_0^t F_s dB_s^{(H)}, 0 \leq t \leq T)$ has a continuous version.

Property 1 is obvious. To show property 2 let $Y_t = \int_0^t F_s\, dB_s^{(H)}, 0 \leq t \leq T$. By the equality (3.41) it follows that

$$E\left[|Y_t - Y_s|^2\right] = E\left[|\int_s^t F_u\, dB_u^{(H)}|^2\right]$$
$$\leq E\left[\int_s^t \int_s^t |D_u^\phi F_v|^2\, du\, dv + \int_s^t \int_s^t F_u F_v \phi(u,v)\, du\, dv\right]$$
$$\leq C(t-s)^2 + E\left[\left\{\sup_{0\leq s\leq T} F_s\right\}^2\right]\int_s^t \int_s^t \phi(u,v)\, du\, dv$$
$$\leq (t-s)^2 + C(t-s)^{2H}\,.$$

By the Kolmogorov's lemma [218], the property 2 is satisfied.

3.7 An Itô formula

Now an analogue of the Itô formula is established, that is, a chain rule for the integral introduced in the last section.

At this purpose we need to introduce first the following theorem that shows how to compute the ϕ-derivative of a stochastic integral of Itô type. It can be verified from the product rule and the Riemann sum approximations to the stochastic integral.

Theorem 3.7.1. Let $(F_t, t \in [0,T])$ be a stochastic process in $\mathcal{L}_\phi(0,T)$ and $\sup_{0\leq s\leq T} E\left[|D_s^\phi F_s|^2\right] < \infty$, and let $\eta_t = \int_0^t F_u\, dB_u^{(H)}$ for $t \in [0,T]$. Then for $s,t \in [0,T]$,

$$D_s^\phi \eta_t = \int_0^t D_s^\phi F_u\, dB_u^{(H)} + \int_0^t F_u \phi(s,u)\, du, \quad a.s.$$

Now a general Itô formula is given. For further comments about Itô formulas for fBm, we refer to Section 6.3.

Theorem 3.7.2. Let $\eta_t = \int_0^t F_u\, dB_u^{(H)}$, where $(F_u, 0 \leq u \leq T)$ is a stochastic process in $\mathcal{L}_\phi(0,T)$. Assume that there is an $\alpha > 1 - H$ such that

$$E\left[|F_u - F_v|^2\right] \leq C|u-v|^{2\alpha},$$

where $|u - v| \leq \delta$ for some $\delta > 0$ and

$$\lim_{0\leq u,v\leq t, |u-v|\to 0} E\left[|D_u^\phi(F_u - F_v)|^2\right] = 0\,.$$

Let $f : \mathbb{R}_+ \times \mathbb{R} \to \mathbb{R}$ be a function having the first continuous derivative in its first variable and the second continuous derivative in its second variable. Assume that these derivatives are bounded. Moreover, it is assumed that $E\left[\int_0^T |F_s D_s^\phi \eta_s| \, ds\right] < \infty$ and $(f'(s, \eta_s) F_s, s \in [0, T])$ is in $\mathcal{L}_\phi(0, T)$. Then for $0 \le t \le T$,

$$f(t, \eta_t) = f(0,0) + \int_0^t \frac{\partial f}{\partial s}(s, \eta_s) \, ds + \int_0^t \frac{\partial f}{\partial x}(s, \eta_s) F_s \, dB_s^{(H)}$$
$$+ \int_0^t \frac{\partial^2 f}{\partial x^2}(s, \eta_s) F_s D_s^\phi \eta_s \, ds \quad a.s.$$

Proof. Let π be a partition defined as above by replacing T by t. Then

$$f(t, \eta_t) - f(0, 0)$$
$$= \sum_{k=0}^{n-1} \left[f(t_{k+1}, \eta_{t_{k+1}}) - f(t_k, \eta_{t_k}) \right]$$
$$= \sum_{k=0}^{n-1} \left[f(t_{k+1}, \eta_{t_{k+1}}) - f(t_k, \eta_{t_{k+1}}) \right] + \sum_{k=0}^{n-1} \left[f(t_k, \eta_{t_{k+1}}) - f(t_k, \eta_{t_k}) \right]$$

by the mean value theorem. It is easy to see that the first sum converges to $\int_0^t \partial f(s, \eta_s)/\partial s \, ds$ in $L^2(\mathbb{P}^H)$. Now consider the second sum. Using Taylor's formula, it follows that

$$f(t_k, \eta_{t_{k+1}}) - f(t_k, \eta_{t_k}) = \frac{\partial f}{\partial x}(t_k, \eta_{t_k}) \left(\eta_{t_{k+1}} - \eta_{t_k} \right)$$
$$+ \frac{1}{2} \frac{\partial^2 f}{\partial x^2}(t_k, \tilde{\eta}_{t_k}) \left(\eta_{t_{k+1}} - \eta_{t_k} \right)^2,$$

where $\tilde{\eta}_{t_k} \in (\eta_{t_k}, \eta_{t_{k+1}})$. An upper bound is obtained for $E\left[\left(\eta_{t_{k+1}} - \eta_{t_k}\right)^2\right]$ as follows

$$E\left[\left(\eta_{t_{k+1}} - \eta_{t_k}\right)^2\right] = E\left[\int_{t_k}^{t_{k+1}} \int_{t_k}^{t_{k+1}} D_s^\phi F_t D_t^\phi F_s \, ds \, dt\right]$$
$$+ E\left[\int_{t_k}^{t_{k+1}} \int_{t_k}^{t_{k+1}} F_u F_v \phi(u, v) \, du \, dv\right]$$
$$\le C(t_{k+1} - t_k)^2$$
$$+ \int_{t_k}^{t_{k+1}} \int_{t_k}^{t_{k+1}} (E\left[F_u^2\right])^{1/2} (E\left[F_v^2\right])^{1/2} \phi(u, v) \, du \, dv$$
$$\le C\left[(t_{k+1} - t_k)^2 + \int_{t_k}^{t_{k+1}} \int_{t_k}^{t_{k+1}} \phi(u, v) \, du \, dv\right]$$
$$\le C(t_{k+1} - t_k)^2 + C(t_{k+1} - t_k)^{2H}$$
$$\le C(t_{k+1} - t_k)^{2H},$$

where $t_{i+1} - t_i < 1$ and C is a constant independent of the partition π that may differ from line to line in this proof. Since

$$E\left[\sum_{k=0}^{n-1} \frac{\partial^2 f}{\partial x^2}(t_k, \tilde{\eta}_{t_k})\left(\eta_{t_{k+1}} - \eta_{t_k}\right)^2\right] \leq C\sum_{k=0}^{n-1} E\left[\left(\eta_{t_{k+1}} - \eta_{t_k}\right)^2\right]$$

$$\leq C\sum_{k=0}^{n-1}(t_{k+1} - t_k)^{2H},$$

then $E\left[\sum_{k=0}^{n-1} \partial^2 f(t_k, \tilde{\eta}_{t_k})/\partial x^2(\eta_{t_{k+1}} - \eta_{t_k})^2\right] \to 0$ as $|\pi| \to 0$. On the other hand,

$$\frac{\partial f}{\partial x}(t_k, \eta_{t_k})\left(\eta_{t_{k+1}} - \eta_{t_k}\right) = \frac{\partial f}{\partial x}(t_k, \eta_{t_k})\left(F_{t_k} \diamond (B^{(H)}_{t_{k+1}} - B^{(H)}_{t_k})\right)$$

$$+ \frac{\partial f}{\partial x}(t_k, \eta_{t_k})\left(\int_{t_k}^{t_{k+1}} (F_s - F_{t_k})\, dB^{(H)}_s\right).$$

The first term on the right-hand side can be expressed as

$$\frac{\partial f}{\partial x}(t_k, \eta_{t_k})\left(F_{t_k} \diamond (B^{(H)}_{t_{k+1}} - B^{(H)}_{t_k})\right)$$

$$= \frac{\partial f}{\partial x}(t_k, \eta_{t_k})\left(F_{t_k}(B^{(H)}_{t_{k+1}} - B^{(H)}_{t_k}) - \int_{t_k}^{t_{k+1}} D^\phi_s F_{t_k}\, ds\right)$$

$$= \frac{\partial f}{\partial x}(t_k, \eta_{t_k})F_{t_k}(B^{(H)}_{t_{k+1}} - B^{(H)}_{t_k}) - \frac{\partial f}{\partial x}(t_k, \eta_{t_k})\int_{t_k}^{t_{k+1}} D^\phi_s F_{t_k}\, ds$$

$$= \left[\frac{\partial f}{\partial x}(t_k, \eta_{t_k})F_{t_k}\right] \diamond (B^{(H)}_{t_{k+1}} - B^{(H)}_{t_k})$$

$$+ \int_{t_k}^{t_{k+1}} D^\phi_s \left(\frac{\partial f}{\partial x}(t_k, \eta_{t_k})F_{t_k}\right)\, ds - \frac{\partial f}{\partial x}(t_k, \eta_{t_k})\int_{t_k}^{t_{k+1}} D^\phi_s F_{t_k}\, ds$$

$$= \left[\frac{\partial f}{\partial x}(t_k, \eta_{t_k})F_{t_k}\right] \diamond (B^{(H)}_{t_{k+1}} - B^{(H)}_{t_k}) + \int_{t_k}^{t_{k+1}} F_{t_k} D^\phi_s \frac{\partial f}{\partial x}(t_k, \eta_{t_k})\, ds.$$

Thus

$$\sum_{k=0}^{n-1} \frac{\partial f}{\partial x}(t_k, \eta_{t_k})\left(\eta_{t_{k+1}} - \eta_{t_k}\right) = \sum_{k=0}^{n-1} \left[\frac{\partial f}{\partial x}(t_k, \eta_{t_k})F_{t_k}\right] \diamond (B^{(H)}_{t_{k+1}} - B^{(H)}_{t_k})$$

$$+ \sum_{k=0}^{n-1} \int_{t_k}^{t_{k+1}} F_{t_k} D^\phi_s \frac{\partial f}{\partial x}(t_k, \eta_{t_k})\, ds.$$

As $|\pi| \to 0$, the first term converges to $\int_0^t F_s \partial f(s, \eta_s)/\partial x\, dB^{(H)}_s$ and the second term converges to $\int_0^t \partial^2 f(s, \eta_s)/\partial x^2\, D^\phi_s \eta_s F_s\, ds$ in L^2. To prove the theorem, it is only necessary to show that

$$\sum_{k=0}^{n-1} E\left[|\frac{\partial f}{\partial x}(t_k, \eta_{t_k}) \int_{t_k}^{t_{k+1}} (F_s - F_{t_k})\, dB_s^{(H)}|\right]$$

converges to 0 as $|\pi| \to 0$. Since f has a bounded second derivative, it follows that
$$|\frac{\partial f}{\partial x}(t_k, \eta_{t_k})| \leq C(1 + |\eta_{t_k}|).$$

Thus
$$E\left[|\frac{\partial f}{\partial x}(t_k, \eta_{t_k})|^2\right] \leq C.$$

Furthermore
$$\sum_{k=0}^{n-1} E\left[|\frac{\partial f}{\partial x}(t_k, \eta_{t_k}) \int_{t_k}^{t_{k+1}} (F_s - F_{t_k})\, dB_s^{(H)}|\right]$$
$$\leq C \sum_{k=0}^{n-1} E\left[|\int_{t_k}^{t_{k+1}} (F_s - F_{t_k})\, dB_s^{(H)}|^2\right]^{1/2}$$
$$= C \sum_{k=0}^{n-1} E\left[\left\{\int_{t_k}^{t_{k+1}} [D_s^\phi (F_s - F_{t_k})]\, ds\right\}^2\right.$$
$$\left. + E\left[\int_{t_k}^{t_{k+1}} \int_{t_k}^{t_{k+1}} (F_u - F_{t_k})(F_v - F_{t_k})\phi(u,v)\, du\, dv\right]\right]^{1/2}$$
$$\leq C \sum_{k=0}^{n-1} \left\{(t_{k+1} - t_k) \int_{t_k}^{t_{k+1}} E\left[(D_s^\phi(F_s - F_{t_k}))^2\right] ds\right.$$
$$\left.+ \int_{t_k}^{t_{k+1}} \int_{t_k}^{t_{k+1}} \sqrt{E[(F_u - F_{t_k})^2] E[(F_v - F_{t_k})^2]} \phi(u,v)\, du\, dv\right\}^{1/2}$$
$$\leq C \sum_{k=0}^{n-1} \left\{\sup_{t_k \leq s \leq t_{k+1}} E\left[|D_s^\phi(F_s - F_{t_k})|^2\right](t_{k+1} - t_k)^2\right.$$
$$\left.+(t_{k+1} - t_k)^{2H} \sup_{t_k \leq s \leq t_{k+1}} E\left[(F_s - F_{t_k})^2\right]\right\}^{1/2}$$
$$\leq C \sup_{t_k \leq s \leq t_{k+1}} E\left[|D_s^\phi(F_s - F_{t_k})|^2\right]^{1/2} + C|\pi|^{H+\alpha-1}.$$

The last term tends to 0 as $|\pi| \to 0$. This proves the theorem. $\square$

The equality (3.7.2) can be formally expressed as
$$df(t, \eta_t) = \frac{\partial f}{\partial t}(t, \eta_t)\, dt + \frac{\partial f}{\partial x}(t, \eta_t) F_t\, dB_t^{(H)} + \frac{\partial^2 f}{\partial x^2}(t, \eta_t) F_t D_t^\phi \eta_t\, dt.$$

If $F(s) = a(s)$ is a deterministic function, then (3.7.2) simplifies as follows.

Corollary 3.7.3. Let $\eta_t = \int_0^t a_u \, dB_u^{(H)}$, where $a \in L_H^2(\mathbb{R})$ and $f : \mathbb{R}_+ \times \mathbb{R} \to \mathbb{R}$ satisfies the conditions in Theorem 3.7.2. Let $(\partial f(s, \eta_s)/\partial x \, a_s, s \in [0, T])$ be in $\mathcal{L}_\phi(0, T)$. Then

$$f(t, \eta_t) = f(0, 0) + \int_0^t \frac{\partial f}{\partial s}(s, \eta_s) \, ds + \int_0^t \frac{\partial f}{\partial x}(s, \eta_s) a_s \, dB_s^{(H)}$$
$$+ \int_0^t \frac{\partial^2 f}{\partial x^2}(s, \eta_s) a_s \int_0^s \phi(s, v) a_v \, dv \, ds$$

alomst surely, or formally,

$$df(t, \eta_t) = \frac{\partial f}{\partial t}(t, \eta_t) \, dt + \frac{\partial f}{\partial x}(t, \eta_t) a_t \, dB_t^{(H)} + \frac{\partial^2 f}{\partial x^2}(t, \eta_t) a_t \int_0^t \phi(t, v) a_v \, dv \, dt.$$

If $a_s \equiv 1$, then we find the same result as in Theorem 4.2.6 in the special case $H > 1/2$.

In the classical stochastic analysis, the stochastic integral can be defined for general semimartingales and an Itô formula can be given. By the Doob–Meyer decomposition [79], a semimartingale can be expressed as the sum of a martingale and a bounded variation process. A semimartingale $(X_t, t \geq 0)$ with respect to a Brownian motion can often be expressed as $X_t = X_0 + \int_0^t f_s \, dB_s + \int_0^t g_s \, ds$. An Itô formula in the analogous form with respect to fBm is given. This generalization of the Itô formula is useful in applications.

Theorem 3.7.4. Let $(F_u, u \in [0, T])$ satisfy the conditions of Theorem 3.7.2, and let $E\left[\sup_{0 \leq s \leq T} |G_s|\right] < \infty$. Denote $\eta_t = \xi + \int_0^t G_u \, du + \int_0^t F_u \, dB_u^{(H)}$, $\xi \in \mathbb{R}$, for $t \in [0, T]$. Let $(\partial f(s, \eta_s)/\partial x F_s, s \in [0, T])$, be in $\mathcal{L}_\phi(0, T)$. Then for $t \in [0, T]$,

$$f(t, \eta_t) = f(0, \xi) + \int_0^t \frac{\partial f}{\partial s}(s, \eta_s) \, ds + \int_0^t \frac{\partial f}{\partial x}(s, \eta_s) G_s \, ds$$
$$+ \int_0^t \frac{\partial f}{\partial x}(s, \eta_s) F_s \, dB_s^{(H)} + \int_0^t \frac{\partial^2 f}{\partial x^2}(s, \eta_s) F_s D_s^\phi \eta_s \, ds \quad \text{a.s.}.$$

Proof. The proof is the same as for Theorem 3.7.2. □

We now present two applications of the Itô formula. First, we provide an L^p estimate of the fWIS integral. Second, we extend the so-called homogeneous chaos to the fBm.

3.8 L^p estimate for the fWIS integral

Let $h_n(x)$ be the Hermite polynomial of degree n, that is,

$$e^{tx - 1/2 t^2} = \sum_{n=0}^\infty t^n h_n(x).$$

Let
$$\|f\|_{H,t} := \left[\int_0^t \int_0^t \phi(u,v) f_u f_v \, du \, dv\right]^{1/2}.$$

Define
$$\tilde{f}_t := \|f\|_{H,t}^{-1} \int_0^t f_s \, dB_s^{(H)},$$

where $f \in L_H^2(\mathbb{R})$ and
$$h_n^{H,f}(t) := \|f\|_{H,t}^n h_n\left(\tilde{f}_t\right). \tag{3.42}$$

Theorem 3.8.1. *If $f \in L_H^2(\mathbb{R})$, then the following equality is satisfied*
$$dh_n^{H,f}(t) = n h_{n-1}^{H,f}(t) f_t \, dB_t^{(H)},$$

where $t \in [0,T]$.

Proof. Fix n and denote $X_t = h_n^{H,f}(t)$ for $t \in [0,T]$. Using the Itô formula (3.7.2), it follows that

$$dX_t = n\|f\|_{H,t}^{n-2} f_t \int_0^t \phi(u,t) f_u \, du \, h_n\left(\tilde{f}_t\right) dt$$
$$- \|f\|_{H,t}^n f_t \int_0^t \phi(u,t) f_u \, du \, h_n'\left(\tilde{f}_t\right) \|f\|_{H,t}^{-3} \left(\int_0^t f_s \, dB_s^{(H)}\right) dt$$
$$+ \|f\|_{H,t}^n h_n'\left(\tilde{f}_t\right) \|f\|_{H,t}^{-1} f_t \, dB_t^{(H)}$$
$$+ \|f\|_{H,t}^n f_t \int_0^t \phi(u,t) f_u \, du \, h_n''\left(\tilde{f}_t\right) \|f\|_{H,t}^{-2} dt$$
$$= n\|f\|_{H,t}^{n-2} h_{n-1}\left(\tilde{f}_t\right) f_t \, dB_t^{(H)}$$
$$+ \|f\|_{H,t}^{n-2} f_t \int_0^t \phi(u,t) f_u \, du \cdot \left\{ n h_n(\tilde{f}_t) - \tilde{f}_t h_n'(\tilde{f}_t) + h_n''(\tilde{f}_t) \right\} dt.$$

It is well known that for each $n \in \mathbb{N}$ the Hermite polynomial satisfies
$$n h_n(x) - x h_n'(x) + h_n''(x) = 0$$

for each $x \in \mathbb{R}$. Thus $n h_n(\tilde{f}_t) - \tilde{f}_t h_n'(\tilde{f}_t) + h_n''(\tilde{f}_t) = 0$. The first term is
$$n h_{n-1}^{H,f}(t) f_t \, dB_t^{(H)}.$$

Thus,
$$dh_n^{H,f}(t) = n h_{n-1}^{H,f}(t) f_t \, dB_t^{(H)}.$$

This proves the theorem. □

3.8 L^p estimate for the fWIS integral

The following estimate for the L^p norm is of Meyer inequality type and is useful in some applications.

Theorem 3.8.2. *Let $(g_s, s \in [0, t])$ be a stochastic process satisfying the assumptions of Theorem 3.7.2. Let $F_t := \int_0^t g_s \, dB_s^{(H)}$ and $p \geq 2$. If $E\left[\int_0^t |g_s|^p \, ds\right] < \infty$, $\int_0^t E\left[|D_s^\phi F_s|^p \, ds\right] < \infty$ and $F^{p-1} g \in \mathcal{L}_\phi(0, t)$, then*

$$E[F_t^p] \leq p^p \left\{ \int_0^t \left(E\left[|g_s D_s^\phi F_s|^{p/2}\right] \right)^{2/p} ds \right\}^{p/2}.$$

Proof. Applying the Itô formula (Theorem 3.7.2) to F_t^p (by the assumption that $F^{p-1} g \in \mathcal{L}_\phi(0, t)$, the restriction on the boundedness of f to its second derivatives in Theorem 3.7.2 can be removed), it follows that

$$F_t^p = p \int_0^t F_s^{p-1} g_s \, dB_s^{(H)} + p(p-1) \int_0^t F_s^{p-2} g_s D_s^\phi F_s \, ds.$$

Thus,

$$E[F_t^p] = p(p-1) \int_0^t E\left[F_s^{p-2} g_s D_s^\phi F_s\right] ds.$$

and

$$E[F_t^p] \leq p(p-1) \int_0^t E\left[|F_s^{p-2} g_s D_s^\phi F_s|\right] ds$$

$$\leq p^2 \int_0^t E[F_s^p]^{(p-2)/p} E\left[|g_s D_s^\phi F_s|^{p/2}\right]^{2/p} ds$$

By an inequality of Langenhop (e.g., [13]), we have

$$E[F_t^p] \leq p^p \left\{ \int_0^t \left(E\left[|g_s D_s^\phi F_s|^{p/2}\right] \right)^{2/p} ds \right\}^{p/2}.$$

This completes the proof of the theorem. □

Corollary 3.8.3. *Let the conditions of Theorem 3.8.2 be satisfied and let $p \geq 2$. Then*

$$E[F_t^p] \leq p^p \left\{ \int_0^t (E[|g_s|^p])^{2/p} \, ds + \int_0^t \left(E\left[|D_s^\phi F_s|^p\right] \right)^{2/p} ds \right\}^{p/2}.$$

Proof. From $|ab| \leq a^2 + b^2$, it follows that

$$E\left[|g_s D_s^\phi F_s|^{p/2}\right] \leq E[|g_s|^p] + E\left[|D_s^\phi F_s|^p\right].$$

Thus

$$\left(E\left[|g_s D_s^\phi F_s|^{p/2}\right] \right)^{2/p} \leq \left(E[|g_s|^p] + E\left[|D_s^\phi F_s|^p\right] \right)^{2/p}$$

$$\leq (E[|g_s|^p])^{2/p} + \left(E\left[|D_s^\phi F_s|^p\right] \right)^{2/p}.$$

This verifies the corollary. □

3.9 Iterated integrals and chaos expansion

As a second application of Itô formula, we now provide a Wiener Itô chaos expansion theorem in terms of iterated integrals.

We recall that $L_H^2(\mathbb{R}_+)$ denotes the subspace generated by the deterministic functions in $L_\phi^2(\mathbb{R}_+)$. In the sequel we identify $f \in L_H^2(\mathbb{R}_+)$ with $fI_{[0,+\infty)} \in L_H^2(\mathbb{R})$ to keep simple the notation concerning the norms. Let $f \in L_H^2(\mathbb{R}_+)$ be such that $\|f\|_H = 1$. Similar to [109], define $(\int_0^\infty f_s dB_s^{(H)})^{\diamond n}$ as the nth Wick power of $\int_0^\infty f_s dB_s^{(H)}$, i.e.,

$$\left(\int_0^\infty f_s dB_s^{(H)}\right)^{\diamond n} := \left(\int_0^\infty f_s dB_s^{(H)}\right)^{\diamond(n-1)} \diamond \int_0^\infty f_s dB_s^{(H)},$$

$$\exp^\diamond\left(\int_0^\infty f_s dB_s^{(H)}\right) := \sum_{n=0}^\infty \frac{1}{n!} \left(\int_0^\infty f_s dB_s^{(H)}\right)^{\diamond n},$$

$$\log^\diamond\left(1 + \int_0^\infty f_s dB_s^{(H)}\right) := \sum_{n=1}^\infty \frac{(-1)^{n-1}}{n} \left(\int_0^\infty f_s dB_s^{(H)}\right)^{\diamond n}$$

for $n = 2, 3, \ldots$.

Lemma 3.9.1. *If $f \in L_H^2(\mathbb{R}_+)$ with $\|f\|_H = 1$, then for each $n \in \mathbb{N}$, $\left(\int_0^\infty f_s dB_s^{(H)}\right)^{\diamond n}$ is well-defined and*

$$\left(\int_0^\infty f_s dB_s^{(H)}\right)^{\diamond n} = h_n\left(\int_0^\infty f_s dB_s^{(H)}\right), \tag{3.43}$$

where h_n denotes the Hermite polynomial of degree n (see (3.8)).

Proof. The equality (3.43) is verified by induction. It is easy to see that (3.43) is true for $n = 1$. Suppose that (3.43) is true for $1, 2, \ldots, n-1$. Then

$$\left(\int_0^\infty f_s dB_s^{(H)}\right)^{\diamond n} = h_{n-1}(\int_0^\infty f_s dB_s^{(H)}) \diamond \int_0^\infty f_s dB_s^{(H)}$$

$$= h_{n-1}(\int_0^\infty f_s dB_s^{(H)}) \int_0^\infty f_s dB_s^{(H)} - D_{\Phi f}\left\{h_{n-1}(\int_0^\infty f_s dB_s^{(H)})\right\}$$

$$= h_{n-1}(\int_0^\infty f_s dB_s^{(H)}) \int_0^\infty f_s dB_s^{(H)} - h'_{n-1}(\int_0^\infty f_s dB_s^{(H)})\|f\|_H^2$$

$$= h_{n-1}(\int_0^\infty f_s dB_s^{(H)}) \int_0^\infty f_s dB_s^{(H)} - h'_{n-1}(\int_0^\infty f_s dB_s^{(H)})$$

$$= h_n(\int_0^\infty f_s dB_s^{(H)})$$

by an identity for Hermite polynomials. This verifies the equation (3.43). □

For an arbitrary, nonzero $f \in L_H^2(\mathbb{R}_+)$, the product defined in (3.43) is extended as follows:

$$\left(\int_0^\infty f_s \, dB_s^{(H)}\right)^{\diamond n} = \|f\|_H^n \left(\frac{\int_0^\infty f_s \, dB_s^{(H)}}{\|f\|_H}\right)^{\diamond n}$$

$$= \|f\|_H^n h_n \left(\int_0^\infty \frac{\int_0^\infty f_s \, dB_s^{(H)}}{\|f\|_H}\right).$$

Lemma 3.9.2. *If $f \in L_H^2(\mathbb{R}_+)$, then $\left(\int_0^\infty f_s \, dB_s^{(H)}\right)^{\diamond n}$ is well-defined for each $n \in \mathbb{N}$ and*

$$\left(\int_0^t f_s \, dB_s^{(H)}\right)^{\diamond n} = h_n^{H,f}(t),$$

where $h_n^{H,f}(t)$ is defined in (3.42).

Since $\int_0^\infty f_s \, dB_s^{(H)}$ is a Gaussian random variable, it is easy to estimate its moments and to show that the series defining $\exp^\diamond \left(\int_0^\infty f_s \, dB_s^{(H)}\right)$ is convergent in $L^2(\mathbb{P}^H)$. Moreover, there is the following corollary.

Corollary 3.9.3. *If $f \in L_H^2(\mathbb{R}_+)$, then*

$$\exp^\diamond \left(\int_0^\infty f_s \, dB_s^{(H)}\right) = \varepsilon(f) = \exp\left(\int_0^\infty f_s \, dB_s^{(H)} - \frac{1}{2}\|f\|_H^2\right).$$

Proof. It follows that

$$\exp^\diamond \left(\int_0^\infty f_s \, dB_s^{(H)}\right) = \sum_{n=0}^\infty \frac{1}{n!} \left(\int_0^\infty f_s \, dB_s^{(H)}\right)^{\diamond n}$$

$$= \sum_{n=0}^\infty \frac{1}{n!} \|f\|_H^n h_n \left(\frac{\int_0^\infty f_s \, dB_s^{(H)}}{\|f\|_H}\right)$$

$$= \exp\left(\|f\|_H \frac{\int_0^\infty f_s \, dB_s^{(H)}}{\|f\|_H} - \frac{1}{2}\|f\|_H^2\right)$$

$$= \exp\left(\int_0^\infty f_s \, dB_s^{(H)} - \frac{1}{2}\|f\|_H^2\right).$$

This completes the proof of the lemma. □

The following lemma is also easy to prove.

Lemma 3.9.4. *For any two functions f and g in $L_H^2(\mathbb{R}_+)$ with $\langle f, g \rangle_H = 0$, the following equality is satisfied*

$$\left(\int_0^\infty f_s \, dB_s^{(H)}\right)^{\diamond n} \diamond \left(\int_0^\infty g_s \, dB_s^{(H)}\right)^{\diamond m}$$
$$= \left(\int_0^\infty f_s \, dB_s^{(H)}\right)^{\diamond n} \left(\int_0^\infty g_s \, dB_s^{(H)}\right)^{\diamond m} = h_n^{H,f}(\infty) h_m^{H,g}(\infty).$$

Since $\int_0^\infty f_s \, dB_s^{(H)}/\|f\|_H$ and $\int_0^\infty g_s \, dB_s^{(H)}/\|g\|_H$ are Gaussian random variables with mean 0 and variance 1, their covariance is

$$E\left[\left(\int_0^\infty \frac{f_s}{\|f\|_H} \, dB_s^{(H)}\right)\left(\int_0^\infty \frac{g_s}{\|g\|_H} \, dB_s^{(H)}\right)\right] = \left\langle \frac{f}{\|f\|_H}, \frac{g}{\|g\|_H}\right\rangle_H,$$

It follows that

$$E\left[\left(\int_0^\infty f_s \, dB_s^{(H)}\right)^{\diamond n}\left(\int_0^\infty g_s \, dB_s^{(H)}\right)^{\diamond m}\right]$$
$$= E\left[\|f\|_H^n \|g\|_H^m h_n\left(\int_0^\infty \frac{f_s}{\|f\|_H} \, dB_s^{(H)}\right) h_m\left(\int_0^\infty \frac{g_s}{\|g\|_H} \, dB_s^{(H)}\right)\right]$$
$$= \begin{cases} 0 & \text{if } m \neq n \\ n! \|f\|_H^n \|g\|_H^n \left\langle \frac{f}{\|f\|_H}, \frac{g}{\|g\|_H}\right\rangle_\phi^n & \text{if } m = n \end{cases}$$
$$= \begin{cases} 0 & \text{if } m \neq n \\ n! \langle f, g\rangle_H^n & \text{if } m = n \end{cases}$$

By a polarization technique [160] it is easy to verify the following lemma.

Lemma 3.9.5. *Let* $f^1, \ldots, f^n, g^1, \ldots, g^m \in L_H^2(\mathbb{R}_+)$. *The following equality is satisfied*

$$E\left[\left(\int_0^\infty f_s^1 dB_s^{(H)} \diamond \cdots \diamond \int_0^\infty f_s^n dB_s^{(H)}\right)\right.$$
$$\left.\cdot \left(\int_0^\infty g_s^1 dB_s^{(H)} \diamond \cdots \diamond \int_0^\infty g_s^m dB_s^{(H)}\right)\right]$$
$$= \begin{cases} 0 & \text{if } n \neq m \\ \sum_\sigma \langle f^1, g^{\sigma(1)}\rangle_H \langle f^2, g^{\sigma(2)}\rangle_H \cdots \langle f^n, g^{\sigma(n)}\rangle_H & \text{if } n = m, \end{cases}$$

where $\sum_\sigma$ *denotes the sum over all permutations* σ *of* $\{1, 2, \ldots, n\}$.

Let $e_1, e_2, \ldots, e_n, \ldots$ be a complete orthonormal basis of $L_H^2(\mathbb{R}_+)$ as defined in (3.10). Consider the nth symmetric tensor product $\hat{L}_H^2(\mathbb{R}_+^n)$ of $L_H^2(\mathbb{R}_+)$:

$$\hat{L}_H^2(\mathbb{R}_+^n) := L_H^2(\mathbb{R}_+) \hat{\otimes} \cdots \hat{\otimes} L_H^2(\mathbb{R}_+).$$

We denote by $\mathcal{L}_n$ the set of all functions of n variables of the following form:

$$f(s_1, \ldots, s_n) = \sum_{1 \leq k_1, \ldots, k_n \leq k} a_{k_1 \cdots k_n} e_{k_1}(s_1) e_{k_2}(s_2) \cdots e_{k_n}(s_n), \quad (3.44)$$

where f is a symmetric function of its variables $s_1, \ldots, s_n$ and k is a positive integer. For an element of the form (3.44), its multiple integral is defined by

$$I_n(f) = \sum_{1 \leq k_1, \ldots, k_n \leq k} a_{k_1 \cdots k_n} \int_0^\infty e_{k_1}(s) \, dB_s^{(H)}$$
$$\diamond \int_0^\infty e_{k_2}(s) \, dB_s^{(H)} \diamond \cdots \diamond \int_0^\infty e_{k_n}(s) \, dB_s^{(H)}.$$

By Lemma 3.9.5, we have

$$E\left[|I_n(f)|^2\right] = n! \int_{\mathbb{R}_+^{2n}} \phi(u_1, v_1) \phi(u_2, v_2) \cdots \phi(u_n, v_n) f(u_1, u_2, \ldots, u_n)$$
$$\cdot f(v_1, v_2, \ldots, v_n) \, du_1 \, du_2 \cdots du_n \, dv_1 \, dv_2 \cdots dv_n \quad (3.45)$$
$$= n! \sum_{\substack{1 \leq k_1, \ldots, k_n \leq k \\ 1 \leq l_1, \ldots, l_n \leq k}} a_{k_1 \cdots k_n} a_{l_1 \cdots l_n} \langle e_{k_1}, e_{l_1} \rangle \cdots \langle e_{k_n}, e_{l_n} \rangle$$
$$= n! \sum_{1 \leq k_1, \ldots, k_n \leq k} a_{k_1 \cdots k_n}^2. \quad (3.46)$$

Given $f \in \hat{L}_\phi^2(\mathbb{R}_+^n)$, we define

$$\|f\|_{L_\phi^2(\mathbb{R}_+^n)} := n! \int_{\mathbb{R}_+^{2n}} \phi(u_1, v_1) \phi(u_2, v_2) \cdots \phi(u_n, v_n) f(u_1, u_2, \ldots, u_n)$$
$$\cdot f(v_1, v_2, \ldots, v_n) \, du_1 \, du_2 \cdots du_n \, dv_1 \, dv_2 \cdots dv_n.$$

We note that the completion of $\hat{L}_H^2(\mathbb{R}_+^n)$ with respect to the norm (3.9) is $L_\phi^2(\mathbb{R}_+^n)$.

Since every element f in $\hat{L}_H^2(\mathbb{R}_+^n)$ is a limit of elements in $\mathcal{L}_n$, the multiple integral $I_n(f)$ for $f \in \hat{L}_H^2(\mathbb{R}_+^n)$ can be defined by a limit from elements in $\mathcal{L}_n$ and it follows that

$$E\left[(|I_n(f)|^2)\right] = n! \|f\|_H^2.$$

The following lemma can also be shown by the polarization technique.

Lemma 3.9.6. If $f \in \hat{L}_H^2(\mathbb{R}_+^n)$ and $g \in \hat{L}_H^2(\mathbb{R}_+^m)$, then

$$E\left[I_n(f) I_m(g)\right] = \begin{cases} n! \langle f, g \rangle_H & \text{if } n = m, \\ 0 & \text{if } n \neq m. \end{cases}$$

Let $f \in \hat{L}_H^2(\mathbb{R}_+^n)$. The iterated integral can be defined by the recursive formula

$$\int_{0 \leq s_1 < \cdots < s_n \leq t} f(s_1, \ldots, s_n) \, dB_{s_1}^{(H)} \cdots dB_{s_n}^{(H)}$$
$$= \int_0^t \left(\int_{0 \leq s_1 < \cdots < s_n} f(s_1, \ldots, s_n) \, dB_{s_1}^{(H)} \cdots dB_{s_{n-1}}^{(H)} \right) dB_{s_n}^{(H)} \quad (3.47)$$

Theorem 3.9.7. If $f \in \hat{L}_H^2(\mathbb{R}_+^n)$, then the iterated integral (3.47) exists and

$$I_n(f) = n! \int_{0 \leq s_1 < \cdots < s_n \leq t} f(s_1, \ldots, s_n) \, dB_{s_1}^{(H)} \cdots dB_{s_n}^{(H)}. \qquad (3.48)$$

Proof. First let f have the special form $f = g^{\otimes n}$, that is, $f(s_1, s_2, \ldots, s_n) = g(s_1)g_2(s_2) \cdots g(s_n)$. Then

$$I_n(f) = h_n^{H,g}(t)$$

and

$$\begin{aligned} dI_n(f) &= dh_n^{H,g}(t) \\ &= nh_{n-1}^{H,g}(t)g(t)\,dB_t^{(H)} \\ &= nI_{n-1}(g^{\otimes(n-1)})g(t)\,dB_t^{(H)}. \end{aligned}$$

This verifies (3.48) for the case where $f = g^{\otimes n}$. By the polarization technique [160], the theorem follows easily. □

Remark 3.9.8. For Brownian motion, a multiple integral was originally introduced by Wiener [235]; Wiener's original multiple integral is in fact a multiple integral of Stratonovich type. The multiple integral of Itô type was introduced in [131].

Theorem 3.9.9. Let $F \in L^2(\mathbb{P}^H)$. Then there exist $f_n \in \hat{L}_H^2(\mathbb{R}_+^n)$ for $n = 0, 1, 2, \ldots$ such that

$$F(\omega) = \sum_{n=0}^{\infty} I_n(f_n),$$

where $I_0(f_0) := E[F]$. Moreover,

$$\|F\|_{L^2(\mathbb{P}^H)}^2 = \sum_{n=0}^{\infty} n! \|f_n\|_{L^2_\phi(\mathbb{R}_+^n)}^2.$$

Proof. The result can be deduced from Theorem 3.1.8 by using the identity

$$\widetilde{\mathcal{H}}_\alpha(\omega) = \int_{\mathbb{R}^n} e_1^{\otimes \alpha_1} \hat{\otimes} \cdots \hat{\otimes} e_m^{\otimes \alpha_m} \, d(B^{(H)})^{\otimes n} \qquad (3.49)$$

if $\alpha = (\alpha_1, \ldots, \alpha_m)$ and $n = |\alpha| = \alpha_1 + \cdots + \alpha_m$, where $\hat{\otimes}$ denotes the symmetric tensor product. The identity (3.49) follows by Theorem 3.1.8, (3.12), and (3.48). □

3.10 Fractional Clark Hausmann Ocone theorem

In this section we prove a generalized Clark Hausmann Ocone formula in the fractional case. For this purpose we extend the differentiation operator to a space of random variables containing $L^2(\mathbb{P}^H)$. A convenient pair of spaces to work with is the following:

Definition 3.10.1. *([1], [192])*

1. *Let $k \in \mathbb{N}$. We say that a function*

$$\psi(\omega) = \sum_{n=0}^{\infty} \int_{\mathbb{R}_+^n} f_n \, d(B^{(H)})^{\otimes n}(t) \in L^2(\mathbb{P}^H), \quad f_n \in \hat{L}_H^2(\mathbb{R}_+^n),$$

belongs to the space $\mathcal{G}_k = \mathcal{G}_k(\mathbb{P}^H)$ if

$$\|\psi\|_{\mathcal{G}_k}^2 := \sum_{n=0}^{\infty} n! \|f_n\|_{L_\phi^2(\mathbb{R}_+^n)}^2 e^{2kn} < \infty,$$

where $\|f_n\|_{\hat{L}_\phi^2(\mathbb{R}_+^n)}$ is defined in (3.9). We put

$$\mathcal{G} = \mathcal{G}(\mathbb{P}^H) = \cap_{k=1}^{\infty} \mathcal{G}_k(\mathbb{P}^H)$$

and equip $\mathcal{G}$ with the projective topology.

2. *Let $q \in \mathbb{N}$. We say that a formal expansion*

$$G = \sum_{n=0}^{\infty} \int_{\mathbb{R}_+^n} g_n \, d(B^{(H)})^{\otimes n}(t), \quad g_n \in \hat{L}_H^2(\mathbb{R}_+^n),$$

belongs to the space $\mathcal{G}_{-q} = \mathcal{G}_{-q}(\mathbb{P}^H)$ if

$$\|G\|_{\mathcal{G}_{-q}}^2 = \sum_{n=0}^{\infty} n! \|g_n\|_{L_\phi^2(\mathbb{R}_+^n)}^2 e^{-2qn} < \infty.$$

We define

$$\mathcal{G}^* = \mathcal{G}^*(\mathbb{P}^H) = \cup_{q \in \mathbb{N}} \mathcal{G}_{-q}(\mathbb{P}^H)$$

and equip $\mathcal{G}^*$ with the inductive topology. Then $\mathcal{G}^*$ is the dual of $\mathcal{G}$, and the action of $G \in \mathcal{G}^*$ on $\psi \in \mathcal{G}$ is given by

$$\langle\!\langle G, \psi \rangle\!\rangle = \sum_{n=0}^{\infty} n! (g_n, f_n)_{L_\phi^2(\mathbb{R}_+^n)}.$$

Remark 3.10.2. Note that by Theorem 3.9.9 we have

$$(\mathcal{S})_H \subset \mathcal{G}(\mathbb{P}^H) \subset L^2(\mathbb{P}^H) = \left(L^2(\mathbb{P}^H)\right)^* \subset \mathcal{G}^*(\mathbb{P}^H) \subset (\mathcal{S})_H^*.$$

Let $\mathcal{F}_t^{(H)}$ be the σ-algebra generated by $B^{(H)}(s, \cdot); 0 \leq s \leq t$. The following operator is useful:

Definition 3.10.3. 1. Let $G = \sum_{n=0}^{\infty} \int_{\mathbb{R}_+^n} g_n(s) \, d(B^{(H)})^{\otimes n}(s) \in \mathcal{G}^*$. Then we define the fractional (or quasi-) conditional expectation of G with respect to $\mathcal{F}_t^{(H)}$ by

$$\tilde{E}\left[G|\mathcal{F}_t^{(H)}\right] := \sum_{n=0}^{\infty} \int_{\mathbb{R}_+^n} g_n(s) \cdot I_{\{0 \leq s \leq t\}} \, d(B^{(H)})^{\otimes n}(s), \tag{3.50}$$

for $t \geq 0$.

2. We say that $G \in \mathcal{G}^*$ is $\mathcal{F}_t^{(H)}$-measurable if $\tilde{E}\left[G|\mathcal{F}_t^{(H)}\right] = G$, $t \geq 0$.

Remark 3.10.4. The fractional conditional expectation $\tilde{E}$ is different from the ordinary conditional expectation. For example, it is easy to check that $\tilde{E}\left[B_t^{(H)}|\mathcal{F}_s^{(H)}\right] = B_s^{(H)}$ for $0 \leq s \leq t$. But the computation of $E\left[B_t^{(H)}|\mathcal{F}_s^{(H)}\right]$ is much more complicated. See, for example, [103].

Note, however, that if $G \in L^2(\mathbb{P}^H)$ then Definition , 2 coincides with the usual definition of being $\mathcal{F}_t^{(H)}$-measurable, i.e., we have

$$\tilde{E}\left[G|\mathcal{F}_t^{(H)}\right] = G \quad \text{a.s.} \Leftrightarrow E\left[G|\mathcal{F}_t^{(H)}\right] = G \quad \text{a.s..}$$

This equivalence follows from the prediction formula for *fBm*. See [103], and also [172] and [227] for related results.

As in Lemma 2.8 of [1], we can get

Lemma 3.10.5. 1. $F \in \mathcal{G}_r$ implies $\|\tilde{E}\left[F|\mathcal{F}_t^{(H)}\right]\|_{\mathcal{G}_r} \leq \|F\|_{\mathcal{G}_r}$.

2. $F \in \mathcal{G}^*$ implies $\tilde{E}\left[F|\mathcal{F}_t^{(H)}\right] \in \mathcal{G}^*$.

3. $F, G \in \mathcal{G}^*$ implies $\tilde{E}\left[F \diamond G|\mathcal{F}_t^{(H)}\right] = \tilde{E}\left[F|\mathcal{F}_t^{(H)}\right] \diamond \tilde{E}\left[G|\mathcal{F}_t^{(H)}\right]$.

Motivated by Lemma 3.3.5 we now extend Definition 3.3.3 to elements in $\mathcal{G}^*$.

Definition 3.10.6. Let $F = \sum_\alpha c_\alpha \widetilde{\mathcal{H}}_\alpha(\omega) \in \mathcal{G}^*$. Then we define the stochastic gradient of F at t by

$$D_t^{(H)} F(\omega) = \frac{dF}{d\omega}(t, \omega) := \sum_\alpha c_\alpha \sum_i \alpha_i \widetilde{\mathcal{H}}_{\alpha-\varepsilon^{(i)}} e_i(t)$$

$$= \sum_\beta \left(\sum_i c_{\beta+\varepsilon^{(i)}} (\beta_i + 1) e_i(t) \right) \widetilde{\mathcal{H}}_\beta(\omega). \tag{3.51}$$

3.10 Fractional Clark Hausmann Ocone theorem

Lemma 3.10.7 (Fractional Clark Hausmann Ocone formula for polynomials). *Let $F(\omega)$ be $\mathcal{F}_T^{(H)}$-measurable and suppose $F(\omega) = P^\diamond(Y)$ for some polynomial $P(y) = \sum_\alpha c_\alpha y^\alpha$, where $Y = (Y_1, \ldots, Y_n)$ with $Y_j = \langle \omega, f_j \rangle$ as in Lemma 3.3.5, $1 \leq j \leq n$. Then*

$$F(\omega) = P^\diamond(Y^{(T)}), \quad \text{where} \quad Y_j^{(T)} = \langle \omega, f_j I_{[0,T]} \rangle,$$

and

$$F(\omega) = E[F] + \int_0^T \tilde{E}\left[D_t^{(H)} F | \mathcal{F}_t^{(H)}\right] dB_t^{(H)}.$$

Proof. The proof of Lemma 3.8 in [1] applies, with only conceptual modifications. □

We now state the main result of this section:

Theorem 3.10.8 (Fractional Clark Hausmann Ocone theorem).

1. *Let $G(\omega) \in \mathcal{G}^*$ be $\mathcal{F}_T^{(H)}$-measurable. Then $D_t^{(H)} G$ and $\tilde{E}\left[D_t^{(H)} G | \mathcal{F}_t^{(H)}\right]$ belong to $\mathcal{G}^*$ for almost all t. Moreover, $\tilde{E}\left[D_t^{(H)} G | \mathcal{F}_t^{(H)}\right] \diamond W^{(H)}(t)$ is integrable in $(\mathcal{S})_H^*$ and*

$$G(\omega) = E[G] + \int_0^T \tilde{E}\left[D_t^{(H)} G | \mathcal{F}_t^{(H)}\right] \diamond W^{(H)}(t) \, dt.$$

2. *Suppose $G(\omega) \in L^2(\mathbb{P}^H)$ is $\mathcal{F}_T^{(H)}$-measurable. Define*

$$\Psi(t, \omega) := \tilde{E}\left[D_t^{(H)} G | \mathcal{F}_t^{(H)}\right], \quad t \in [0, T].$$

Then Ψ belongs to the space $\mathcal{L}_\phi(0, T)$ and

$$G(\omega) = E[G] + \int_0^T \tilde{E}\left[D_t^{(H)} G | \mathcal{F}_t^{(H)}\right] dB_t^{(H)}.$$

Proof. The proof of 1 is similar to the proof of Theorem 3.15 in [1] and is omitted. The proof of 2 is similar to the proof of Theorem 3.11 in [1], but with minor modifications. For completeness we include a proof of 2.

Let $G(\omega) \in L^2(\mathbb{P}^H)$ have the expansion $G(\omega) = \sum_{\alpha \in \mathcal{J}} c_\alpha \tilde{\mathcal{H}}_\alpha(\omega)$. Define

$$\mathcal{J}_n = \{\alpha \in \mathcal{J} : |\alpha| \leq n \quad \text{and} \quad \text{length}(\alpha) \leq n\}$$

and put

$$G_n(\omega) = \sum_{\alpha \in \mathcal{J}_n} c_\alpha \tilde{\mathcal{H}}_\alpha(\omega).$$

Then

$$G_n(\omega) = \sum_{\alpha \in \mathcal{I}_n} c_\alpha X^{\diamond n},$$

where
$$X = (X_1, \ldots, X_n); \quad X_i = \langle \omega, e_i \rangle.$$

Hence, by Lemma 3.10.7 we have
$$G_n(\omega) = E[G_n] + \int_0^T \tilde{E}\left[D_t^{(H)} G_n | \mathcal{F}_t^{(H)}\right] dB_t^{(H)}$$

for all n. Since $G_n \to G$ in $L^2(\mathbb{P}^H)$ as $n \to \infty$, we deduce that
$$G(\omega) = E[G] + \lim_{n \to \infty} \int_0^T \tilde{E}\left[D_t^{(H)} G_n | \mathcal{F}_t^{(H)}\right] dB_t^{(H)}.$$

The sequence $\tilde{E}\left[D_t^{(H)} G_n | \mathcal{F}_t^{(H)}\right]$ is a Cauchy sequence and hence convergent in $\mathcal{L}_\phi(0,T)$. By 1 we know that this limit can be represented as
$$\tilde{E}\left[D_t^{(H)} G | \mathcal{F}_t^{(H)}\right].$$

This completes the proof of 2. $\square$

We shall call $\tilde{E}\left[D_t^{(H)} G | \mathcal{F}_t^{(H)}\right]$ the *fractional Clark derivative* of G in analogy with the classical Brownian motion case. We will use the notation
$$\nabla_t^\phi G = \tilde{E}\left[D_t^{(H)} G | \mathcal{F}_t^{(H)}\right].$$

Example 3.10.9. Let $\xi \in \mathbb{R}$ and let
$$X(t) = \exp\left(i\xi B_t^{(H)} + \frac{1}{2}\xi^2 t^{2H}\right), \quad t \geq 0.$$

Then from Example 3.4.4 it follows that
$$X(T) = 1 + i\xi \int_0^T X(s) \, dB_s^{(H)}$$

Thus we have
$$\nabla_t^\phi X(T) = i\xi X(t).$$

Consequently,
$$\nabla_t^\phi e^{i\xi B_T^{(H)}} = i\xi e^{i\xi B_t^{(H)} + 1/2 \xi^2 (t^{2H} - T^{2H})}. \tag{3.52}$$

Let $f \in \mathcal{S}(\mathbb{R})$ and let $\hat{f}$ be the Fourier transform of f, i.e. $\hat{f}(\xi) = \int_\mathbb{R} e^{-ix\xi} f(x) \, dx$. Then $f(x) = 1/(2\pi) \int_\mathbb{R} e^{ix\xi} \hat{f}(\xi) \, d\xi$. Consequently, we have
$$f(B_T^{(H)}) = \frac{1}{2\pi} \int_\mathbb{R} e^{i B_T^{(H)} \xi} \hat{f}(\xi) \, d\xi.$$

Therefore, by (3.52) we obtain

$$\nabla_t^\phi f(B_T^{(H)}) = \frac{1}{2\pi} \int_{\mathbb{R}} \nabla_t^\phi e^{i B_T^{(H)} \xi} \hat{f}(\xi)\, d\xi$$

$$= \frac{1}{2\pi} \int_{\mathbb{R}} i\xi \exp\left(i\xi B_t^{(H)} - \frac{1}{2}\xi^2\left(T^{2H} - t^{2H}\right)\right) \hat{f}(\xi)\, d\xi$$

$$= g(B_t^{(H)}),$$

where g is the inverse Fourier transform of the product of the following two functions: $\hat{f}(\xi)$ and

$$Q(\xi) = i\xi e^{-1/2\xi^2(T^{2H}-t^{2H})}.$$

However, $Q(\xi)$ is the Fourier transform of $dP_{t,T}(x)/dx$, where

$$P_{t,T}(x) = \frac{1}{\sqrt{2\pi(T^{2H} - t^{2H})}} \exp\left(-\frac{x^2}{2(T^{2H} - t^{2H})}\right),$$

which is the heat kernel at time $T^{2H} - t^{2H}$. Thus we have obtained

$$g(x) = \int_{\mathbb{R}} q_{t,T}(x - y) f(y)\, dy,$$

where $q_{t,T}(x) = dP_{t,T}(x)/dx$.

In general, we can obtain the following

Proposition 3.10.10. *Let f be a function such that $E\left[|f(B_T^{(H)})|\right] < \infty$. Then*

$$\nabla_t^\phi f(B_T^{(H)}) = \int_{\mathbb{R}} q_{t,T}(B_t^{(H)} - y) f(y)\, dy, \tag{3.53}$$

where $q_{t,T}(x) = dP_{t,T}(x)/dx$ with

$$P_{t,T}(x) = \frac{1}{\sqrt{2\pi(T^{2H} - t^{2H})}} \exp\left(-\frac{x^2}{2(T^{2H} - t^{2H})}\right) \tag{3.54}$$

and $0 \leq t \leq T$.

Remark 3.10.11. When $H = 1/2$, (3.53) and (3.54) reduce to known formulas. (See [111].)

3.11 Multidimensional fWIS integral

We wish to extend the results of the first part of this chapter to the multidimensional case. Let $B^{(H)}(t)$ be an m-*dimensional* fBm *with Hurst parameter* $H = (H_1, \ldots, H_m) \in (0,1)^m$ with respect to the probability measure $\mathbb{P}^H$ defined on $\Omega := (\mathcal{S}'(\mathbb{R}))^d$ by

$$\mathbb{P}^H := \mathbb{P}^{H_1} \otimes \cdots \otimes \mathbb{P}^{H_m}. \tag{3.55}$$

Note that for the sake of simplicity we are using the symbol $\mathbb{P}^H$ also to denote the product measure (3.55). This is the (version of the) *fBm* we will work with from now on. Similar to the one-dimensional case, we can define the multidimensional fWIS integral over $[0,T]$:

$$\int_0^T f(t,\omega)\,dB^{(H)}(t) = \sum_{k=1}^m \int_0^T f_k(t,\omega)\,dB_k^{(H)}(t) \in L^2(\mathbb{P}^H)$$

for $f = (f_1,\ldots,f_m)$ in

$$\mathcal{L}_\phi^{(m)}(0,T) := \mathcal{L}_{\phi_1}(0,T) \times \cdots \times \mathcal{L}_{\phi_m}(0,T).$$

In particular, we have

$$\left\| \int_\mathbb{R} f\,dB^{(H)} \right\|_{L^2(\mathbb{P}^H)} \leq \sum_{k=1}^m \|f_k\|_{\mathcal{L}_{\phi_k}(0,T)}. \tag{3.56}$$

It is useful to have an explicit expression for the norm on the left-hand side of (3.56).

Theorem 3.11.1 (Multidimensional fWIS isometry I). *Let* $f,g \in \mathcal{L}_\phi^{(m)}(0,T)$. *Then*

$$E\left[\left(\int_0^T f\,dB^{(H)}\right) \cdot \left(\int_0^T g\,dB^{(H)}\right)\right] = (f,g)_{\mathcal{L}_\phi^{(m)}(0,T)}$$

where

$$(f,g)_{\mathcal{L}_\phi^{(m)}(0,T)} = E\left[\sum_{k=1}^m \int_0^T \int_0^T f_k(s)g_k(t)\phi_k(s,t)\,ds\,dt \right. \\ \left. + \sum_{k,\ell=1}^m \left(\int_0^T \int_0^T D_{\ell,t}^\phi f_k(s) D_{k,s}^\phi g_\ell(t)\,ds\,dt\right)\right]. \tag{3.57}$$

Remark 3.11.2. Note the crossing of the indices ℓ, k of the derivatives and the components f_k, g_ℓ in the last terms of the right-hand side of (3.57).

To prove Theorem 3.11.1, we proceed as in Section 3.6, but with the appropriate modifications. In the sequel we put

$$L_H^{2,(m)}(\mathbb{R}) := L_{H_1}^2(\mathbb{R}) \times \cdots \times L_{H_m}^2(\mathbb{R}).$$

If $\alpha = (\alpha_1,\ldots,\alpha_m) \in L_H^{2,(m)}(\mathbb{R})$ we define the corresponding Wick exponential

$$\mathcal{E}(\alpha) = \exp^\diamond\left(\int_\mathbb{R} \alpha(t)\,dB^{(H)}(t)\right) = \exp^\diamond\left(\sum_{k=1}^m \int_\mathbb{R} \alpha_k(t)\,dB_k^{(H)}(t)\right)$$

$$= \exp\Big(\sum_{k=1}^{m}\int_{\mathbb{R}}\alpha_k(t)\,dB_k^{(H)}(t) - \frac{1}{2}\|\alpha\|_H^2\Big), \tag{3.58}$$

where

$$\|\alpha\|_H^2 = \sum_{k=1}^{m}\int_{\mathbb{R}}\alpha_k(s)\alpha_k(t)\phi_k(s,t)\,ds\,dt = \sum_{k=1}^{m}\|\alpha\|_{H_k}^2. \tag{3.59}$$

Let $\mathcal{E}$ be the linear span of all $\mathcal{E}(\alpha)$; $\alpha \in L_H^{2,(m)}(\mathbb{R})$. By Theorem 3.1.4 we have that $\mathcal{E}$ is a dense subset of $L^p(\mathbb{P}^H)$, for all $p \geq 1$, and we can reformulate Theorem 3.1.5 for the multidimensional case as follows.

Theorem 3.11.3. *Let $g_i = (g_{i_1},\ldots,g_{i_m}) \in L_H^{2,(m)}(\mathbb{R})$ for $i = 1, 2, \ldots, n$ such that*

$$\|g_{i_k} - g_{j_k}\|_{H_k} \neq 0 \quad \text{if } i \neq j, \ k = 1,\ldots,m.$$

Then $\mathcal{E}(g_1),\ldots,\mathcal{E}(g_n)$ are linearly independent in $L^2(\mathbb{P}^H)$.

If $F \in L^2(\mathbb{P}^H)$ and $g_k \in L_{H_k}^2(\mathbb{R})$, we put, as in [83],

$$D_{k,\Phi(g_k)}F = \int_{\mathbb{R}} D_{k,t}^\phi F \cdot g_k(t)\,dt.$$

We list some useful differentiation and Wick product rules. The proofs are similar to the one-dimensional case and are omitted.

Lemma 3.11.4. *Let $f = (f_1,\ldots,f_m) \in L_H^{2,(m)}(\mathbb{R})$, $g = (g_1,\ldots,g_m) \in L_H^{2,(m)}(\mathbb{R})$. Then*

1. $D_{k,\Phi(g_k)}\Big(\sum_{i=1}^{m}\int_{\mathbb{R}}f_i\,dB_i^{(H)}\Big) = \langle f, g\rangle_{H_k}$, $k = 1,\ldots,m$, where

$$\langle f, g\rangle_{H_k} = \int_{\mathbb{R}}\int_{\mathbb{R}} f_k(s)g_k(t)\phi_k(s,t)\,ds\,dt, \quad k = 1,\ldots,m.$$

2. $D_{k,s}^\phi\Big(\sum_{i=1}^{m}\int_{\mathbb{R}}f_i\,dB_i^{(H)}\Big) = \int_{\mathbb{R}} f_k(u)\phi_k(s,u)\,du$; $k = 1,\ldots,m$.
3. $D_{k,\Phi(g_k)}\mathcal{E}(f) = \mathcal{E}(f)\cdot \langle f, g\rangle_{H_k}$; $k = 1,\ldots,m$.
4. $D_{k,s}^\phi \mathcal{E}(f) = \mathcal{E}(f)\cdot \int_{\mathbb{R}} f_k(u)\phi_k(s,u)\,du$; $k = 1,\ldots,m$.
5. $\mathcal{E}(f)\diamond\mathcal{E}(g) = \mathcal{E}(f+g)$.
6. $F \diamond \int_{\mathbb{R}} g_k\,dB_k^{(H)} = F\cdot \int_{\mathbb{R}} g_k\,dB_k^{(H)} - D_{k,\Phi(g_k)}F$, $k = 1,\ldots,m$, provided that $F \in L^2(\mathbb{P}^H)$ and $D_{k,\Phi(g_k)}F \in L^2(\mathbb{P}^H)$.
7. $E[\mathcal{E}(f)\cdot \mathcal{E}(g)] = \exp\langle f, g\rangle_H$.

Lemma 3.11.5. *Suppose $\alpha_k \in L_{H_k}^2(\mathbb{R})$, $\beta_\ell \in L_{H_\ell}^2(\mathbb{R})$, $D_{\ell,\Phi(\beta_\ell)}F \in L^2(\mathbb{P}^H)$ and $D_{k,\Phi(\alpha_k)}G \in L^2(\mathbb{P}^H)$. Then*

$$E\Big[\Big(F\diamond\int_{\mathbb{R}}\alpha_k\,dB_k^{(H)}\Big)\cdot\Big(G\diamond\int_{\mathbb{R}}\beta_\ell\,dB_\ell^{(H)}\Big)\Big] \tag{3.60}$$
$$= E\big[(D_{\ell,\Phi(\beta_\ell)}F)\cdot(D_{k,\Phi(\alpha_k)}G) + \delta_{k\ell}FG\langle\alpha_k,\beta_k\rangle_{H_k}\big],$$

where
$$\delta_{k\ell} = \begin{cases} 1 & \text{if } k = \ell, \\ 0 & \text{otherwise.} \end{cases}$$

Proof. We adapt the argument in Theorem 3.5.5 to the multidimensional case. First note that by a density argument we may assume that

$$F = \mathcal{E}(f) = \exp\left(\int_{\mathbb{R}} f(t)\, dB^{(H)}(t) - \frac{1}{2}\|f\|_H^2\right)$$

and

$$G = \mathcal{E}(g) = \exp\left(\int_{\mathbb{R}} g(t)\, dB^{(H)}(t) - \frac{1}{2}\|g\|_H^2\right),$$

for some $f \in L_H^{2,(m)}(\mathbb{R})$, $g \in L_H^{2,(m)}(\mathbb{R})$. Choose $\delta = (\delta_1, \ldots, \delta_m) \in \mathbb{R}^m$, and $\gamma = (\gamma_1, \ldots, \gamma_m) \in \mathbb{R}^m$, and put $\delta \times f = (\delta_1 f_1, \ldots, \delta_m f_m)$ and $\gamma \times g = (\gamma_1 g_1, \ldots, \gamma_m g_m)$. Then by Lemma 3.11.4

$$E\left[\mathcal{E}(f) \diamond \mathcal{E}(\delta \times \alpha)) \cdot (\mathcal{E}(g) \diamond \mathcal{E}(\gamma \times \beta))\right] \tag{3.61}$$
$$= E\left[\mathcal{E}(f + \delta \times \alpha) \cdot \mathcal{E}(g + \gamma \times \beta)\right] = \exp(\langle f + \delta \times \alpha, g + \gamma \times \beta\rangle_H)$$
$$= \exp\left\{\sum_{i=1}^{m} \int_{\mathbb{R}} \int_{\mathbb{R}} (f_i + \delta_i \alpha_i)(s)(g_i + \gamma_i \beta_i)(t) \phi_i(s,t)\, ds\, dt\right\}. \tag{3.62}$$

We now compute the double derivatives

$$\frac{\partial^2}{\partial \delta_k \partial \gamma_\ell}$$

of (3.61) and (3.62) at $\delta = \gamma = 0$. We distinguish between two cases:

1. $k \neq \ell$: Then if we differentiate (3.61), we get

$$\frac{\partial^2}{\partial \delta_k \partial \gamma_\ell} E\left[(\mathcal{E}(f) \diamond \mathcal{E}(\delta \times \alpha)) \cdot (\mathcal{E}(g) \diamond \mathcal{E}(\gamma \times \beta))\right]_{\delta=\gamma=0}$$
$$= \frac{\partial}{\partial \gamma_\ell} E\left[\left(\mathcal{E}(f) \diamond \mathcal{E}(\delta \times \alpha) \diamond \int_{\mathbb{R}} \alpha_k\, dB_k^{(H)}\right) \cdot (\mathcal{E}(g) \diamond \mathcal{E}(\gamma \times \beta))\right]_{\delta=\gamma=0}$$
$$= E\left[\left(\mathcal{E}(f) \diamond \int_{\mathbb{R}} \alpha_k\, dB_k^{(H)}\right) \cdot \left(\mathcal{E}(g) \diamond \int_{\mathbb{R}} \beta_\ell\, dB_\ell^{(H)}\right)\right]. \tag{3.63}$$

On the other hand, if we differentiate (3.62) we get

$$\frac{\partial^2}{\partial \delta_k \partial \gamma_\ell}\left[\exp\langle f + \delta \times \alpha, g + \gamma \times \beta\rangle_H\right]_{\delta=\gamma=0}$$
$$= \frac{\partial}{\partial \gamma_\ell}\left[\exp\langle f + \delta \times \alpha, g + \gamma \times \beta\rangle_H\right.$$

$$\cdot \iint_{\mathbb{R}\,\mathbb{R}} \alpha_k(s)(g_k + \gamma_k \beta_k)(t)\phi_k(s,t)\,ds\,dt\bigg]_{\delta=\gamma=0}$$

$$= \exp\langle f,g\rangle_H \cdot \iint_{\mathbb{R}\,\mathbb{R}} \alpha_k(s)g_k(t)\phi_k(s,t)\,ds\,dt \cdot \iint_{\mathbb{R}\,\mathbb{R}} \beta_\ell(s)f_\ell(t)\phi_\ell(s,t)\,ds\,dt$$

$$= \exp\langle f,g\rangle_H \cdot \langle \alpha_k, g_k\rangle_{H_k} \cdot \langle \beta_\ell, f_\ell\rangle_{H_\ell}$$

$$= E\left[\mathcal{E}(f) \cdot \langle \beta_\ell, f_\ell\rangle_{H_\ell} \cdot \mathcal{E}(g) \cdot \langle \alpha_k, g_k\rangle_{H_k}\right]$$

$$= E\left[D_{\ell, \Phi(\beta_\ell)} \mathcal{E}(f) \cdot D_{k, \Phi(\alpha_k)} \mathcal{E}(g)\right]. \tag{3.64}$$

This proves (3.60) in this case.

2. $k = \ell$: In this case, if we differentiate (3.61), we get

$$\frac{\partial^2}{\partial \delta_k \partial \gamma_k} E\left[(\mathcal{E}(f) \diamond \mathcal{E}(\delta \times \alpha)) \cdot (\mathcal{E}(g) \diamond \mathcal{E}(\gamma \times \beta))\right]_{\delta=\gamma=0}$$

$$= \frac{\partial}{\partial \gamma_k} E\left[\left(\mathcal{E}(f) \diamond \mathcal{E}(\delta \times \alpha) \diamond \int_{\mathbb{R}} \alpha_k\, dB_k^{(H)}\right) \cdot (\mathcal{E}(g) \diamond \mathcal{E}(\gamma \times \beta))\right]_{\delta=\gamma=0}$$

$$= E\left[\left(\mathcal{E}(f) \diamond \int_{\mathbb{R}} \alpha_k\, dB_k^{(H)}\right) \cdot \left(\mathcal{E}(g) \diamond \int_{\mathbb{R}} \beta_k\, dB_k^{(H)}\right)\right]. \tag{3.65}$$

On the other hand, if we differentiate (3.62), we get

$$\frac{\partial^2}{\partial \delta_k \partial \gamma_k}\left[\exp\langle f + \delta \times \alpha, g + \gamma \times \beta\rangle_H\right]_{\delta=\gamma=0}$$

$$= \frac{\partial}{\partial \gamma_k}\bigg[\exp\langle f + \delta \times \alpha, g + \gamma \times \beta\rangle_H$$

$$\cdot \iint_{\mathbb{R}\,\mathbb{R}} \alpha_k(s)(g_k + \gamma_k \beta_k)(t)\phi_k(s,t)\,ds\,dt\bigg]_{\delta=\gamma=0}$$

$$= \exp\langle f,g\rangle_H \cdot \left[\langle \alpha_k, g_k\rangle_{H_k} \cdot (\beta_k, f_k)_{H_k} + \iint_{\mathbb{R}\,\mathbb{R}} \alpha_k(s)\beta_k(t)\phi_k(s,t)\,ds\,dt\right]$$

$$= E\left[D_{k,\Phi(\beta_k)} \mathcal{E}(f) \cdot D_{k,\Phi(\alpha_k)} \mathcal{E}(g) + \mathcal{E}(f)\mathcal{E}(g)\langle \alpha_k, \beta_k\rangle_{H_k}\right]. \tag{3.66}$$

This proves (3.60) also for this case. Hence, the proof of the lemma is complete. □

We are now ready to prove Theorem 3.11.1.

Proof. Put $\tilde{f}(t) = f(t)I_{[0,T]}(t)$. Then $\int_0^T f(t)\,dB_k^{(H)}(t) = \int_{\mathbb{R}} \tilde{f}(t)\,dB_k^{(H)}(t)$. We may consider $\int_{\mathbb{R}} \tilde{f}_k(t)\,dB_k^{(H)}(t)$ as the limit of sums of the form

$$\sum_{i=1}^N \tilde{f}_k(t_i) \diamond \left[B_k^{(H)}(t_{i+1}) - B_k^{(H)}(t_i)\right]$$

when $\Delta t_i = t_{i+1} - t_i \to 0$, $t_1 < t_2 < \cdots < t_N$, $N = 2, 3, \ldots$. Hence, $E\left[\left(\int_{\mathbb{R}} f\, dB^{(H)}\right)^2\right] = E\left[\left(\sum_{k=1}^{m} \int_{\mathbb{R}} \tilde{f}_k\, dB_k^{(H)}\right)^2\right]$ is the limit of sums of the form

$$\sum_{i,j,k,\ell} E\left[(\tilde{f}_k(t_i) \diamond (B_k^{(H)}(t_{i+1}) - B_k^{(H)}(t_i))) \cdot (\tilde{f}_\ell(t_j) \diamond (B_\ell^{(H)}(t_{j+1}) - B_\ell^{(H)}(t_j)))\right],$$

which by Lemma 3.11.5 is equal to

$$\sum_{i,j,k,\ell} E\left[\int_{t_i}^{t_{i+1}} \int_{t_j}^{t_{j+1}} D_{\ell,s}^\phi \tilde{f}_k(t_j) D_{k,t}^\phi \tilde{f}_\ell(t_i)\, ds\, dt\right]$$

$$+ E\left[\delta_{k\ell} \int_{t_i}^{t_{i+1}} \int_{t_j}^{t_{j+1}} \tilde{f}_k(t_i)\tilde{f}_k(t_j)\phi_k(s,t)\, ds\, dt\right].$$

When $\Delta t_i \to 0$, this converges to

$$E\left[\sum_{k,\ell=1}^{m} \left(\int_{\mathbb{R}} \int_{\mathbb{R}} D_{\ell,t}^\phi \tilde{f}_k(s) D_{k,s}^\phi \tilde{f}_\ell(t)\, ds\, dt\right)\right.$$

$$\left. + \sum_{k=1}^{m} \int_{\mathbb{R}} \int_{\mathbb{R}} \tilde{f}_k(s)\tilde{f}_k(t)\phi_k(s,t)\, ds\, dt\right].$$

This proves (3.57) when $f = g$. By polarization the proof of Theorem 3.11.1 is complete. □

Using Theorem 3.11.1, we can now proceed as in the one-dimensional case with appropriate modifications, and obtain a fractional multidimensional Itô formula. We omit the proof (for a proof in $n = 2$, see also [83]).

Theorem 3.11.6 (The fractional multidimensional Itô formula). *Let* $X(t) = (X_1(t), \ldots, X_n(t))$, *with*

$$dX_i(t) = \sum_{j=1}^{m} \sigma_{ij}(t,\omega) dB_j^{(H)}(t),$$

where $\sigma_i = (\sigma_{i1}, \ldots, \sigma_{im}) \in \mathcal{L}_\phi^{(m)}(0,T)$, $1 \leq i \leq n$. *Suppose that for all* $j = 1, \ldots, m$ *there exists* $\theta_j > 1 - H_j$ *such that*

$$\sup_i E\left[(\sigma_{ij}(u) - \sigma_{ij}(v))^2\right] \leq C|u-v|^{\theta_j} \quad \text{if } |u-v| < \delta$$

where $\delta > 0$ *is a constant. Moreover, suppose that*

$$\lim_{\substack{0 \leq u,v \leq t \\ |u-v| \to 0}} \sup_{i,j,k} E\left[(D_{k,u}^\phi \{\sigma_{ij}(u) - \sigma_{ij}(v)\})^2\right] = 0.$$

Let $f \in C^{1,2}(\mathbb{R} \times \mathbb{R}^n)$ with bounded second-order derivatives with respect to x. Then for $t \in [0, T]$,

$$f(t, X(t)) = f(0, X(0)) + \int_0^t \frac{\partial f}{\partial s}(s, X(s))\, ds + \int_0^t \sum_{i=1}^n \frac{\partial f}{\partial x_i}(s, X(s))\, dX_i(s)$$

$$+ \int_0^t \Big\{ \sum_{i,j=1}^m \frac{\partial^2 f}{\partial x_i \partial x_j}(s, X(s)) \sum_{k=1}^m \sigma_{ik}(s) D_{k,s}^\phi (X_j(s)) \Big\}\, ds$$

$$= f(0, X(0)) + \int_0^t \frac{\partial f}{\partial s}(s, X(s))\, ds$$

$$+ \sum_{j=1}^m \int_0^t \Big[\sum_{i=1}^n \frac{\partial f}{\partial x_i}(s, X(s)) \sigma_{ij}(s, \omega)\Big] dB_j^{(H)}(s)$$

$$+ \int_0^t \mathrm{Tr}\left[\Lambda^T(s) f_{xx}(s, X(s)) \right] ds, \qquad (3.67)$$

where $(\cdot)^T$ denotes matrix transposed, $\mathrm{Tr}[\cdot]$ denotes matrix trace, and $\Lambda = [\Lambda_{ij}] \in \mathbb{R}^{n \times m}$ with

$$\Lambda_{ij}(s) = \sum_{k=1}^m \sigma_{ik} D_{k,s}^\phi(X_j(s)), \qquad 1 \le i \le n, \quad 1 \le j \le m, \qquad (3.68)$$

$$f_{xx} = \Big[\frac{\partial^2 f}{\partial x_i \partial x_j}\Big]_{1 \le i,j \le n} \in \mathbb{R}^{n \times n}. \qquad (3.69)$$

If we combine Theorem 3.11.6 with Theorem 3.11.1, we get the following result, which also may be regarded as a fractional Itô isometry:

Theorem 3.11.7 (Fractional Itô isometry II). *Suppose $f = (f_1, \ldots, f_m) \in \mathcal{L}_\phi^{(m)}(0, T)$. Then, for $T > 0$,*

$$E\Big[\Big(\int_0^T \int_0^T D_{\ell,t}^\phi f_k(s) D_{k,s}^\phi f_\ell(t)\, ds\, dt\Big)\Big]$$

$$= E\Big[\int_0^T \Big\{f_k(t) \int_0^t D_{k,t}^\phi f_\ell(s)\, dB_\ell^{(H)}(s)$$

$$+ f_\ell(t) \int_0^T D_{\ell,t}^\phi f_k(s)\, dB_k^{(H)}(s)\Big\} dt\Big]$$

Proof. By the Itô formula (Theorem 3.11.6) we have

$$E\Big[\Big(\int_0^T f_k\, dB_k^{(H)}\Big) \cdot \Big(\int_0^T f_\ell\, dB_\ell^{(H)}\Big)\Big]$$

$$= E\Big[\int_0^T f_k(t) D_{k,t}^\phi \Big(\int_0^t f_\ell(s)\, dB_\ell^{(H)}(s)\Big) dt\Big]$$

$$+ E\Big[\int_0^T f_k(t) D^\phi_{\ell,t}\Big(\int_0^t f_k(s)\,dB_k^{(H)}(s)\Big)\,dt\Big]$$

$$= E\Big[\int_0^T \Big\{ f_k(t) \int_0^t D^\phi_{k,t} f_\ell(s)\,dB_\ell^{(H)}(s)$$

$$+ f_\ell(t) \int_0^t D^\phi_{\ell,t} f_k(s)\,dB_k^{(H)}(s) \Big\}\,dt\Big]$$

$$+ \delta_{k\ell} E\Big[\int_0^T \int_0^t \{f_k(t)f_k(s) + f_\ell(t)f_k(s)\}\phi_k(s,t)\,ds\,dt\Big], \quad (3.70)$$

where we have used that, for $u > 0$,

$$D^\phi_{k,t}\Big[\int_0^u f_\ell(s)\,dB_\ell^{(H)}(s)\Big]$$

$$= \int_0^u D^\phi_{k,t} f_\ell(s)\,dB_\ell^{(H)}(s) + \delta_{k\ell} \int_0^u f_k(s)\phi_k(t,s)\,ds.$$

On the other hand, the Itô isometry (Theorem 3.11.1) gives that

$$E\Big[\Big(\int_0^T f_k\,dB_k^{(H)}\Big) \cdot \Big(\int_0^T f_\ell\,dB_\ell^{(H)}\Big)\Big]$$
$$= E\Big[\Big(\int_0^T \int_0^T D^\phi_{\ell,t} f_k(s) D^\phi_{k,s} f_\ell(t)\,ds\,dt\Big) + \delta_{k\ell}\|f_k\|^2_{H_k}\Big]. \quad (3.71)$$

Comparing (3.70) and (3.71), we get the theorem. □

We end this section by proving a fractional integration by parts formula. First we recall the following:

Theorem 3.11.8 (Fractional Girsanov formula). *Let* $\gamma = (\gamma_1, \ldots, \gamma_m)$ *be in* $\in (L^2(\mathbb{R}))^m$ *and* $\hat\gamma = (\hat\gamma_1, \ldots, \hat\gamma_m) \in L_H^{2,(m)}(\mathbb{R}) \cap C(\mathbb{R}, \mathbb{R}^m)$ *are related by*

$$\gamma_k(t) = \int_\mathbb{R} \hat\gamma_k(s)\phi_k(s,t)\,ds, \qquad t \in \mathbb{R}, \quad k = 1, \ldots, m. \quad (3.72)$$

Let $G \in L^2(\mathbb{P}^H)$. *Then*

$$E[G(\omega + \gamma)] = E[G(\omega)\exp^\circ(\langle\omega,\hat\gamma\rangle)] = E\Big[G(\omega)\mathcal{E}\Big(\int_\mathbb{R} \hat\gamma\,dB^{(H)}\Big)\Big].$$

The proof in the multidimensional case is similar to the one-dimensional case (Theorem 3.2.2).

Definition 3.11.9. *If* $F \in L^2(\mathbb{P}^H)$ *and* $\gamma = (\gamma_1, \ldots, \gamma_m) \in (L^2(\mathbb{R}))^m$, *the directional derivative of* F *in the direction* γ *is defined by*

$$D_\gamma^{(H)} F(\omega) = \lim_{\varepsilon \to 0} \frac{F(\omega + \varepsilon \gamma) - F(\omega)}{\varepsilon},$$

provided the limit exists in $L^2(\mathbb{P}^H)$. We say that F is differentiable *if there exists a process* $D_t^{(H)} F(\omega) = (D_{1,t}^{(H)} F(\omega), \ldots, D_{m,t}^{(H)} F(\omega))$ such that $D_{k,t}^{(H)} F(\omega) \in L^2(d\mathbb{P}^H \times dt)$ for all $k = 1, \ldots, m$ and

$$D_\gamma^{(H)} F(\omega) = \int_\mathbb{R} D_t^{(H)} F(\omega) \cdot \gamma(t) \, dt \qquad \forall \gamma \in (L^2(\mathbb{R}))^m.$$

Theorem 3.11.10 (Fractional integration by parts I). *Consider $F, G \in L^2(\mathbb{P}^H)$ and $\gamma \in (L^2(\mathbb{R}))^m$, and assume that the directional derivatives $D_\gamma^{(H)} F$, $D_\gamma^{(H)} G$ exist. Then*

$$E\left[D_\gamma^{(H)} F \cdot G\right] = E\left[F \cdot G \cdot \int_\mathbb{R} \hat\gamma \, dB^{(H)}\right] - E\left[F \cdot D_\gamma^{(H)} G\right],$$

where $\hat\gamma$ is defined in (3.72).

Proof. By Theorem 3.11.8 we have, for all $\varepsilon > 0$,

$$E\left[F(\omega + \varepsilon\gamma) G(\omega)\right] = E\left[F(\omega) G(\omega - \varepsilon\gamma) \exp^\circ(\varepsilon\langle\omega, \hat\gamma\rangle)\right].$$

Hence

$$
\begin{aligned}
E\left[D_\gamma^{(H)} F \cdot G\right] &= E\left[\lim_{\varepsilon \to 0} \frac{1}{\varepsilon} \{F(\omega + \varepsilon\gamma) - F(\omega)\} G(\omega)\right] \\
&= E\left[\lim_{\varepsilon \to 0} \frac{1}{\varepsilon} \{F(\omega)[G(\omega - \varepsilon\gamma) \exp^\circ(\varepsilon\langle\omega, \hat\gamma\rangle) - G(\omega)]\}\right] \\
&= E\left[F(\omega) \frac{d}{d\varepsilon} \left\{G(\omega - \varepsilon\gamma) \exp\left(\varepsilon \int_\mathbb{R} \hat\gamma \, dB^{(H)} - \frac{1}{2}\varepsilon^2 \|\hat\gamma\|_H^2\right)\right\}_{\varepsilon=0}\right] \\
&= E\left[F(\omega) G(\omega) \int_\mathbb{R} \hat\gamma \, dB^{(H)}\right] - E\left[F(\omega) D_\gamma^{(H)} G(\omega)\right].
\end{aligned}
$$

$\square$

We now apply the above to the fractional gradient

$$D_t^\phi F = \int_\mathbb{R} D_s^{(H)} F \cdot \phi(s,t) \, ds = \sum_{k=1}^m \int_\mathbb{R} D_{k,s}^{(H)} F \cdot \phi_k(s,t) \, ds.$$

Theorem 3.11.11 (Fractional integration by parts II). *Suppose $F, G \in L^2(\mathbb{P}^H)$ are differentiable with fractional gradients $D_t^\phi F$, $D_t^\phi G$. Then for each $t \in \mathbb{R}$, $k \in \{1, \ldots, m\}$, we have*

$$E\left[D_{k,t}^\phi F \cdot G\right] = E\left[F \cdot G \cdot B_k^{(H)}(t)\right] - E\left[F \cdot D_{k,t}^\phi G\right].$$

Proof. Choose a sequence $\hat{\gamma}_k^{(j)} \in L_{H_k}^2(\mathbb{R})$, $j = 1, 2, \ldots$, such that $\lim_{j\to\infty} \hat{\gamma}_k^{(j)} = \delta_t(\cdot)$ (the point mass at t) in the sense that if we define

$$\phi_k^{(j)}(s) = \int_{\mathbb{R}} \hat{\gamma}_k^{(j)}(r) \phi_k(s,r) \, dr$$

then $\phi_k^{(j)}(\cdot) \to \phi_k(\cdot, t)$ in $L^2(\mathbb{R})$. Then by Theorem 3.11.10

$$\begin{aligned} E\left[D_{k,t}^{\phi} F \cdot G\right] &= E\left[\lim_{j\to\infty} D_{\phi_k^{(j)}} F \cdot G\right] = \lim_{j\to\infty} E\left[D_{\phi_k^{(j)}} F \cdot G\right] \\ &= \lim_{j\to\infty} E\left[F \cdot G \cdot \int_{\mathbb{R}} \hat{\gamma}^{(j)} dB^{(H)}\right] - E\left[F \cdot D_{\phi_k^{(j)}} G\right] \\ &= E\left[F \cdot G \cdot B_k^{(H)}(t)\right] - E\left[F \cdot D_{k,t}^{\phi} G\right]. \end{aligned}$$

$\square$

3.12 Relation between the fWIS integral and the divergence-type integral for $H > 1/2$

We now investigate the relation between the stochastic integral and the divergence-type integral introduced in Chapter 2 in the case when $H > 1/2$. In the following we refer to the notation used in Chapter 2 and recall that $\mathcal{H} = \overline{(L^2([0,T]), <,>_H)}$.

First of all, by Remark 3.1.1 we have $L_\phi^2(\mathbb{R}) \supseteq L_\phi^2([0,T]) = \mathcal{H}$ if we identify $\psi \in L_\phi^2([0,T])$ with $\psi I_{[0,T]}$. Hence, for every deterministic $\psi \in \mathcal{H}$ $B^{(H)}(\psi)$ coincides with the fWIS integral of ψ. Next we note that for any $\psi \in \mathcal{H}$ and any random variable $F \in L^2(\mathbb{P}^H)$, the Wick product $F \diamond B^{(H)}(\psi)$ exists in $L^2(\mathbb{P}^H)$ if and only if ψF belongs to dom δ_H. Hence, the equality

$$F \diamond B^{(H)}(\psi) = \delta_H(F\psi) \tag{3.73}$$

follows by Proposition 3.5.4 and equations (2.33) to (2.34). The relation (3.73) can also be seen as a consequence of the characterization of dom δ_H in terms of the Wiener chaos expansion.

Proposition 3.12.1. *Consider $H > 1/2$. Let u be a process that is λ-Hölder continuous in the norm $\mathbb{D}^{1,2}(\mathcal{H})$ with $\lambda > H - 1/2$. Then*

$$\lim_{|\pi|\to 0^+} \sum_{i=1}^n u_{t_i} \diamond (B^{(H)}(t_{i+1}) - B^{(H)}(t_i)) = \delta_H(u),$$

where $\pi = \{0 = s_0 < s_1 < \cdots < s_{n+1} = T\}$ is a partition of $[0,T]$ with mesh size $|\pi| = \max_{i=1,\ldots,n} |t_{i+1} - t_i|$ and the convergence holds in $L^2(\mathbb{P}^H)$.

3.12 Relation between the fWIS and the divergence-type integral

Proof. We sketch here the proof of Proposition 4 of [6]. Consider $u^\pi(t) = \sum_{i=1}^{n} u_{t_i} I_{[t_i,t_{i+1})}$. By (3.73) we obtain

$$\delta_H(u^\pi) = \sum_{i=1}^{n} u_{t_i} \diamond \left[B^{(H)}(t_{i+1}) - B^{(H)}(t_i) \right].$$

The thesis follows since u^π converges to u in the norm $\mathbb{D}_H^{1,2}$. □

Hence we can conclude that under the hypotheses of Proposition 3.12.1, the fWIS integral coincides with the divergence operator for $H > 1/2$. Moreover, in Chapter 5 we will investigate the relation between the divergence operator (respectively, the fWIS integral) and the forward integral.

4
Wick Itô Skorohod (WIS) integrals for fractional Brownian motion

In this chapter we again study the stochastic integral for the *fBm* following the white noise approach. However, the integral is defined here as an element in the *classical* Hida distribution space by using the *white noise theory* and *Malliavin calculus for standard Brownian motion* introduced in Appendix A. The main advantage of this method with respect to the one presented in Chapter 3 is that it permits to define the stochastic integral for any $H \in (0,1)$. In addition, it doesn't require the introduction of the fractional white noise theory since it uses the well-established one for the standard case.

On the other side, the approach of Chapter 3 can be seen as more intrinsic. For a further discussion of the relation among these two types of integrals we refer to Chapter 6. The main references for this chapter are [34] and [89].

4.1 The M operator

Let $\mathcal{S}(\mathbb{R})$ denote the Schwartz space of rapidly decreasing smooth functions on $\mathbb{R}$, and let $\Omega := \mathcal{S}'(\mathbb{R})$ be its dual, the *space of tempered distributions*. Let $\mathbb{P}$ be the *white noise probability measure* on the Borel sets $\mathcal{B}(\mathcal{S}'(\mathbb{R}))$ defined by the property that

$$\int_{\mathcal{S}'(\mathbb{R})} \exp(i<\omega,f>)\,d\mathbb{P}(\omega) = \exp\left(-\frac{1}{2}\|f\|^2_{L^2(\mathbb{R})}\right), \quad f \in \mathcal{S}(\mathbb{R}),$$

where $i^2 = -1$ and $<\omega,f> = \omega(f)$ is the action of $\omega \in \Omega = \mathcal{S}'(\mathbb{R})$ on $f \in \mathcal{S}(\mathbb{R})$.

Remark 4.1.1. Note that the underlying probability measure $\mathbb{P}$ is here *the same as for the standard Brownian motion* $B(t)$. For further details about the white noise theory and Malliavin calculus for standard Brownian motion applied in this chapter to the *fBm* case, we refer to Appendix A.

The main idea of this approach is to relate the fBm $B^{(H)}(t)$ with Hurst parameter $H \in (0,1)$ to classical Brownian motion $B(t)$ (which corresponds to $H = 1/2$) via the following operator M:

Definition 4.1.2. *Let $0 < H < 1$. The operator $M = M_H$ is defined on functions $f \in \mathcal{S}(\mathbb{R})$ by*

$$\widehat{Mf}(y) = |y|^{1/2-H} \hat{f}(y), \quad y \in \mathbb{R} \tag{4.1}$$

where

$$\hat{g}(y) := \int_{\mathbb{R}} e^{-ixy} g(x) \, dx$$

denotes the Fourier transform.

We remark that in the sequel we will usually use the notation M instead of M_H, unless we need to specify the associated Hurst parameter H. Equivalently, for every $0 < H < 1$ the operator M can be defined as

$$Mf(x) = -\frac{d}{dx} \frac{C_H}{(H-1/2)} \int_{\mathbb{R}} (t-x)|t-x|^{H-3/2} f(t) \, dt \tag{4.2}$$

where $f \in \mathcal{S}(\mathbb{R})$ and where

$$C_H = \left\{ 2\Gamma\left(H - \frac{1}{2}\right) \cos\left[\frac{\pi}{2}\left(H - \frac{1}{2}\right)\right] \right\}^{-1} [\Gamma(2H+1)\sin(\pi H)]^{1/2},$$

with $\Gamma(\cdot)$ denoting the classical gamma function. This can be restated as follows. Let $f \in \mathcal{S}(\mathbb{R})$. For $0 < H < 1/2$ we have

$$Mf(x) = C_H \int_{\mathbb{R}} \frac{f(x-t) - f(x)}{|t|^{3/2-H}} \, dt.$$

For $H = 1/2$ we have
$$Mf(x) = f(x).$$

For $1/2 < H < 1$ we have

$$Mf(x) = C_H \int_{\mathbb{R}} \frac{f(t)}{|t-x|^{3/2-H}} \, dt. \tag{4.3}$$

See [89]. The exact domains and ranges of M have been characterized in [206]. The operator M extends in a natural way from $\mathcal{S}(\mathbb{R})$ to the space

$$\begin{aligned} L_H^2(\mathbb{R}) :&= \{f: \mathbb{R} \to \mathbb{R} \text{ (deterministic)} : |y|^{1/2-H} \hat{f}(y) \in L^2(\mathbb{R})\} \\ &= \{f: \mathbb{R} \to \mathbb{R} : Mf(x) \in L^2(\mathbb{R})\} \\ &= \{f: \mathbb{R} \to \mathbb{R} : \|f\|_H < \infty\}, \end{aligned}$$

where
$$\|f\|_H := \|Mf\|_{L^2(\mathbb{R})}.$$
The inner product on this space is
$$\langle f, g \rangle_H = \langle Mf, Mg \rangle_{L^2(\mathbb{R})}. \qquad (4.4)$$

We remark that $L_H^2(\mathbb{R})$ is not closed with respect the inner product (4.4), as we have seen in Chapter 2 (see also [188]).

In particular, the indicator function $I_{[0,t]}(\cdot)$ is easily seen to belong to $L_H^2(\mathbb{R})$ for fixed $t \in \mathbb{R}$, and we write
$$MI_{[0,t]}(x) := M[0,t](x).$$
Note that if $f, g \in L^2(\mathbb{R}) \cap L_H^2(\mathbb{R})$, then
$$\langle f, Mg \rangle_{L^2(\mathbb{R})} = \langle \hat{f}, \widehat{Mg} \rangle_{L^2(\mathbb{R})} =$$
$$\int_{\mathbb{R}} |y|^{1/2-H} \hat{f}(y) \hat{g}(y) \, dy = \langle \widehat{Mf}, \hat{g} \rangle_{L^2(\mathbb{R})} = \langle Mf, g \rangle_{L^2(\mathbb{R})}. \qquad (4.5)$$

Remark 4.1.3. Note that by (4.1) it follows that for $H \in (0,1)$ we have
$$M_H(M_{1-H}f) = f, \quad f \in \mathcal{S}(\mathbb{R}).$$

Example 4.1.4. We now compute $M[a,b] := MI_{[a,b]}(x)$, i.e., Mf when f is the indicator function of an interval $[a,b]$ with $a < b$. By [89], $M[a,b]$ can be calculated explicitly as
$$M[a,b](x) = \frac{[\Gamma(2H+1)\sin(\pi H)]^{1/2}}{2\Gamma(H+1/2)\cos[\pi/2(H+1/2)]} \left[\frac{b-x}{|b-x|^{3/2-H}} - \frac{a-x}{|a-x|^{3/2-H}} \right]$$

This can be computed directly by using the characterization (4.2). We also remark that $Mf \in L^2(\mathbb{R})$ for this choice of f.

Moreover, by using (4.1) we obtain by Parceval's Theorem, for $0 < H < 1$,
$$\int_{\mathbb{R}} [M[a,b](x)]^2 \, dx = \frac{1}{2\pi} \int_{\mathbb{R}} [\widehat{M[a,b]}(\xi)]^2 \, d\xi$$
$$= \frac{1}{2\pi} \int_{\mathbb{R}} |\xi|^{1-2H} \frac{|e^{-ib\xi} - e^{-ia\xi}|^2}{|\xi|^2} \, d\xi$$
$$= (b-a)^{2H}, \qquad (4.6)$$
where we have used the fact that
$$\widehat{I_{[a,b]}}(\xi) = \frac{[e^{-ib\xi} - e^{-ia\xi}]}{-i\xi}.$$

Since $M[s,t] = M[0,t] - M[0,s]$ for $s < t$, by polar identity arguments we obtain

$$\int_{\mathbb{R}} M[0,t](x)M[0,s](x)\,dx = \frac{1}{2}(|t|^{2H} + |s|^{2H} - |t-s|^{2H}), \tag{4.7}$$

which holds for arbitrary $s,t \in \mathbb{R}$ (see [89, (A.10)]). When $t < 0$, $M[0,t] = -M[0,-t]$. Moreover, we also have, for $0 < H, L < 1$,

$$\int_{\mathbb{R}} M_H[a,b](x)M_L[a,b](x)\,dx = |a-b|^{H+L}, \tag{4.8}$$

and

$$\int_{\mathbb{R}} M_H[0,t](x)M_L[0,s](x)\,dx = \frac{1}{2}(|t|^{H+L} + |s|^{H+L} - |t-s|^{H+L}). \tag{4.9}$$

If $H = L$, (4.8) and (4.9) coincide with (4.6) and (4.7), respectively.

Remark 4.1.5. By (4.7) we have that for $s,t > 0$,

$$\int_{\mathbb{R}} M[0,t](x)M[0,s](x)\,dx = R_H(t,s) = \langle I_{[0,t]}, I_{[0,s]} \rangle_H$$

as in (2.7) and (2.25). This justifies the use of the same notation for the two inner products.

Moreover, comparing (3.2) and (4.7), by limit argument we obtain that

$$\langle Mf, Mg \rangle_{L^2(\mathbb{R})} = \int_{\mathbb{R}} \int_{\mathbb{R}} f(s)g(t)\phi(s,t)\,ds\,dt, \tag{4.10}$$

where ϕ is defined in (3.1). Hence, for $H > 1/2$, $L_H^2(\mathbb{R})$ is the subspace of deterministic functions of the space $L_\phi^2(\mathbb{R})$ defined in Section 3.1.

We now define, for $t \in \mathbb{R}$,

$$\tilde{B}^{(H)}(t) := \tilde{B}^{(H)}(t,\omega) := <\omega, M[0,t](\cdot)>, \tag{4.11}$$

where $<\omega, f> = \omega(f)$ is the action of $\omega \in \Omega = \mathcal{S}'(\mathbb{R})$ on $f \in \mathcal{S}(\mathbb{R})$. Then $\tilde{B}^{(H)}(t)$ is Gaussian, $\tilde{B}^{(H)}(0) = E[\tilde{B}^{(H)}(t)] = 0$ for all $t \in \mathbb{R}$, and by (4.7)

$$E[\tilde{B}^{(H)}(s)\tilde{B}^{(H)}(t)] = \int_{\mathbb{R}} M[0,s](x)M[0,t](x)\,dx$$

$$= \frac{1}{2}(|t|^{2H} + |s|^{2H} - |s-t|^{2H}).$$

Therefore, the continuous version $B^{(H)}(t)$ of $\tilde{B}^{(H)}(t)$ is a *fBm*, as defined in Chapter 1.

Let $f(x) = \sum_j a_j I_{[t_j,t_{j+1}]}(x)$ be a step function. Then by (4.11) and linearity

$$<\omega, Mf> = \sum_j a_j <\omega, M[t_j, t_{j+1}]>$$

$$= \sum_j a_j (B^{(H)}(t_{j+1}) - B^{(H)}(t_j))$$

$$=: \int_{\mathbb{R}} f(t)\, dB^{(H)}(t). \tag{4.12}$$

Since

$$\| <\omega, Mf> \|_{L^2(\mathbb{P})} = \|Mf\|_{L^2(\mathbb{R})} = \|f\|_H,$$

we see that definition (4.12) extends to all $f \in L_H^2(\mathbb{R})$. Comparing with (A.4), we obtain

$$\int_{\mathbb{R}} f(t)\, dB^{(H)}(t) = \int_{\mathbb{R}} Mf(t)\, dB(t), \quad f \in L_H^2(\mathbb{R}). \tag{4.13}$$

Since $Mf \in L^2(\mathbb{R})$ for all $f \in \mathcal{S}(\mathbb{R})$, we can, by (A.2), define $M : \mathcal{S}'(\mathbb{R}) \to \mathcal{S}'(\mathbb{R})$ by

$$<M\omega, f> = <\omega, Mf>, \quad f \in \mathcal{S}(\mathbb{R}) \text{ for } \mathbb{P}\text{-a.e. } \omega \in \Omega = \mathcal{S}'(\mathbb{R}). \tag{4.14}$$

Remark 4.1.6. By (4.11) and (4.13) we obtain

$$B^{(H)}(t) = \int_{\mathbb{R}} M_H[0,t](u)\, dB(u),$$

and because of the properties of M_H [see, e.g., (4.6)] we also have

$$B(t) = \int_{\mathbb{R}} M_{1-H}[0,t](u)\, dB^{(H)}(u).$$

4.2 The Wick Itô Skorohod (WIS) integral

Let $\{\xi_k\}_{k=1}^{\infty}$ be the Hermite functions as defined in (A.6) of Appendix A. Define

$$e_k(x) = M^{-1}\xi_k(x), \quad k = 1, 2, \ldots$$

Then $\{e_k\}_{k=1}^{\infty}$ is orthonormal in $L_H^2(\mathbb{R})$, and the closed linear span of $\{e_k\}_{k=1}^{\infty}$ contains $L_H^2(\mathbb{R})$. Note, however, that this closed linear span of also contains distributions that do not belong to $L_H^2(\mathbb{R})$. See Chapter 2 for further details.

Example 4.2.1 (Fractional white noise). From (4.11) we see that for each t the random variable $B^{(H)}(t, \omega)$ belongs to $L^2(\mathbb{P})$. Let $\mathcal{H}_\alpha(\omega)$ be defined as in (A.8) and denote by $\varepsilon^{(k)}$ the unit vectors

$$\varepsilon^{(k)} = (0, 0, \ldots, 0, 1)$$

with 1 on the kth entry and 0 otherwise, $k = 1, 2, \ldots$ By Theorem A.1.3 the chaos expansion of $B^{(H)}(t)$ can be found as follows:

$$B^{(H)}(t) = <\omega, M[0,t](\cdot)> = <M\omega, I_{[0,t]}(\cdot)>$$

$$= <M\omega, \sum_{k=1}^{\infty} \langle I_{[0,t]}, e_k \rangle_H e_k(\cdot)>$$

$$= <M\omega, \sum_{k=1}^{\infty} \langle M[0,t], Me_k \rangle_{L^2(\mathbb{R})} e_k(\cdot)>$$

$$= \sum_{k=1}^{\infty} \langle M[0,t], \xi_k \rangle_{L^2(\mathbb{R})} <M\omega, e_k>$$

$$= \sum_{k=1}^{\infty} \langle I_{[0,t]}, M\xi_k \rangle_{L^2(\mathbb{R})} <\omega, Me_k>$$

$$= \sum_{k=1}^{\infty} \int_0^t M\xi_k(s), ds\ \mathcal{H}_{\varepsilon(k)}(\omega), \qquad (4.15)$$

where we have used (4.11), (4.14), and (4.5) and the fact that $\xi_n \in L^2(\mathbb{R}) \cap L^2_H(\mathbb{R})$.

Now define the *fractional white noise* $W^{(H)}(t)$ by the expansion

$$W^{(H)}(t) = \sum_{k=1}^{\infty} M\xi_k(t) \mathcal{H}_{\varepsilon(k)}(\omega). \qquad (4.16)$$

Then it can be shown (see Example 8.2.3) that $W^{(H)}(t) \in (\mathcal{S})^*$ for all t and

$$\frac{dB^{(H)}(t)}{dt} = W^{(H)}(t) \text{ in } (\mathcal{S})^*. \qquad (4.17)$$

For the definition of the Hida space $(\mathcal{S})^*$ stochastic distributions see Definition A.1.4.

In view of Theorem A.2.2 the following definition is natural.

Definition 4.2.2 (The Wick Itô Skorohod integral).
Let $Y : \mathbb{R} \to (\mathcal{S})^*$ be such that $Y(t) \diamond W^{(H)}(t)$ is dt-integrable in $(\mathcal{S})^*$ *(Definition A.1.5)*. Then we say that Y is Wick Itô Skorohod integrable *(WIS integrable)* and we define the Wick Itô Skorohod integral *(WIS integral)* of $Y(t) = Y(t,\omega)$ with respect to $B^{(H)}(t)$ by

$$\int_{\mathbb{R}} Y(t,\omega)\, dB^{(H)}(t) := \int_{\mathbb{R}} Y(t) \diamond W^{(H)}(t)\, dt,$$

where the Wick product $\diamond$ is introduced in Definition A.1.7 and $W^{(H)}(t)$ is the fractional white noise defined in (4.16).

Note that by (4.16) this definition coincides with (4.13) if $Y = f \in L^2_H(\mathbb{R})$ since in that case, by (4.5),

$$\int_{\mathbb{R}} f(t) \diamond W^{(H)}(t)\, dt = \sum_{k=1}^{\infty} \left[\int_{\mathbb{R}} f(t) M\xi_k(t)\, dt \right] \mathcal{H}_{\varepsilon^{(k)}}(\omega)$$

$$= \sum_{k=1}^{\infty} \langle f, M\xi_k \rangle_{L^2(\mathbb{R})} \mathcal{H}_{\varepsilon^{(k)}}(\omega) = \sum_{k=1}^{\infty} \langle Mf, \xi_k \rangle_{L^2(\mathbb{R})} \mathcal{H}_{\varepsilon^{(k)}}(\omega)$$

$$= \int_{\mathbb{R}} Mf \diamond W(t)\, dt = \int_{\mathbb{R}} Mf(t)\, dB(t),$$

where $W(t) = dB(t)/dt$ is the white noise of the standard Brownian motion [see (A.11)]. In particular, we define the stochastic integral over a finite interval $[0,T]$ as

$$\int_0^T Y(t,\omega)\, dB^{(H)}(t) := \int_{\mathbb{R}} Y(t) I_{[0,T]} \diamond W^{(H)}(t)\, dt.$$

Example 4.2.3. Using Wick calculus in $(\mathcal{S})^*$ (see Definition A.1.7), we get

$$\int_0^T B^{(H)}(t)\, dB^{(H)}(t) = \int_0^T B^{(H)}(t) \diamond W^{(H)}(t)\, dt$$

$$= \int_0^T B^{(H)}(t) \diamond \frac{dB^{(H)}(t)}{dt}\, dt = \frac{1}{2}[(B^{(H)}(t))^{\diamond 2}]_0^T = \frac{1}{2}(B^{(H)}(T))^{\diamond 2}$$

$$= \frac{1}{2}(<\omega, M[0,T]>)^{\diamond 2} = \frac{1}{2}[(<\omega,M[0,T]>)^2 - \langle M[0,T], M[0,T] \rangle_{L^2(\mathbb{R})}]$$

$$= \frac{1}{2}(B^{(H)}(T))^2 - \frac{1}{2}\|M[0,T]\|_{L^2(\mathbb{R})}^2 = \frac{1}{2}(B^{(H)}(T))^2 - \frac{1}{2}T^{2H},$$

where we have used (A.12) and (4.6).

Example 4.2.4 (The WIS exponential).
Consider the fractional stochastic differential equation

$$dX(t) = \alpha(t)X(t)\, dt + \beta(t)X(t)\, dB^{(H)}(t), \quad t \geq 0,$$

which is just a shorthand notation for

$$X(t) = X(0) + \int_0^t \alpha(s)X(s)\, ds + \int_0^t \beta(s)X(s)\, dB^{(H)}(s).$$

Here $\alpha(\cdot), \beta(\cdot)$ are locally bounded deterministic functions. To solve this equation, we use (4.17) to rewrite it as a differential equation in $(\mathcal{S})^*$:

$$\frac{dX(t)}{dt} = \alpha(t)X(t) + \beta(t)X(t) \diamond W^{(H)}(t) = X(t) \diamond [\alpha(t) + \beta(t)W^{(H)}(t)]. \quad (4.18)$$

This is the familiar differential equation for the exponential, but with ordinary product replaced by Wick product. Thus, by analogy we guess that the solution is

$$X(t) = X(0) \diamond \exp^\diamond(\int_0^t \alpha(s)\,ds + \int_0^t \beta(s)\,dB^{(H)}(s)), \quad (4.19)$$

where

$$\int_0^t \beta(s)\,dB^{(H)}(s) = \int_{\mathbb{R}} \beta(s)I_{[0,t]}(s)\,dB^{(H)}(s)$$

and, in general,

$$\exp^\diamond F = \sum_{n=0}^\infty \frac{1}{n!} F^{\diamond n}$$

if convergent in $(\mathcal{S})^*$. By Wick calculus (Appendix A) we see that (4.19) is indeed the (unique) solution of (4.18).

In general we have (see [89, (3.5)] or [121, (3.15)])

$$\exp^\diamond[<\omega, Mf>] = \exp\left(<\omega, Mf> - \frac{1}{2}\|Mf\|_{L^2(\mathbb{R})}^2\right).$$

Therefore, the solution can also be written

$$X(t) = X(0) \diamond \exp(\int_0^t \beta(s)\,dB^{(H)}(s) + \int_0^t \alpha(s)\,ds$$
$$- \frac{1}{2}\int_{\mathbb{R}} (M_s\beta(s)I_{[0,t]}(s))^2\,ds),$$

where M_s is the operator M acting on the variable s. If $X(0) = x$ is deterministic, this becomes

$$X(t) = x\exp(\int_0^t \beta(s)\,dB^{(H)}(s) + \int_0^t \alpha(s)\,ds$$
$$- \frac{1}{2}\int_{\mathbb{R}} (M_s(\beta(s)I_{[0,t]}(s)))^2\,ds).$$

In particular, if $\beta(s) = \beta, \alpha(s) = \alpha$ are constants, we get, by using (4.7),

$$X(t) = x\exp(\beta B^{(H)}(t) + \alpha t - \frac{1}{2}\beta^2 t^{2H}), \quad t \geq 0.$$

4.2 The Wick Itô Skorohod (WIS) integral

Remark 4.2.5. Note that if the expansion of the process $Y(s)$ is

$$Y(s) = \sum_{\alpha \in \mathcal{J}} c_\alpha(s) \mathcal{H}_\alpha(\omega) \quad \text{for each } s \in \mathbb{R},$$

then the expansion of its WIS integral is

$$\int_{\mathbb{R}} Y(s) \, dB^{(H)}(s) = \int_{\mathbb{R}} [\sum_{\alpha \in \mathcal{J}} c_\alpha(s) \mathcal{H}_\alpha(\omega)] \diamond [\sum_{k=1}^{\infty} M\xi_k(s) \mathcal{H}_{\varepsilon^{(k)}}(\omega)] \, ds$$

$$= \int_{\mathbb{R}} [\sum_{\alpha \in \mathcal{J}, k \in \mathbb{N}} c_\alpha(s) M\xi_k(s) \mathcal{H}_{\alpha + \varepsilon^{(k)}}(\omega)] \, ds$$

$$= \sum_{\alpha \in \mathcal{J}, k \in \mathbb{N}} \langle c_\alpha, M\xi_k \rangle_{L^2(\mathbb{R})} \mathcal{H}_{\alpha + \varepsilon^{(k)}}(\omega). \tag{4.20}$$

Note that this expansion is not necessarily orthogonal since it may happen that $\alpha + \varepsilon^{(k)} = \beta + \varepsilon^{(j)}$ for some $\alpha, \beta, j, k, \alpha \neq \beta$. In particular, if $\int_{\mathbb{R}} Y(s) \, dB^{(H)}(s) \in L^2(\mathbb{P})$, then

$$E[\int_{\mathbb{R}} Y(s) \, dB^{(H)}(s)] = 0. \tag{4.21}$$

Indeed, the WIS integral shares many of the properties of the Skorohod integral for classical Brownian motion (see Section A.2).

We end this section by presenting an Itô formula for fBm valid for *all* H in $(0, 1)$. A formula for general $H \in (0, 1)$ similar to ours has been obtained independently in [14] and [90]. Our proof, given below, is different from the proofs of these authors. A more general form for $H > 1/2$ is proved in Theorem 3.7.2 and Theorem 6.3.6. The relations among these Itô formulas will be clarified in Chapter 6.

Theorem 4.2.6 (A fractional Itô formula). *Let $H \in (0,1)$. Assume that $f(s,x) : \mathbb{R} \times \mathbb{R} \to \mathbb{R}$ belongs to $C^{1,2}(\mathbb{R} \times \mathbb{R})$, and assume that the random variables*

$$f(t, B^{(H)}(t)), \quad \int_0^t \frac{\partial f}{\partial s}(s, B^{(H)}(s)) \, ds \quad \text{and} \quad \int_0^t \frac{\partial^2 f}{\partial x^2}(s, B^{(H)}(s)) s^{2H-1} \, ds$$

all belong to $L^2(\mathbb{P})$. Then

$$f(t, B^{(H)}(t)) = f(0,0) + \int_0^t \frac{\partial f}{\partial s}(s, B^{(H)}(s)) \, ds$$
$$+ \int_0^t \frac{\partial f}{\partial x}(s, B^{(H)}(s)) \, dB^{(H)}(s) + H \int_0^t \frac{\partial^2 f}{\partial x^2}(s, B^{(H)}(s)) s^{2H-1} \, ds. \tag{4.22}$$

Proof. Let $\alpha \in \mathbb{R}$ be constant, and let $\beta : \mathbb{R} \to \mathbb{R}$ be a deterministic differentiable function. Define

$$g(t,x) = \exp(\alpha x + \beta(t)) \qquad (4.23)$$

and put

$$Y(t) = g(t, B^{(H)}(t)).$$

Then

$$\begin{aligned} Y(t) &= \exp(\alpha B^{(H)}(t))\exp(\beta(t)) \\ &= \exp^\diamond(\alpha B^{(H)}(t) + \frac{1}{2}\alpha^2 t^{2H})\exp(\beta(t)). \end{aligned}$$

Therefore, by Wick calculus in $(\mathcal{S})^*$,

$$\begin{aligned} \frac{d}{dt}Y(t) &= \exp^\diamond(\alpha B^{(H)}(t) + \frac{1}{2}\alpha^2 t^{2H}) \diamond (\alpha W^{(H)}(t) + H\alpha^2 t^{2H-1})\exp(\beta(t)) \\ &\quad + \exp^\diamond(\alpha B^{(H)}(t) + \frac{1}{2}\alpha^2 t^{2H})\exp(\beta(t))\beta'(t) \\ &= Y(t)\beta'(t) + Y(t) \diamond (\alpha W^{(H)}(t)) + Y(t)H\alpha^2 t^{2H-1}. \end{aligned}$$

Hence,

$$Y(t) = Y(0) + \int_0^t Y(s)\beta'(s)\,ds + \int_0^t Y(s)\alpha\,dB^{(H)}(s) + H\int_0^t Y(s)\alpha^2 s^{2H-1}\,ds.$$

This can be written

$$\begin{aligned} g(t, B^{(H)}(t)) &= g(0,0) + \int_0^t \frac{\partial g}{\partial s}(s, B^{(H)}(s))\,ds \\ &\quad + \int_0^t \frac{\partial g}{\partial x}(s, B^{(H)}(s))\,dB^{(H)}(s) + H\int_0^t \frac{\partial^2 g}{\partial x^2}(s, B^{(H)}(s))s^{2H-1}\,ds, \end{aligned} \qquad (4.24)$$

which is (4.22).

Now let $f(t,x)$ be as in Theorem 4.2.6. Then we can find a sequence $f_n(t,x)$ of linear combinations of functions $g(t,x)$ of the form (4.23) such that

$$f_n(t,x) \to f(t,x), \quad \frac{\partial f_n}{\partial t}(t,x) \to \frac{\partial f}{\partial t}(t,x), \quad \frac{\partial f_n}{\partial x}(t,x) \to \frac{\partial f}{\partial x}(t,x)$$

and $\partial^2 f_n(t,x)/\partial x^2 \to \partial^2 f(t,x)/\partial x^2$ pointwise dominatedly as $n \to \infty$. By (4.24) we have for all n

$$f_n(t, B^{(H)}(t)) = f_n(0,0) + \int_0^t \frac{\partial f_n}{\partial s}(s, B^{(H)}(s))\,ds$$
$$+ \int_0^t \frac{\partial f_n}{\partial x}(s, B^{(H)}(s))\,dB^{(H)}(s) + H\int_0^t \frac{\partial^2 f_n}{\partial x^2}(s, B^{(H)}(s))s^{2H-1}\,ds. \quad (4.25)$$

Taking the limit of (4.25) in $L^2(\mathbb{P})$ [and hence also in $(\mathcal{S})^*$], we get

$$f(t, B^{(H)}(t)) = f(0,0) + \int_0^t \frac{\partial f}{\partial s}(s, B^{(H)}(s))\,ds$$
$$+ \lim_{n\to\infty} \int_0^t \frac{\partial f_n}{\partial x}(s, B^{(H)}(s))\,dB^{(H)}(s)$$
$$+ H\int_0^t \frac{\partial^2 f}{\partial x^2}(s, B^{(H)}(s))s^{2H-1}\,ds. \quad (4.26)$$

Since $s \to \partial f_n(s, B^{(H)}(s))/\partial x$ is continuous in $(\mathcal{S})^*$ we have

$$\int_0^t \frac{\partial f_n}{\partial x}(s, B^{(H)}(s))\,dB^{(H)}(s) = \int_0^t \frac{\partial f_n}{\partial x}(s, B^{(H)}(s)) \diamond W^{(H)}(s)\,ds. \quad (4.27)$$

Hence, (4.27) converges to

$$\int_0^t \frac{\partial f}{\partial x}(s, B^{(H)}(s)) \diamond W^{(H)}(s)\,ds \quad (4.28)$$

in $(\mathcal{S})^*$ as $n \to \infty$. Comparing (4.26) and (4.28), we see that $\int_0^t \partial f(s, B^{(H)}(s))/\partial x \diamond W^{(H)}(s)\,ds \in L^2(\mathbb{P})$ and (4.22) follows. □

4.3 Girsanov theorem

The first Girsanov-type theorem for *fBm* was obtained by [168]. Using the approach of Section 5 of [89], we obtain the following version of the Girsanov theorem for *fBm* valid for every Hurst index $H \in (0,1)$. Let $\psi \in L^2(\mathbb{R})$ and consider the new probability measure $\hat{\mathbb{P}}$ on $(\Omega, \mathcal{F})$ with Radon–Nikodym density

$$\frac{d\hat{\mathbb{P}}}{d\mathbb{P}} = \exp\left(\int_\mathbb{R} \psi(s)\,dB(s) - \frac{1}{2}\|\psi\|^2_{L^2(\mathbb{R})}\right).$$

The classical Girsanov theorem then states that under $\hat{\mathbb{P}}$ the process

$$\hat{B}(t) := B(t) - \int_0^t \psi(s)\,ds$$

is a standard Brownian motion. Hence,

$$\hat{B}^{(H)}(t) := \int_{\mathbb{R}} M(0,t)(s)\,d\hat{B}(s) = \int_{\mathbb{R}} M(0,t)(s)\,dB(s) - \int_{\mathbb{R}} M(0,t)(s)\psi(s)\,ds$$

is a *fBm* under $\hat{\mathbb{P}}$. To eliminate the drift, we have to solve equations of the form

$$\int_{\mathbb{R}} M(0,t)(s)\psi(s)\,ds = g(t),$$

or equivalently,

$$\int_0^t M\psi(s)\,ds = g(t).$$

In the particular case when $g(t) = At$ for $t \in [0,T]$ for some constant A and $g(t) = 0$ otherwise, by [89] we obtain for $0 \le t \le T$,

$$\psi(t) = \frac{A[(T-t)^{1/2-H} + t^{1/2-H}]}{2\Gamma(3/2-H)\cos(\pi/2\,(1/2-H))}.$$

4.4 Differentiation

We now recall the approach in [121] to differentiation, as modified and extended by [89].

Definition 4.4.1. *Let $F : \Omega \to \mathbb{R}$ and choose $\gamma \in \Omega$. Then we say F has a directional derivative in the direction γ if*

$$D_\gamma^{(H)} F(\omega) := \lim_{\varepsilon \to 0} \tfrac{1}{\varepsilon}[F(\omega + \varepsilon M\gamma) - F(\omega)]$$

exists in $(\mathcal{S})^*$. *In that case we call $D_\gamma^{(H)} F$ the directional derivative of F in the direction γ.*

Example 4.4.2. 1. Suppose $F(\omega) = <\omega, Mf>^n$ for some $f \in L_H^2(\mathbb{R})$, $n \in \mathbb{N}$. Since

$$\frac{1}{\varepsilon}[F(\omega + \varepsilon M\gamma) - F(\omega)]$$
$$= \frac{1}{\varepsilon}[<\omega + \varepsilon M\gamma, Mf>^n - <\omega, Mf>^n]$$
$$= \frac{1}{\varepsilon}[(<\omega, Mf> + \varepsilon <M\gamma, Mf>)^n - <\omega, Mf>^n],$$

we get the *chain rule*

$$D_\gamma^{(H)}(<\omega, Mf>^n) = n<\omega, Mf>^{n-1}<M\gamma, Mf>. \quad (4.29)$$

In particular, choosing $n=1$ and $\gamma \in L_H^2(\mathbb{R})$, we get

$$D_\gamma^{(H)}(\int_\mathbb{R} f(t)\, dB^{(H)}(t)) = \langle M\gamma, Mf\rangle_{L^2(\mathbb{R})} = \langle \gamma, f\rangle_{L_H^2(\mathbb{R})}.$$

2. Suppose $G(\omega) = <\omega, Mg>^{\diamond n}$ for some $g \in L_H^2(\mathbb{R})$, $n \in \mathbb{N}$. Assume $\|Mg\|_{L^2(\mathbb{R})} = 1$. By Example A.1.8 we have $<\omega, Mg>^{\diamond n} = h_n(<\omega, Mg>)$ with h_n as in (A.5); hence, we get

$$\lim_{\varepsilon \to 0} \frac{1}{\varepsilon}[G(\omega + \varepsilon M\gamma) - G(\omega)]$$
$$= \lim_{\varepsilon \to 0} \frac{1}{\varepsilon}[<\omega + \varepsilon M\gamma, Mg>^{\diamond n} - <\omega, Mg>^{\diamond n}]$$
$$= \lim_{\varepsilon \to 0} \frac{1}{\varepsilon}[h_n(<\omega + \varepsilon M\gamma, Mg>) - h_n(<\omega, Mg>)]$$
$$= \lim_{\varepsilon \to 0} \frac{1}{\varepsilon}[h_n(<\omega, Mg> + \varepsilon <M\gamma, Mg>) - h_n(<\omega, Mg>)]$$
$$= h_n'(<\omega, Mg>)<M\gamma, Mg> = nh_{n-1}(\langle\omega, Mg\rangle)\langle M\gamma, Mg\rangle,$$

by a well-known property of the Hermite polynomials $\{h_n\}_{n=1}^\infty$. Hence, the following *Wick chain rule* holds:

$$D_\gamma^{(H)}(<\omega, Mg>^{\diamond n}) = n<\omega, Mg>^{\diamond(n-1)}<M\gamma, Mg>. \quad (4.30)$$

By linearity this holds also if $\|Mg\|_{L^2(\mathbb{R})} \neq 1$.

Definition 4.4.3. *A process* $Y(t) = \sum_{\alpha \in \mathcal{J}} c_\alpha(t)\mathcal{H}_\alpha(\omega) \in (\mathcal{S})^*$ *belongs to* $\mathcal{M}$ *if* $c_\alpha(\cdot) \in L_H^2(\mathbb{R})$ *and* $\sum_{\alpha \in \mathcal{J}} Mc_\alpha(t)\mathcal{H}_\alpha(\omega)$ *converges in* $(\mathcal{S})^*$ *for all* t. *If* $Y \in \mathcal{M}$, *we define* $MY(t)$ *by the expansion*

$$MY(t) := \sum_{\alpha \in \mathcal{J}} Mc_\alpha(t)\mathcal{H}_\alpha(\omega). \quad (4.31)$$

In particular, by (4.16) we see that the relation between fractional and classical white noise is given by

$$W^{(H)}(t) = MW(t).$$

Definition 4.4.4. *We say that* $F : \Omega \to \mathbb{R}$ *is differentiable if there exists a function* $\Psi : \mathbb{R} \to (\mathcal{S})^*$ *in* $\mathcal{M}$ *such that*

$$D_\gamma^{(H)} F(\omega) = \int_\mathbb{R} M\Psi(t) M\gamma(t)\, dt \quad \forall\, \gamma \in L_H^2(\mathbb{R}).$$

Then we write

$$D_t^{(H)} F := \frac{\partial^{(H)}}{\partial \omega} F(t,\omega) = \Psi(t),$$

and we call $D_t^{(H)} F$ *the Hida Malliavin derivative or the stochastic gradient of* F *at* t.

Example 4.4.5. From (4.29) to (4.30) we get, for $f \in L^2_H(\mathbb{R})$,

$$D_t^{(H)}(<\omega, Mf>^n) = n<\omega, Mf>^{n-1} f(t) \quad \text{for a.a. } t$$

$$D_t^{(H)}(\int_\mathbb{R} f(s)dB^{(H)}(s)) = f(t) \quad \text{for a.a. } t$$

$$D_t^{(H)}(<\omega, Mf>^{\diamond n}) = n<\omega, Mf>^{\diamond(n-1)} f(t) \quad \text{for a.a. } t \quad (4.32)$$

These examples illustrate that the stochastic gradient satisfies a *chain rule* both with respect to ordinary products and with respect to Wick products. Note that for $\alpha = (\alpha_1, \ldots, \alpha_m) \in \mathcal{J}$ we have

$$\mathcal{H}_\alpha(\omega) = h_{\alpha_1}(<\omega, \xi_1>) \cdots h_{\alpha_m}(<\omega, \xi_m>)$$
$$= <\omega, \xi_1>^{\diamond\alpha_1} \cdots <\omega, \xi_m>^{\diamond\alpha_m}$$
$$= <\omega, Me_1>^{\diamond\alpha_1} \cdots <\omega, Me_m>^{\diamond\alpha_m}$$

Therefore, by (4.32),

$$D_t^{(H)} \mathcal{H}_\alpha(\omega) = \sum_{i=1}^\infty \alpha_i h_{\alpha_i - 1}(<\omega, Me_i>) \Pi_{j \neq i} h_{\alpha_j}(<\omega, \xi_j>) e_i(t)$$
$$= \sum_{i=1}^\infty \alpha_i \mathcal{H}_{\alpha - \varepsilon^{(i)}}(\omega) e_i(t).$$

This motivates the following:

Definition 4.4.6 (Fractional stochastic Sobolev spaces). Let $\mathbb{D}_H^{1,2}$ be the set of all $F \in L^2(\mathbb{P})$ whose chaos expansion

$$F(\omega) = \sum_{\alpha \in \mathcal{J}} c_\alpha \mathcal{H}_\alpha(\omega)$$

satisfies

$$\sum_{\alpha \in \mathcal{J}} \sum_{i=1}^\infty c_\alpha^2 \alpha_i \alpha! < \infty.$$

If $F \in \mathbb{D}_H^{1,2}$, we define the fractional stochastic derivative of F by

$$D_t^{(H)} F(\omega) = \frac{\partial^{(H)} F}{\partial \omega}(t, \omega) = \sum_{\alpha \in \mathcal{J}} \sum_{i=1}^\infty c_\alpha \alpha_i \mathcal{H}_{\alpha - \varepsilon^{(i)}}(\omega) e_i(t).$$

Note that if $F \in \mathbb{D}_H^{1,2}$, then $E\left[\|D_t^{(H)} F(\omega)\|_H^2\right] = \sum_{\alpha,i} c_\alpha^2 \alpha_i \alpha! < +\infty$. Next, we extend this to $(\mathcal{S})^*$:

Definition 4.4.7 (The general fractional stochastic gradient). Let $F \in (\mathcal{S})^*$ with chaos expansion

$$F(\omega) = \sum_{\alpha \in \mathcal{J}} c_\alpha \mathcal{H}_\alpha(\omega).$$

Then we define the fractional stochastic gradient of F by the expansion

$$D_t^{(H)} F(\omega) := \frac{\partial^{(H)} F}{\partial \omega}(t, \omega) := \sum_{\alpha \in \mathcal{J}} \sum_{i=1}^\infty c_\alpha \alpha_i \mathcal{H}_{\alpha - \varepsilon^{(i)}}(\omega) e_i(t) \quad (4.33)$$

$$= \sum_{\gamma \in \mathcal{J}} [\sum_{\alpha, i : \alpha - \varepsilon^{(i)} = \gamma} c_\alpha \alpha_i e_i(t)] \mathcal{H}_\gamma(\omega)$$

$$= \sum_{\gamma \in \mathcal{J}} [\sum_{i=1}^{l(\gamma)} c_{\gamma + \varepsilon^{(i)}} (\gamma_i + 1) e_i(t)] \mathcal{H}_\gamma(\omega), \quad (4.34)$$

where $l(\gamma) = \max\{i : \gamma_i \neq 0\}$ is the length of γ. One can show that $D_t^{(H)} F \in (\mathcal{S})^*$ for almost all $t \in \mathbb{R}$ (see [89]).

Remark 4.4.8. We note that for the fractional gradient we are using the same notation as in Definition 3.3.3. We remark that here the gradient is defined as an element in the *classical* Hida space $(\mathcal{S})^*$ of stochastic distributions (Definition A.1.4), while in Chapter 3 it belongs to the *fractional* Hida distribution space $(\mathcal{S})_H^*$ (Definition 3.1.10). The relation between the two stochastic gradients is clarified in Proposition 6.3.8.

The following result gives a relation between the Wick product and the ordinary product. It has been obtained with a different proof in Lemma 2 of [155].

Lemma 4.4.9. *Let $g \in L_H^2(\mathbb{R})$ and $F \in \mathbb{D}_H^{1,2}$. Then*

$$F \diamond \int_\mathbb{R} g(t)\, dB^{(H)}(t) = F \cdot \int_\mathbb{R} g(t)\, dB^{(H)}(t) - \langle g, D^{(H)} F \rangle_H$$

Proof. For $y \in \mathbb{R}$ define

$$G_y = \exp^\diamond(y \int_\mathbb{R} g(t) dB^{(H)}(t)) = \exp\left(y \int_\mathbb{R} g(t) dB^{(H)}(t) - \frac{1}{2} y^2 \|g\|_H^2 \right)$$

Choose $F = \exp^\diamond(\int_\mathbb{R} f(t)\, dB^{(H)}(t)) = \exp\left(\int_\mathbb{R} f(t)\, dB^{(H)}(t) - 1/2 \|f\|_H^2\right)$, where $f \in L_H^2(\mathbb{R})$. Then

$$F \diamond G_y = \exp^\diamond \left(\int_\mathbb{R} f(t)\, dB^{(H)}(t) \right) \diamond \exp^\diamond \left(y \int_\mathbb{R} g(t)\, dB^{(H)}(t) \right)$$

$$= \exp^\diamond \left(\int_\mathbb{R} (f(t) + y g(t))\, dB^{(H)}(t) \right)$$

$$= \exp\left(\int_{\mathbb{R}} (f(t) + yg(t))\, dB^{(H)}(t) - \frac{1}{2}\|f(t) + yg(t)\|_H^2\right)$$

$$= \exp^{\diamond}\left(\int_{\mathbb{R}} f(t)\, dB^{(H)}(t)\right) \exp^{\diamond}\left(y\int_{\mathbb{R}} g(t)\, dB^{(H)}(t)\right) \exp(-y\langle g, f\rangle_H)$$

$$= F \cdot G_y \cdot \exp(-y\langle g, f\rangle_H). \qquad (4.35)$$

Now differentiate with respect to y. We have

$$\frac{d}{dy}(F \diamond G_y) = F \diamond \frac{d}{dy}(G_y) = F \diamond \left\{ G_y \cdot \left[\int_{\mathbb{R}} g(t)\, dB^{(H)}(t) - y\|g\|_H\right]\right\} \qquad (4.36)$$

and

$$\frac{d}{dy}(F \cdot G_y \cdot \exp(-y\langle g, f\rangle_H))$$
$$= F \cdot G_y \cdot \left[\int_{\mathbb{R}} g(t)\, dB^{(H)}(t) - y\|g\|_H\right] \cdot \exp(-y\langle g, f\rangle_H)$$
$$- \langle g, f\rangle_H F \cdot G_y \exp(-y\langle g, f\rangle_H). \qquad (4.37)$$

Comparing (4.36) and (4.37) and using (4.35), we get

$$F \diamond \left(G_y \cdot \left(\int_{\mathbb{R}} g(t)\, dB^{(H)}(t) - y\|g\|_H\right)\right) = \frac{d}{dy}(F \diamond G_y) =$$
$$= F \cdot G_y \cdot \left(\int_{\mathbb{R}} g(t)\, dB^{(H)}(t) - y\|g\|_H\right) \exp(-y\langle g, f\rangle_H)$$
$$- \langle g, f\rangle_H F \cdot G_y \exp(-y\langle g, f\rangle_H) \qquad (4.38)$$

In particular, if we put $y = 0$, we get

$$F \diamond \int_{\mathbb{R}} g(t)\, dB^{(H)}(t) = F \cdot \int_{\mathbb{R}} g(t)\, dB^{(H)}(t)) - F\langle g, f\rangle_H$$
$$= F \cdot \int_{\mathbb{R}} g(t)\, dB^{(H)}(t)) - F \int_{\mathbb{R}} Mg(t) Mf(t)\, dt$$

By the chain rule $D_t^{(H)} F = Ff(t)$, and hence

$$F \int_{\mathbb{R}} Mg(t) Mf(t)\, dt = \int_{\mathbb{R}} Ff(t) M^2 g(t)\, dt$$
$$= \int_{\mathbb{R}} D_t^{(H)} F M^2 g(t)\, dt = \langle g, D_t^{(H)} F\rangle_H.$$

Combined with (4.38), this proves the theorem in the case when $F = \exp^{\diamond}(\int_{\mathbb{R}} f(t)\, dB^{(H)}(t))$, where $f \in L_H^2(\mathbb{R})$. Since the linear combinations of such F is dense in $\mathbb{D}_H^{1,2}$, the result follows. $\square$

4.5 Relation with the standard Malliavin calculus

In this section we study the relation between the WIS integration for *fBm* and the classical stochastic calculus. As before, $D_t^{(H)}$ denotes the Hida Malliavin derivative with respect to $B^{(H)}(\cdot)$. In the classical case ($H = 1/2$) we use the notation D_t for the corresponding Hida Malliavin derivative (for further details, see Appendix A).

Proposition 4.5.1 (Differentiation). *Let* $F \in (\mathcal{S})^*$. *Then*

$$D_t F = M D_t^{(H)} F \quad \text{for a.a. } t \in \mathbb{R}.$$

Proof. Let F have the expansion

$$F(\omega) = \sum_{\alpha \in \mathcal{J}} c_\alpha \mathcal{H}_\alpha(\omega).$$

Then by (4.34) and (4.31) we get

$$M D_t^{(H)} F = M\left(\sum_{\gamma \in \mathcal{J}} \left[\sum_{i=1}^{l(\gamma)} c_{\gamma + \varepsilon^{(i)}} (\gamma_i + 1) e_i(t)\right] \mathcal{H}_\gamma(\omega)\right)$$

$$= \sum_{\gamma \in \mathcal{J}} \left[\sum_{i=1}^{l(\gamma)} c_{\gamma + \varepsilon^{(i)}} (\gamma_i + 1) \xi_i(t)\right] \mathcal{H}_\gamma(\omega)$$

$$= D_t F.$$

$\square$

Proposition 4.5.2 (Integration). *Suppose* $Y : \mathbb{R} \to (\mathcal{S})^*$ *is WIS integrable (Definition 4.2.2) and* $Y \in \mathcal{M}$. *Then*

$$\int_\mathbb{R} Y(t) \, dB^{(H)}(t) = \int_\mathbb{R} MY(t) \delta B(t),$$

where $\delta B(t)$ *denotes the Skorohod integral with respect to the standard Brownian motion* $B(t)$ *defined in Definition A.2.1.*

Proof. Suppose $Y(t) = \sum_{\alpha \in \mathcal{J}} c_\alpha(t) \mathcal{H}_\alpha(\omega)$. Then by (4.20) and (4.5) we have

$$\int_\mathbb{R} Y(t) \, dB^{(H)}(t) = \sum_{\alpha \in \mathcal{J}, k \in \mathbb{N}} \langle c_\alpha, M\xi_k \rangle_{L^2(\mathbb{R})} \mathcal{H}_{\alpha + \varepsilon^{(k)}}(\omega)$$

$$= \sum_{\alpha \in \mathcal{J}, k \in \mathbb{N}} \langle Mc_\alpha, \xi_k \rangle_{L^2(\mathbb{R})} \mathcal{H}_{\alpha + \varepsilon^{(k)}}(\omega)$$

$$= \int_\mathbb{R} MY(t) \delta B(t).$$

$\square$

Note that

$$\int_0^T Y(t)\,dB^{(H)}(t) = \int_{\mathbb{R}} Y(t) I_{[0,T]}(t)\,dB^{(H)}(t)$$

$$= \int_{\mathbb{R}} M(Y I_{[0,T]}) \delta B(t) \neq \int_0^T MY(t)\delta B(t).$$

Theorem 4.5.3. *Let* $Y \in \mathcal{M}$. *Suppose* $Y : \mathbb{R} \to (\mathcal{S})^*$ *and* $D_t^{(H)} Y(\cdot) : \mathbb{R} \to (\mathcal{S})^*$ *are WIS integrable. Then*

$$D_t^{(H)} \Big(\int_{\mathbb{R}} Y(s)\,dB^{(H)}(s) \Big) = \int_{\mathbb{R}} D_t^{(H)} Y(s)\,dB^{(H)}(s) + Y(t) \qquad (4.39)$$

Proof. If $Y(s) = \sum_{\alpha \in \mathcal{J}} c_\alpha(s) \mathcal{H}_\alpha(\omega)$ then by (4.20) and (4.33) we get

$$D_t^{(H)} \Big(\int_{\mathbb{R}} Y(s) dB^{(H)}(s) \Big)$$

$$= D_t^{(H)} \Big(\sum_{\alpha \in \mathcal{J}, k \in \mathbb{N}} \langle c_\alpha, M\xi_k \rangle_{L^2(\mathbb{R})} \mathcal{H}_{\alpha+\varepsilon^{(k)}}(\omega) \Big)$$

$$= \sum_{\alpha \in \mathcal{J}, k \in \mathbb{N}} \langle c_\alpha, M\xi_k \rangle_{L^2(\mathbb{R})} \sum_{i \in \mathbb{N}} (\alpha + \varepsilon^{(k)})_i \mathcal{H}_{\alpha+\varepsilon^{(k)}-\varepsilon^{(i)}}(\omega) e_i(t)$$

$$= \sum_{\alpha \in \mathcal{J}, k \in \mathbb{N}, i \in \mathbb{N}} \langle c_\alpha, M\xi_k \rangle_{L^2(\mathbb{R})} \alpha_i \mathcal{H}_{\alpha+\varepsilon^{(k)}-\varepsilon^{(i)}} e_i(t)$$

$$+ \sum_{\alpha \in \mathcal{J}, k \in \mathbb{N}} \langle c_\alpha, M\xi_k \rangle_{L^2(\mathbb{R})} \mathcal{H}_\alpha(\omega) e_k(t). \qquad (4.40)$$

Applying (4.20) to the integrand $D_t^{(H)} Y(s)$, we see that the right-hand side of (4.39) is

$$\sum_{\alpha \in \mathcal{J}, k, i \in \mathbb{N}} \langle c_\alpha, M\xi_k \rangle_{L^2(\mathbb{R})} \alpha_i \mathcal{H}_{\alpha-\varepsilon^{(i)}+\varepsilon^{(k)}}(\omega) e_i(t) + \sum_{\alpha \in \mathcal{J}, k \in \mathbb{N}} \langle c_\alpha, e_k \rangle_H e_k(t) \mathcal{H}_\alpha(\omega),$$

which coincides with (4.40) since by (4.4) and (4.5)

$$\langle c_\alpha, e_k \rangle_H = \langle Mc_\alpha, Me_k \rangle_{L^2(\mathbb{R})} = \langle Mc_\alpha, \xi_k \rangle_{L^2(\mathbb{R})} = \langle c_\alpha, M\xi_k \rangle_{L^2(\mathbb{R})}.$$

$\square$

Theorem 4.5.4 (Integration by parts). *Consider* $F \in L^2(\mathbb{P})$ *and let* $Y : \mathbb{R} \to (\mathcal{S})^*$ *be in* $\mathcal{M}$. *Then*

$$F \int_{\mathbb{R}} Y(s)\,dB^{(H)}(s) = \int_{\mathbb{R}} FY(s)\,dB^{(H)}(s) + \int_{\mathbb{R}} MY(s) M D_s^{(H)} F\,ds$$

provided that at least two of the terms are well-defined and belong to $L^2(\mathbb{P})$.

Proof. The classical ($H = 1/2$) integration by parts formula states that

$$F \int_{\mathbb{R}} Y(s)\delta B(s) = \int_{\mathbb{R}} FY(s)\delta B(s) + \int_{\mathbb{R}} Y(s) D_s F \, ds.$$

See, e.g., [176, (1.49)]. Combining this with Proposition 4.5.2 and Proposition 4.5.1, we get

$$F \int_{\mathbb{R}} Y(s) \, dB^{(H)}(s) = F \int_{\mathbb{R}} MY(s)\delta B(s)$$

$$= \int_{\mathbb{R}} FMY(s)\delta B(s) + \int_{\mathbb{R}} MY(s) D_s F \, ds$$

$$= \int_{\mathbb{R}} M_s(FY(s))\delta B(s) + \int_{\mathbb{R}} MY(s) M_s D_s^{(H)} F \, ds$$

$$= \int_{\mathbb{R}} FY(s) \, dB^{(H)}(s) + \int_{\mathbb{R}} MY(s) M_s D_s^{(H)} F \, ds.$$

□

Since WIS integrals have expectation 0 (see (4.21)), we deduce the following:

Corollary 4.5.5. *Let $F, Y(s)$ be such that $F \in L^2(\mathbb{P})$, $E\left[(\int_{\mathbb{R}} Y_s dB^{(H)}(s))^2\right] < \infty$, $E\left[(\int_{\mathbb{R}} FY_s dB^{(H)}(s))^2\right] < \infty$. Then*

$$E[F \int_{\mathbb{R}} Y(s) \, dB^{(H)}(s)] = E[\int_{\mathbb{R}} MY(s) MD_s^{(H)} F \, ds].$$

The following WIS isometry was first proved by [89]. We give a different proof, based on the results above.

Theorem 4.5.6. (The WIS isometry)
Let $Y : \mathbb{R} \to (\mathcal{S})^*$ be WIS integrable and belong to $\mathcal{M}$. Then

$$E[(\int_{\mathbb{R}} Y(t) \, dB^{(H)}(t))^2] = E[\int_{\mathbb{R}} (MY(t))^2 \, dt]$$

$$+ E[\int_{\mathbb{R}} \int_{\mathbb{R}} D_t^{(H)} M_s^2 Y(s) \cdot D_s^{(H)} M_t^2 Y(t) \, ds \, dt]$$

provided that at least two of these terms are well-defined. Here M_t indicates that the operator M acts on the variable t, and similarly with M_s.

Proof. By (4.5), Corollary 4.5.5, Theorem 4.5.3, and Corollary 4.5.5 again, we get

$$E[\int_\mathbb{R}\int_\mathbb{R} D_t^{(H)} M_s^2 Y(s) \cdot D_s^{(H)} M_t^2 Y(t)\, ds\, dt]$$

$$= E[\int_\mathbb{R} (\int_\mathbb{R} M_s D_t^{(H)} Y(s) \cdot M_s(D_s^{(H)} M_t^2 Y(t))\, ds)\, dt]$$

$$= E[\int_\mathbb{R} (M_t^2 Y(t) \cdot \int_\mathbb{R} D_t^{(H)} Y(s)\, dB^{(H)}(s)\, dt]$$

$$= E[\int_\mathbb{R} M_t^2 Y(t) \cdot \{D_t^{(H)}(\int_\mathbb{R} Y(s)\, dB^{(H)}(s)) - Y(t)\}\, dt]$$

$$= E[\int_\mathbb{R} (M_t^2 Y(t) \cdot D_t^{(H)}(\int_\mathbb{R} Y(s)\, dB^{(H)}(s))\, dt] - E[\int_\mathbb{R} M_t^2 Y(t) \cdot Y(t)\, dt]$$

$$= E[(\int_\mathbb{R} Y(s)\, dB^{(H)}(s))^2] - E[\int_\mathbb{R} (M_t Y(t))^2\, dt].$$

$\square$

4.6 The multidimensional case

We now proceed to the multidimensional case. In the following we let $H_1, \ldots, H_N$ be N numbers (Hurst coefficients) in $(0,1)$, and we put

$$H = (H_1, \ldots, H_N) \in (0,1)^N.$$

With $(\Omega, \mathbb{P})$ as in Section 4.1, we let $(\Omega_1, \mathbb{P}_1), \ldots, (\Omega_N, \mathbb{P}_N)$ be N copies of $(\Omega, \mathbb{P})$, and we put

$$\Omega = \Omega_1 \times \cdots \times \Omega_N, \quad \mathbb{P} = \mathbb{P}_1 \otimes \cdots \otimes \mathbb{P}_N. \tag{4.41}$$

Note that here we are denoting the product measure (4.41) simply with $\mathbb{P}$ for notational simplicity. Then the N-dimensional *fBm* with Hurst vector $H = (H_1, \ldots, H_N)$ is defined by

$$B^{(H)}(t) = (B_1^{(H)}(t), \ldots, B_N^{(H)}(t)),$$

where

$$B_k^{(H)}(t) = B_k^{(H)}(t, \omega) = B^{(H_k)}(t, \omega_k), \quad \omega = (\omega_1, \ldots, \omega_N) \in \Omega,$$

is a one-dimensional *fBm* with Hurst coefficient $H_k \in (0,1)$, $k = 1, \ldots, N$. Thus $B^{(H)}(t)$ consists of N independent one-dimensional *fBms* $B^{(H_1)}(t), \ldots,$

$B^{(H_N)}(t)$.

We let $\mathcal{J}^N$ be the set of all N-tuples $\alpha = (\alpha^{(1)}, \ldots, \alpha^{(N)})$ with $\alpha^{(j)} = (\alpha_1^{(j)}, \ldots, \alpha_{l(\alpha^{(j)})}^{(j)}) \in \mathcal{J}$ for all $j = 1, \ldots, N$, and we put

$$\mathcal{H}_\alpha(\omega) = \mathcal{H}_{\alpha^{(1)}}(\omega_1) \cdots \mathcal{H}_{\alpha^{(N)}}(\omega_N) \quad \text{for } \alpha \in \mathcal{J}^N. \tag{4.42}$$

Then the family $\{\mathcal{H}_\alpha\}_{\alpha \in \mathcal{J}^N}$ constitutes an orthogonal basis for $L^2(\mathbb{P})$ and

$$E[(\mathcal{H}_\alpha)^2] = \alpha! := \alpha^{(1)}! \cdots \alpha^{(N)}!.$$

Therefore, every $F \in L^2(\mathbb{P})$ has a chaos expansion

$$F(\omega) = \sum_{\alpha \in \mathcal{J}^N} c_\alpha \mathcal{H}_\alpha(\omega),$$

where $c_\alpha \in \mathbb{R}$ for all $\alpha \in \mathcal{J}^N$ with

$$\|F\|_{L^2(\mathbb{P})}^2 = \sum_{\alpha \in \mathcal{J}^N} c_\alpha^2 \alpha!.$$

The nth component of $B^{(H)}(t)$, $B_n^{(H)}(t)$, has the expansion

$$B_n^{(H)}(t) = B^{(H_n)}(t, \omega_n) = \sum_{k=1}^\infty \int_0^t M^{(H_n)} \xi_k(s)\, ds \mathcal{H}_{\varepsilon^{(k)}}(\omega_n),$$

where $M^{(H_n)}$ is as in Definition 4.1.2 with $H = H_n$. The corresponding expansion for *the nth component of fractional white noise* is

$$W_n^{(H)}(t) = W^{(H_n)}(t, \omega_n) = \sum_{k=1}^\infty M^{(H_n)} \xi_k(t) \mathcal{H}_{\varepsilon^{(k)}}(\omega_n).$$

As in Section 4.4 we have $W_n^{(H)}(t) \in (\mathcal{S})^*$ and

$$W_n^{(H)}(t) = \frac{d}{dt} B_n^{(H)}(t) \quad \text{in } (\mathcal{S})^* \text{ for } n = 1, \ldots, N.$$

[See (4.15) to (4.17).]

The Wick product on $(\mathcal{S})^*$ is defined as in Appendix A. For the multidimensional case we have

$$\sum_{\alpha \in \mathcal{J}^N} a_\alpha \mathcal{H}_\alpha(\omega) \diamond \sum_{\beta \in \mathcal{J}^N} b_\beta \mathcal{H}_\beta(\omega) = \sum_{\alpha, \beta \in \mathcal{J}^N} a_\alpha b_\beta \mathcal{H}_{\alpha+\beta}(\omega).$$

Note in particular that if $m, n \in \{1, \ldots, N\}$ with $m \neq n$, then by (4.42)

$$\mathcal{H}_\alpha(\omega_m) \diamond \mathcal{H}_\beta(\omega_n) = \mathcal{H}_\alpha(\omega_m) \mathcal{H}_\beta(\omega_n).$$

As in Definition 4.2.2 we now define the multidimensional WIS integral.

Definition 4.6.1 (The multidimensional WIS integral).

1. *If* $X : \mathbb{R} \to (\mathcal{S})^*$ *is such that* $X(t) \diamond W_n^{(H)}(t)$ *is dt-integrable in* $(\mathcal{S})^*$, *then we define*

$$\int_\mathbb{R} X(t,\omega)\, dB_n^{(H)}(t) = \int_\mathbb{R} X(t,\omega) \diamond W_n^{(H)}(t)\, dt.$$

2. *If* $Y : \mathbb{R} \to ((\mathcal{S})^*)^N$ *is such that* $Y_n(t) \diamond W_n^{(H)}(t)$ *is dt-integrable in* $(\mathcal{S})^*$ *for all* $n = 1, \ldots, N$, *we say that* Y *is* WIS integrable *with respect to* $B^{(H)}$ *and define the* multidimensional WIS integral *as given by*

$$\int_\mathbb{R} Y(t)\, dB^{(H)}(t) := \sum_{n=1}^N \int_\mathbb{R} Y_n(t)\, dB_n^{(H)}(t) = \sum_{n=1}^N \int_\mathbb{R} Y_n(t)\, dB^{(H_n)}(t).$$

Example 4.6.2. Let $m \neq n$. Then

$$\int_0^T B_m^{(H)}(t)\, dB_n^{(H)}(t) = \int_\mathbb{R} B_m^{(H)}(T) I_{[0,T]}(t) \diamond W_n^{(H)}(t)\, dt$$

$$= B_m^{(H)}(T) \diamond \int_0^T W_n^{(H)}(t)\, dt = B_m^{(H)}(T) \diamond B_n^{(H)}(T)$$

$$= B^{(H_m)}(T, \omega_m) B^{(H_n)}(T, \omega_n).$$

Therefore, if we choose

$$Y_k(t) = \begin{cases} B_m^{(H)}(t) \cdot I_{[0,T]}(t) & \text{if } k = n, \\ -B_n^{(H)}(t) \cdot I_{[0,T]}(t) & \text{if } k = m, \\ 0 & \text{otherwise}, \end{cases}$$

then

$$\int_\mathbb{R} Y(t)\, dB^{(H)}(t) = B^{(H_m)}(T) B^{(H_n)}(T) - B^{(H_n)}(T) B^{(H_m)}(T) = 0,$$

even though $Y \neq 0$.

Proceeding as in Section 4.4, we are led to the following definition of a stochastic derivative in the direction n (see Definition 4.4.6).

Definition 4.6.3. *Let* $\mathbb{D}_H^{1,2}$ *be the set of all* $F \in L^2(\mathbb{P})$ *whose chaos expansion*

$$F(\omega) = \sum_{\alpha \in \mathcal{J}^N} c_\alpha \mathcal{H}_\alpha(\omega)$$

satisfies

$$\sum_{\alpha \in \mathcal{J}^N} \sum_{i=1}^{\infty} c_\alpha^2 \alpha_i^{(n)} \alpha! < \infty \quad \text{for } n = 1, \ldots, N. \tag{4.43}$$

If $F \in \mathbb{D}_H^{1,2}$, we define the fractional stochastic derivative of F in direction n ($n = 1, 2, \ldots, N$) by

$$D_{n,t}^{(H)} F = \frac{\partial^{(H_n)} F}{\partial \omega_n}(t, \omega) = \sum_{\alpha \in \mathcal{J}^N} \sum_{i=1}^{\infty} c_\alpha \alpha_i^{(n)} \mathcal{H}_{\alpha - \varepsilon^{(n,i)}}(\omega) e_{n,i}(t),$$

where

$$\varepsilon^{(n,i)} = (0, \ldots, 0, \varepsilon^{(i)}, 0, \ldots, 0) \in \mathcal{J}^N$$

with $\varepsilon^{(i)}$ on the nth place and $e_{n,i}(t) = (M^{(H_n)})^{(-1)} \xi_i(t)$.
We define the fractional stochastic gradient *of F by*

$$\nabla^{(H)} F(t, \omega) = \left(\frac{\partial^{(H_1)} F}{\partial \omega_1}(t, \omega), \ldots, \frac{\partial^{(H_N)} F}{\partial \omega_N}(t, \omega) \right).$$

Note that by (4.43) we have

$$\nabla^{(H)} F(t, \omega) \in L^2(\lambda \times \mathbb{P}) \quad \text{if } F \in \mathbb{D}_H^{1,2},$$

where as before λ denotes Lebesgue measure on $\mathbb{R}$.

As in the one-dimensional case we extend this to $F \in (\mathcal{S})^*$ by setting

$$D_{n,t}^{(H)} F = \frac{\partial^{(H)} F}{\partial \omega_n}(t, \omega) = \sum_{\alpha \in \mathcal{J}^N} \sum_{i=1}^{\infty} c_\alpha \alpha_i^{(n)} \mathcal{H}_{\alpha - \varepsilon^{(n,i)}}(\omega) e_i(t).$$

(See Definition 4.4.7.)

We now give the multidimensional versions of the results of Section 4.5. These results can either be obtained similarly to those in Section 4.5, or by reducing to the one-dimensional cases, and we therefore omit the proofs.

Proposition 4.6.4 (Differentiation). *Let $F \in (\mathcal{S})^*$. Then*

$$D_{n,t} F = M^{(H_n)} D_{n,t}^{(H)} F \quad \text{for } n = 1, \ldots, N.$$

Proposition 4.6.5 (Integration). *Suppose $Y : \mathbb{R} \to ((\mathcal{S})^*)^N$ is WIS integrable with respect to $B^{(H)}$ (Definition 4.6.1) and belongs to $(\mathcal{M})^N$. Then*

$$\int_{\mathbb{R}} Y(t) \, dB^{(H)}(t) = \int_{\mathbb{R}} M^{(H)} Y(t) \delta B(t),$$

where

$$M^{(H)} Y(t) = (M^{(H_1)} Y_1(t), \ldots, M^{(H_N)} Y_N(t)),$$

$B(t) = (B^1(t), \cdots, B^N(t))$ *is an N-dimensional standard Brownian motion, and*

$$\int_{\mathbb{R}} M^{(H)} Y(t) \delta B(t) = \sum_{i=1}^{N} \int_{\mathbb{R}} M_{H_i} Y^i(t) \delta B^i(t)$$

is a multidimensional Skorohod integral with respect to $B(t)$.

Theorem 4.6.6. *Suppose $Y : \mathbb{R} \to ((\mathcal{S})^*)^N$ is WIS integrable with respect to $B^{(H)}$. Then*

$$D_{n,t}^{(H)}(\int_{\mathbb{R}} Y(s) \, dB^{(H)}(s)) = \sum_{j=1}^{N} \int_{\mathbb{R}} D_{n,t}^{(H_n)} Y_j(s) \, dB_j^{(H)}(s) + Y_n(t).$$

Theorem 4.6.7 (Integration by parts). *Let $F \in L^2(\mathbb{P})$ and $Y : \mathbb{R} \to ((\mathcal{S})^*)^N$. Then*

$$F \int_{\mathbb{R}} Y(s) \, dB^{(H)}(s) = \int_{\mathbb{R}} F Y(s) \, dB^{(H)}(s)$$
$$+ \int_{\mathbb{R}} M^{(H)} Y(s) \cdot M^{(H)} \nabla^{(H)} F(s) \, ds,$$

(where the dot $\cdot$ denotes inner product in $\mathbb{R}^N$), provided that at least two of the terms are well-defined and belong to $L^2(\mathbb{P})$.

Theorem 4.6.8 (The multi-dimensional WIS isometry). *Let $Y : \mathbb{R} \to ((\mathcal{S})^*)^N$. Then*

$$E[(\int_{\mathbb{R}} Y(t) \, dB^{(H)}(t))^2] = E[\sum_{n=1}^{N} \int_{\mathbb{R}} (M_t^{(H_n)} Y_n(t))^2 \, dt]$$
$$+ E[\sum_{m,n=1}^{N} \int_{\mathbb{R}} \int_{\mathbb{R}} D_{m,t}^{(H_m)}((M_s^{(H_n)})^2 Y_n(s))$$
$$\cdot D_{n,s}^{(H_n)}((M_t^{(H_m)})^2 Y_m(t)) \, ds \, dt],$$

provided that at least two of the terms are well-defined.

5
Pathwise integrals for fractional Brownian motion

We conclude our overview of the stochastic integrals for *fBm* with the pathwise ones. A natural way to introduce a stochastic integral with respect to the *fBm* is to consider the so-called Riemann sums:

$$\sum_{i=1}^{n} f(t_i) \left[B^{(H)}(t_{i+1}) - B^{(H)}(t_i) \right],$$

where $0 = t_1 < \cdots < t_n = T$ is a partition of $[0, T]$, and then to investigate the conditions on f under which the convergence of this quantity holds at least in probability. Here we summarize the main approaches and results for pathwise integration with respect to *fBm* and investigate their relation with the other definitions of stochastic integral (divergence, fWIS and WIS integrals) seen in the previous chapters. For further details we also refer to Chapter 6.

We remark that since all definitions are made pathwise, adaptedness of integrands is not required and only sample-path regularity counts. This implies that in general it is not trivial to compute the expectation of any of these integrals. In the following sections, we recall some of the main results of [5], [6], [33], [42], [46], [54], [55], [61],[62], [70], [72], [74], [94], [101], [102], [149], [172], [173], [177], [237], [238], [239], [240].

5.1 Symmetric, forward and backward integrals for *fBm*

The *symmetric* (respectively, *forward* and *backward*) integral for the standard Brownian motion has been introduced in [10], [24] and [145] and developed in [198], [199] and [201]. These definitions have been extended to the *fBm* case in [33], [101], [102],[177], [237], [238], [239], [240].

Here we consider the definition of the *symmetric stochastic integral* in the *fBm* case as in the approach of [177].

Definition 5.1.1. *Let* $H \in (0,1)$. *Let* $(u_t)_{t \in [0,T]}$ *be a process with integrable trajectories. The* symmetric integral *of u with respect to* $B^{(H)}$ *is defined as*

124   5 Pathwise integrals for fractional Brownian motion

$$\lim_{\epsilon \to 0} \frac{1}{2\epsilon} \int_0^T u(s) \left[B^{(H)}(s+\epsilon) - B^{(H)}(s-\epsilon) \right] ds,$$

provided that the limit exists in probability, *and is denoted by* $\int_0^T u(s) d^\circ B^{(H)}(s)$.

We also define the *forward* and *backward* integrals for *fBm* as in the approach of [201].

Definition 5.1.2. *Let* $H \in (0,1)$. *Suppose that* $(u_t)_{t \in [0,T]}$ *is a process with integrable trajectories. The* forward integral *of u with respect to* $B^{(H)}$ *is defined as*

$$\lim_{\epsilon \to 0} \frac{1}{\epsilon} \int_0^T u(s) \frac{B^{(H)}(s+\epsilon) - B^{(H)}(s)}{\epsilon} ds, \qquad (5.1)$$

provided that the limit exists in probability, *and is denoted by* $\int_0^T u(s) d^- B^{(H)}(s)$.
  The backward integral *is defined as*

$$\lim_{\epsilon \to 0} \frac{1}{\epsilon} \int_0^T u(s) \frac{B^{(H)}(s-\epsilon) - B^{(H)}(s)}{\epsilon} ds,$$

provided that the limit exists in probability, *and is denoted by* $\int_0^T u(s) d^+ B^{(H)}(s)$.

For a survey of all the different possible definitions of the forward integral for *fBm*, we refer to [101] and to [240]. Here the "classical" forward integral defined in [201] has been extended to the case of *fBm* in the general context of m-order integrals for nonsemimartingale processes.

If f is a deterministic function with bounded variation, the symmetric integral $\int_0^T f(t) d^\circ B^{(H)}(t)$ can be obtained as the almost sure limit of Riemann sums, as will be made precise in the sequel. The convergence of these Riemann sums is equivalent to the convergence of the Riemann–Stieltjes integral appearing on the right-hand side of the integration by parts formula

$$\int_0^T f(t) d^\circ B^{(H)}(t) = f(T) B^{(H)}(T) - \int_0^T B^{(H)}(s) df(s),$$

which holds because of the continuity of the sample paths of $B^{(H)}$.

For $H = 1/2$, the symmetric integral is a generalization of the Stratonovich integral for standard Brownian motion, while the forward integral extends the Itô integral. In order to clarify the relation between the forward and the symmetric integral for every $H \in (0,1)$, by following [101] and [201] we define the (generalized) covariation.

Definition 5.1.3. *If X, Y are two continuous (respectively, locally bounded) processes, then their* covariation *is defined as the limit*

$$[X,Y]_t = \lim_{\epsilon \to 0} \frac{1}{\epsilon} \int_0^t (X_{u+\epsilon} - X_u)(Y_{u+\epsilon} - Y_u) \, du$$

if the limit exists uniformly on compacts in probability [i.e., in uniform convergence in probability *(ucp)].*

If X is such that $[X, X]$ exists, then X is called finite quadratic variation process. *Moreover, if $[X, X] = 0$, then X is called* zero quadratic variation process.

If X, Y are two continuous (respectively, locally bounded) processes, the following relation among the symmetric integral and the forward integral holds:

$$\int_0^t Y_u \, d^o X_u = \int_0^t Y_u \, d^- X_u + [X, Y]_t,$$

provided that two of these three terms exist. If X_t, Y_t are continuous semi-martingales, the covariation coincides with the usual square bracket. Hence, for $H = 1/2$, $[B, B]_t = t$. Since by Theorem 1.6.1 we have that sample paths of a *fBm* with parameter H are, outside a negligible event, Hölder continuous with every exponent $\beta < H$, it follows that $B^{(H)}$ is a zero quadratic variation process for $H > 1/2$. On the other hand, the quadratic variation is infinite for $H < 1/2$. However, $B^{(H)}$ admits for every $H \in (0,1)$ a *strong α-variation* according to the following definition provided originally in [201].

Definition 5.1.4. *Let X be a continuous process and $\alpha > 0$. The strong α-variation of X is the increasing continuous process given by*

$$[X]_t^{(\alpha)} = \lim_{\epsilon \to 0} \frac{1}{\epsilon} \int_0^t |X_{u+\epsilon} - X_u|^\alpha du,$$

if this limit exists in ucp.

In Proposition 3.14 of [201] it is proved that $B^{(H)}$ has a strong $1/H$-variation for every $H \in (0,1)$ and that

$$[B^{(H)}]_t^{(1/H)} = \rho_H t,$$

where $\rho_H = E\left[|G|^{1/H}\right]$ and G is a centered Gaussian random variable with unit variance. For further details we also refer to [101], [102] and [201].

5.2 On the link between fractional and stochastic calculus

In order to clarify the link between stochastic calculus for *fBm* and fractional calculus presented in Appendix B, we follow the approach of [237] and introduce the following definition:

Definition 5.2.1. *Let $(h_t)_{t \in [0,T]}$ be a stochastic process. We define the* extended forward integral *of h with respect to $B^{(H)}$ as*

$$\int_0^t h(s)\,d^-B^{(H)}(s) \tag{5.2}$$
$$= \lim_{\epsilon \to 0} \frac{1}{\Gamma(\epsilon)} \int_0^T u^{\epsilon-1} \int_0^t h(s) \frac{B^{(H)}(s+u) - B^{(H)}(s)}{s}\,ds\,du,$$

if the limit exists in ucp as a function of $t \in [0,T]$.

Definition 5.2.1 provides an extension of the definition of the forward integral for fBm given in Definition 5.1.2. In fact, the existence of the limit

$$\lim_{u \to 0} \int_0^t h(s) \frac{B^{(H)}(s+u) - B^{(H)}(s)}{u}\,ds$$

in uniform convergence in probability implies the existence of (5.2). This notion is a generalization of the one provided in [199]. This representation is the key in order to describe the link between stochastic and fractional calculus as described in the following.

Consider two deterministic functions f, g on $[0,T]$ satisfying the conditions of Definition B.1.4 and the fractal integral $\int_a^b f\,dg, 0 \le a < b \le T$, as introduced in (B.6) [respectively, in (B.7)]. Then the following *approximation* property of the integral holds:

$$\int_a^b f\,dg = \lim_{\epsilon \to 0} \int_a^b I_{a+}^\epsilon f\,dg.$$

By (B.4) we obtain that

$$\int_a^b I_{a+}^\epsilon f\,dg = \frac{1}{\Gamma(\epsilon)} \int_0^T u^{\epsilon-1} \int_a^b f(s) \frac{g_{b-}(s+u) - g_{b-}(s)}{u}\,ds\,du, \tag{5.3}$$

where $g_{b-}(x) = I_{(a,b)}(g(x) - g(b-))$ (see Definition B.1.4). This formula is valid if the degrees of differentiability of f and g sum up at least to $1 - \epsilon$. By using (5.3), we extend the definition of fractal integral provided in Definition B.1.4 as

$$\int_a^b f\,dg = \lim_{\epsilon \to 0} \frac{1}{\Gamma(\epsilon)} \int_0^T u^{\epsilon-1} \int_a^b f(s) \frac{g_{b-}(s+u) - g_{b-}(s)}{u}\,ds\,du,$$

whenever the limit on the right-hand side is determined. Applying this definition to stochastic processes via uniform convergence in probability, we obtain the natural extension of fractional calculus to the stochastic case.

5.3 The case $H < 1/2$

The definition of a pathwise integral for fBm with Hurst index $H < 1/2$ is more delicate since new difficulties appear. For example, the forward integral

$\int_0^T B^{(H)}(s)\, d^- B^{(H)}(s)$ does not exist in the sense of Definition 5.1.2 when the limit (5.1) is meant in the L^2 sense, as it is shown in the following example provided in [177]. Given a partition $t_i = iT/n$ of the interval $[0, T]$, the Riemann sums

$$\sum_{i=1}^n B^{(H)}(t_{i-1})\left[B^{(H)}(t_i) - B^{(H)}(t_{i-1})\right]$$

have the expectation

$$\sum_{i=1}^n E\left[B^{(H)}(t_{i-1})(B^{(H)}(t_i) - B^{(H)}(t_{i-1}))\right] = \frac{1}{2}\sum_{i=1}^n t_i^{2H} - t_{i-1}^{2H} - (t_i - t_{i-1})^{2H}$$

$$= \frac{1}{2}T^{2H}(1 - n^{1-2H})$$

that diverges as n goes to infinity if $H < 1/2$. On the other hand, the expectation of the symmetric sums

$$\sum_{i=1}^n E\left[(B^{(H)}(t_i) + B^{(H)}(t_{i-1}))(B^{(H)}(t_i) - B^{(H)}(t_{i-1}))\right] = \sum_{i=1}^n [t_i^{2H} - t_{i-1}^{2H}]$$

$$= T^{2H}$$

is finite.

There are several approaches introduced in order to avoid these problems. In [6], [42] and [46] the stochastic integral for fBm when $H < 1/2$ is defined by regularizing the kernel $K_H(t, s)$ that appears in the representation $B^{(H)}(t) = \int_0^t K_H t, s)\, dB(s)$ of the fBm with respect to the standard Bm [see (2.30)], in order to obtain a semimartingale. They then use the classical theory of stochastic integration and pass to the limit after a stochastic integration by parts in the sense of Malliavin calculus. Furthermore, they prove that under suitable Hölder continuity assumptions on the integrand, the integral is the limit in probability of the Riemann sums.

In order to establish when the symmetric integral is well-defined for $H < 1/2$, here we follow the approach of [5], [54] and [177]. By Chapter 2, we have that the operator K_H^* defined in (2.26) induces an isometry between the Hilbert space $\mathcal{H} = \overline{(\mathcal{E}, <, >_H)}$ (introduced in Section 2.1.2) and $L^2([0, T])$. We recall that for $H < 1/2$

$$\frac{\partial K_H}{\partial t}(t, s) = c_H\left(H - \frac{1}{2}\right)\left(\frac{t}{s}\right)^{H-1/2}(t-s)^{H-3/2},$$

that can be then estimated as follows:

$$\left|\frac{\partial K_H}{\partial t}(t, s)\right| \le c_H\left(\frac{1}{2} - H\right)(t-s)^{H-3/2} \tag{5.4}$$

if $t > s$. Using (5.4), we can introduce the following space of suitable integrands.

Definition 5.3.1. *Consider on the space $\mathcal{E}$ of step functions on $[0,T]$ the following seminorm:*

$$\|\phi\|_{K_H}^2 = \int_0^T \phi^2(s) K_H(T,s)^2 \, ds$$
$$+ \int_0^T \left[\int_s^T |\phi(t) - \phi(s)|(t-s)^{H-3/2} dt \right]^2 ds.$$

We denote by $\mathcal{H}_{K_H}$ the completion of $\mathcal{E}$ with respect to this seminorm.

The space $\mathcal{H}_{K_H}$ is continuously embedded in the Hilbert space $\mathcal{H}$.

Proposition 5.3.2. *Let $H < 1/2$ and $(u_t)_{t \in [0,T]}$ be a stochastic process in the space $\mathbb{D}^{1,2}(\mathcal{H}_{K_H})$. Suppose that the trace defined as*

$$\operatorname{Tr} D^{(H)} u := \lim_{\epsilon \to 0} \frac{1}{2\epsilon} \int_0^T \langle D^{(H)} u_s, I_{[s-\epsilon, s+\epsilon]} \rangle_H \, ds$$

exists as a limit in probability and that

$$E\left[\int_0^T u^2(s)(s^{2H-1} + (T-s)^{2H-1}) \, ds \right] < \infty,$$

$$E\left[\int_0^T \int_0^T (D_r^{(H)} u_s)^2 (s^{2H-1} + (T-s)^{2H-1}) \, ds \, dr \right] < \infty.$$

Then the symmetric integral $\int_0^T u(s) d^\circ B^{(H)}(s)$ of u with respect to the fBm defined as the limit in probability

$$\lim_{\epsilon \to 0} \int_0^T u_s \frac{B^{(H)}(s+\epsilon) - B^{(H)}(s-\epsilon)}{2\epsilon} \, ds$$

exists and

$$\int_0^T u(s) \, d^\circ B^{(H)}(s) = \delta_H(u) + \operatorname{Tr} D^{(H)} u.$$

Proof. For the proof, we refer to [5]. □

We can compute the trace in the particular case when the process $u_t = f(B_t^{(H)})$, where $f \in C^2(\mathbb{R})$ and satisfies the growth condition

$$\max |f(x)|, |f'(x)|, |f''(x)| \leq c e^{\lambda x^2}, \tag{5.5}$$

with c, λ positive constants such that $\lambda < 1/(4T^{2H})$. If $1/2 > H > 1/4$, then the process $u_t = f(B_t^{(H)})$ belongs to $\mathbb{D}^{1,2}(\mathcal{H}_{K_H})$, $\operatorname{Tr} D^{(H)} u$ exists and

$$\operatorname{Tr} D^{(H)} u = H \int_0^T f'(B^{(H)}(t)) t^{2H-1} \, dt.$$

By Proposition 5.3.2 we get

$$\int_0^T f(B^{(H)}(t))\,d^\circ B^{(H)}(t) = \delta_H(f(B^{(H)})) + H\int_0^T f'(B^{(H)}(t))t^{2H-1}\,dt.$$

where $\delta_H(f(B^{(H)}))$ is the divergence-type integral introduced in Chapter 2.

In order to consider a wider class of integrands with respect to the one considered in (5.5), we use the properties of the extended divergence operator as in the approach of [54]. They have shown that the symmetric integral of a general smooth function of $B^{(H)}$ with respect to $B^{(H)}$ exists in $L^2(\mathbb{P}^H)$ if and only if $H > 1/6$.

Note that if $h : \mathbb{R} \longrightarrow \mathbb{R}$ is a continuous function, then

$$\lim_{\epsilon \to 0} \int_0^T h(s)\frac{h(s+\epsilon) - h(s-\epsilon)}{2\epsilon}\,ds$$

$$= \lim_{\epsilon \to 0} \frac{1}{2\epsilon}\left(\int_0^T h(s)h(s+\epsilon)\,ds - \int_{-\epsilon}^{T-\epsilon} h(s)h(s+\epsilon)\,ds\right)$$

$$= \lim_{\epsilon \to 0} \frac{1}{2\epsilon}\left(\int_{T-\epsilon}^T h(s)h(s+\epsilon)\,ds - \int_{-\epsilon}^0 h(s)h(s+\epsilon)\,ds\right)$$

$$= \frac{1}{2}h^2(T) - \frac{1}{2}h^2(0)$$

Hence it follows that for all $H \in (0,1)$,

$$\lim_{\epsilon \to 0} \int_a^b B^{(H)}(s)\frac{B^{(H)}(s+\epsilon) - B^{(H)}(s-\epsilon)}{2\epsilon}\,ds = \frac{1}{2}\left[B^{(H)}(b)^2 - B^{(H)}(a)^2\right]$$

almost surely. Since for $H > 1/2$, $B^{(H)}$ is a zero quadratic variation process, by Theorem 2.1 of [199] that for all $H > 1/2$ and $g \in C^1(\mathbb{R})$

$$\int_a^b g(B^{(H)}(s))\,d^\circ B^{(H)}(s) = G(B^{(H)}(b)^2) - G(B^{(H)}(a)^2), \qquad (5.6)$$

where $G'(x) := g(x), x \in \mathbb{R}$. In [200], this formula is proved for $H = 1/2$ even in the case when $g \in L^2_{loc}(\mathbb{R})$. For $H < 1/2$, $B^{(H)}$ has infinite quadratic variation, but equation (5.6) holds for $g \in C^1(\mathbb{R})$ if $H \in (1/4, 1/2)$ as shown in [5]. In Theorem 4.1 of [102], (5.6) is proved even for $H = 1/4$, but with $g \in C^3(\mathbb{R})$. The most general result in this direction is contained in Theorem 5.3 of [54], where they show that for $g \in C^3(\mathbb{R})$, $H = 1/6$ is the critical value for the existence of the symmetric integral in (5.6). Here we state the result of [54].

Proposition 5.3.3. *Let $g \in C^3(\mathbb{R})$. Then the following results hold:*

1. *For every $H \in (1/6, 1/2)$,*

$$\int_a^b g(B^{(H)}(s))\, d^\circ B^{(H)}(s) = G(B^{(H)}(b)^2) - G(B^{(H)}(a)^2),$$

where $G(x) = \int_0^x g(s)\, ds$, $x \in \mathbb{R}$.
2. On the other hand, if $H \in (0, 1/6]$, then

$$\int_a^b [B^{(H)}(s)]^2\, d^\circ B^{(H)}(s)$$

does not exist.

Proof. The proof of this result is quite technical and we refer to [54, Theorem 5.3, Lemma 5.4, 5.5, 5.6 and Proposition 5.7] for further details. □

5.4 Relation with the divergence integral

In Proposition 5.3.2 we have already investigated the relation between the symmetric and the divergence-type integrals for $H < 1/2$. Here we study the case $H > 1/2$ following the approach of [8] and [177].

We recall that the space $\mathcal{H} = \overline{(L^2([0,T]), <,>_H)}$ is introduced in Theorem 2.1.6, $|\mathcal{H}|$ in (2.24), the stochastic derivative $D^{(H)}$ in Section 2.2 and that $\mathbb{P}^H$ denotes the law of $B^{(H)}$.

Proposition 5.4.1. *Let $H > 1/2$. Suppose $(u_t)_{t \in [0,T]}$ is a stochastic process in $\mathbb{D}^{1,2}(|\mathcal{H}|)$ and that*

$$\int_0^T \int_s^T |D_s^{(H)} u(t)||t-s|^{2H-2}\, ds\, dt < \infty, \quad a.s. \tag{5.7}$$

Then the symmetric integral exists and the following relation holds:

$$\int_0^T u(t)\, d^\circ B^{(H)}(t)$$
$$= \delta_H(u) + H(2H-1) \int_0^T \int_0^T D_s^{(H)} u(t) |t-s|^{2H-2}\, ds\, dt. \tag{5.8}$$

Under the assumptions of Proposition 5.4.1, the symmetric, backward and forward integrals coincide. For example, a sufficient condition for (5.7) is that

$$\int_0^T \left(\int_0^T |D_s^{(H)} u(t)|^p\, dt \right)^{1/p} ds < \infty$$

for some $p > 1/(2H-1)$.

Proof. Here we sketch a short proof of (5.8) by following [177]. For further details, we refer to the proof of Proposition 3 of [8]. We start by approximating u by

$$u_t^\epsilon = \frac{1}{2\epsilon} \int_{t-\epsilon}^{t+\epsilon} u(s)\, ds.$$

Then
$$\|u^\epsilon\|_{\mathbb{D}^{1,2}(|\mathcal{H}|)}^2 \leq d_H \|u\|_{\mathbb{D}^{1,2}(|\mathcal{H}|)}^2$$

for some positive constant d_H depending on H. Since by (2.33) we have the following integration by part formula
$$\delta_H(Fu) = F\delta_H(u) - \langle D^{(H)}F, u\rangle_H,$$

if $F \in \mathbb{D}_H^{1,2}$, $u \in \text{dom } \delta_H$, and Fu, $F\delta_H(u) + \langle D^{(H)}F, u\rangle_H \in L^2(\mathbb{P}^H)$, we obtain

$$\int_0^T u(s) \frac{B^{(H)}(s+\epsilon) - B^{(H)}(s-\epsilon)}{2\epsilon} ds$$
$$= \int_0^T u(s) \frac{1}{2\epsilon} \int_{s-\epsilon}^{s+\epsilon} dB^{(H)}(u)\, ds$$
$$= \int_0^T \delta_H(u(s) \frac{1}{2\epsilon} I_{[s-\epsilon, s+\epsilon]})\, ds + \frac{1}{2\epsilon} \int_0^T \langle D^{(H)}u(s), I_{[s-\epsilon, s+\epsilon]}\rangle_H\, ds$$
$$= \delta_H(u^\epsilon) + \frac{1}{2\epsilon} \int_0^T \langle D^{(H)}u(s), I_{[s-\epsilon, s+\epsilon]}\rangle_H\, ds.$$

The proposition follows taking the limit as ϵ tends to 0. $\square$

We now investigate when the symmetric integral coincides almost surely with the limit of Riemann sums, i.e., with the pathwise Riemann–Stieltjes integral. Using the representation (5.8), we obtain the following:

Proposition 5.4.2. *Let $H > 1/2$. If u is an adapted process continuous in the norm of $\mathbb{D}^{1,2}(\mathcal{H})$ such that*

$$\lim_{n \to \infty} \int_0^T \sup_{s,s' \in (r, r+1/n) \cap [0,T]} E\left[|D_r^{(H)}u(s) - D_r^{(H)}u(s')|^2\right] dr = 0, \quad (5.9)$$

then we have

$$\lim_{|\pi| \to 0} \sum_{i=1}^n u(t_i)(B^{(H)}(t_{i+1}) - B^{(H)}(t_i)) = \int_0^T u(s)\, d^\circ B^{(H)}(s),$$

where the convergence holds in $L^2(\mathbb{P}^H)$ and $\pi : 0 = t_0 < t_1 < \ldots < t_N = T$ is a partition of $[0,T]$ with mesh size $|\pi| = \sup_{i=0,\ldots,n}(t_{i+1} - t_i)$.

Proof. Here we adapt the proof of Proposition 5 of [6] to the fBm case for $H > 1/2$. By (5.8) we have that

$$\int_0^T u(t)\, d^\circ B^{(H)}(t) = \delta_H(u) + H(2H-1) \int_0^T \int_0^T D_s^{(H)} u(t) |t-s|^{2H-2}\, ds\, dt.$$

Since by Proposition 3.12.1

$$\delta_H(u) = \lim_{|\pi| \to 0} \sum_{i=1}^{n} u(t_i) \diamond (B^{(H)}(t_{i+1}) - B^{(H)}(t_i)),$$

where the convergence is in $L^2(\mathbb{P}^H)$, we have

$$\sum_{i=1}^{n} u(t_i) \diamond (B^{(H)}(t_{i+1}) - B^{(H)}(t_i))$$

$$= \sum_{i=1}^{n} u(t_i)(B^{(H)}(t_{i+1}) - B^{(H)}(t_i)) + \sum_{i=1}^{n} \langle D^{(H)} u(s_i), I_{(s_i, s_{i+1}]} \rangle_H.$$

We need to study the convergence of the second term of the sum. By using the properties of the operator K_H^* defined in (2.15) with respect to $\langle \cdot, \cdot \rangle_H$, we get

$$A_\pi = \sum_{i=1}^{n} \langle D^{(H)} u(s_i), I_{(s_i, s_{i+1}]} \rangle_H = \sum_{i=1}^{n} \langle Du(s_i), K_H^* I_{(s_i, s_{i+1}]} \rangle_{L^2([0,T])}$$

$$= \sum_{i=1}^{n} \int_0^T D_r u(s_i) (K_H(s_{i+1}, r) - K_H(s_i, r)) \, dr,$$

where we have also applied (2.36). Then

$$E\left[(A_\pi - \int_0^T \int_0^T D_s^{(H)} u(t) |t-s|^{2H-2} \, ds \, dt)^2\right]$$

$$\leq TE\left[\left(\int_0^T \sum_{i=1}^{n} \int_{s_i}^{s_{i+1}} |D_r u(s) - D_r u(s_i)| |r-s|^{2H-2} \, ds\right)^2 dr\right]$$

$$\leq T \int_0^T |T-r|^{2H-2} \sum_{i=1}^{n} \int_{s_i}^{s_{i+1}} E\left[|D_r u(s) - D_r u(s_i)|^2\right] |r-s|^{2H-2} \, ds \, dr$$

$$\leq T \int_0^T |T-r|^{2(2H-2)} \, dr$$

$$\cdot \int_0^T \sup_{s, s' \in (t, t+|\pi|) \cap [0, T]} E\left[|D_t^{(H)} u(s) - D_t^{(H)} u(s')|^2\right] dt.$$

The last term converges to zero since (5.9) holds. □

5.5 Relation with the fWIS integral

We now investigate the relation between the fWIS integral introduced in Chapter 3 and the pathwise integral in the case $H > 1/2$. Here $\int_0^T F(s) \, dB^{(H)}(s)$

must be interpreted as the fWIS integral defined in Definition 3.4.1 for $H > 1/2$. Main reference for this section is [83]. For other approaches see also [68] and [149].

Consider the symmetric integral as introduced in Definition 5.1.1 and characterized in Proposition 5.4.2. We recall that in Chapter 3 the space $\mathcal{L}_\phi(0,T)$ of integrands is defined as the family of stochastic processes F on $[0,T]$ with the following properties: $F \in \mathcal{L}_\phi(0,T)$ if $E\left[\|F\|_H^2\right] < \infty$, F is ϕ-differentiable, the trace of $D_s^\phi F_t, 0 \le s \le T, 0 \le t \le T$ exists, and $E\left[\int_0^T \int_0^T (D_t^\phi F_s)^2 \, ds \, dt\right] < \infty$ and for each sequence of partitions $(\pi_n, n \in \mathbb{N})$ such that $|\pi_n| \to 0$ as $n \to \infty$,

$$\sum_{i=0}^{n-1} E\left[\int_{t_i^{(n)}}^{t_{i+1}^{(n)}} \int_{t_j^{(n)}}^{t_{j+1}^{(n)}} |D_s^\phi F_{t_i^{(n)}}^\pi D_t^\phi F_{t_j^{(n)}}^\pi - D_s^\phi F_t D_t^\phi F_s| \, ds \, dt\right]$$

and

$$E\left[\|F^\pi - F\|_H^2\right]$$

tend to 0 as $n \to \infty$, where $\pi_n : 0 = t_0^{(n)} < t_1^{(n)} < \cdots < t_{n-1}^{(n)} < t_n^{(n)} = T$.

Theorem 5.5.1. *Let $H > 1/2$ and $F \in \mathcal{L}_\phi(0,T)$. Then the symmetric integral $\int_0^T F_s \, d^\circ B^{(H)}(s)$ exists and the following equality is satisfied*

$$\int_0^T F_s \, d^\circ B^{(H)}(s) = \int_0^T F_s \, dB^{(H)}(s) + \int_0^T D_s^\phi F_s \, ds \quad a.s.,$$

where $\int_0^T F_s \, dB^{(H)}(s)$ is the fWIS integral defined in Theorem 3.6.1 and $D_s^\phi F$ in Definition 3.5.1.

Proof. Since if $g \in L_H^2(\mathbb{R})$, $F \in L^2(\mathbb{P}^H)$, and $D_{\Phi g} F \in L^2(\mathbb{P}^H)$, by Proposition 3.5.4 we have that

$$F \diamond \int_\mathbb{R} g_s \, dB^{(H)}(s) = F \int_\mathbb{R} g_s \, dB^{(H)}(s) - D_{\Phi g} F.$$

Hence,

$$\sum_{i=0}^{n-1} F_{t_i^{(n)}} (B^{(H)}(t_{i+1}^{(n)}) - B^{(H)}(t_i^{(n)}))$$

$$= \sum_{i=0}^{n-1} F_{t_i^{(n)}} \diamond (B^{(H)}(t_{i+1}^{(n)}) - B^{(H)}(t_i^{(n)})) + \sum_{i=0}^{n-1} D_{\Phi I_{[t_i^{(n)}, t_{i+1}^{(n)}]}} F_{t_i^{(n)}}$$

$$= \sum_{i=0}^{n-1} \left\{ F_{t_i^{(n)}} \diamond \left[B^{(H)}(t_{i+1}^{(n)}) - B^{(H)}(t_i^{(n)}) \right] + \int_{t_i^{(n)}}^{t_{i+1}^{(n)}} D_s^\phi F_{t_i^{(n)}} \, ds \right\}.$$

This equality easily proves the theorem. □

These two types of stochastic integrals are both interesting and present different characteristics. The expectation of $\int_0^t F_s \, dB^{(H)}(s)$ is 0, but the chain rule for this type of integral is more complicated than for the pathwise integral. On the contrary, we have that $E\left[\int_0^t F_s \, d^\circ B^{(H)}(s)\right] \neq 0$ in general, as the following shows.

It is well known that if X is a standard normal random variable, $X \sim N(0,1)$, then

$$E[X^n] = \begin{cases} \dfrac{n!}{(\sqrt{2})^n (n/2)!} & \text{if } n \text{ is even,} \\ 0 & \text{if } n \text{ is odd.} \end{cases}$$

Let $f(x) = x^n$. If n is odd, then

$$E\left[\int_0^t f(B^{(H)}(s)) \, d^\circ B^{(H)}(s)\right] = E\left[\int_0^t D_s^\phi f(B^{(H)}(s)) \, ds\right]$$

$$= E\left[\int_0^t f'(B^{(H)}(s)) D_s^\phi B^{(H)}(s) \, ds\right]$$

$$= E\left[\int_0^t f'(B^{(H)}(s)) \int_0^s \phi(u,s) \, du \, ds\right]$$

$$= H \int_0^t s^{2H-1} E\left[f'(B^{(H)}(s))\right] ds$$

$$= nH \int_0^t s^{2H-1} E\left[(B^{(H)}(s))^{n-1}\right] ds$$

$$= nH \int_0^t s^{2H-1} E\left[\left(\frac{B^{(H)}(s)}{s^H}\right)^{n-1}\right] s^{nH-H} \, ds$$

$$= \frac{n! H t^{(n+1)H}}{\sqrt{2}^{n-1}(n+1)H((n-1)/2)!}$$

which is not 0. If n is even, then by the same computation, we obtain

$$E\left[\int_0^t (B^{(H)}(s))^n \, d^\circ B^{(H)}(s)\right] = 0.$$

We now show another interesting phenomenon. Let π be a partition of the interval $[0,T]$: $0 = t_0 < t_1 < t_2 < \ldots < t_n = T$. If $(f(s), s \geq 0)$ is a continuous stochastic process, the Itô integral with respect to the Brownian motion $(B_t, t \in [0,T])$ can be defined as the limit of the Riemann sums

$$\sum_{i=0}^{n-1} f_{t_i}(B_{t_{i+1}} - B_{t_i})$$

as the partition mesh size $|\pi| \to 0$, where $\pi : 0 = t_0 < t_1 < t_2 < \ldots < t_n = T$. The Stratonovich integral in the standard Brownian motion case is defined as the limit of the Riemann sums

5.5 Relation with the fWIS integral

$$\sum_{i=0}^{n-1} \frac{f_{t_i} + f_{t_{i+1}}}{2} (B_{t_{i+1}} - B_{t_i})$$

as the partition $|\pi| \to 0$. Here we prove that the symmetric integral is a Stratonovich-type integral for fBm $B_t^{(H)}$, $t \geq 0$, with $H > 1/2$ since in the sequel we show that the above two limits are the same for a large class of stochastic processes.

Initially the following lemma is given.

Lemma 5.5.2. *Let p be a positive even integer. Then*

$$E\left[\left(B^{(H)}(t) - B^{(H)}(s)\right)^p\right] = \frac{p!}{2^{p/2}(p/2)!}|t-s|^{pH}.$$

Proof. Since

$$E\left[|B^{(H)}(t) - B^{(H)}(s)|^2\right] = |t-s|^{2H},$$

we have that $(B^{(H)}(t) - B^{(H)}(s))/|t-s|^H$ is a standard Gaussian random variable and

$$E\left[|B^{(H)}(t) - B^{(H)}(s)|^p\right] = |t-s|^{pH} E\left[\left(\frac{B^{(H)}(t) - B^{(H)}(s)}{|t-s|^H}\right)^p\right]$$

$$= \frac{p!}{2^{p/2}(p/2)!}|t-s|^{pH}.$$

□

Corollary 5.5.3. *For each $\alpha > 1$, there is a $C_\alpha < \infty$ such that*

$$E\left[|B^{(H)}(t) - B^{(H)}(s)|^\alpha\right] \leq C_\alpha |t-s|^{\alpha H}.$$

Definition 5.5.4. *The process $(f_s, 0 \leq s \leq T)$ is said to be a bounded quadratic variation process if there are constants $p \geq 1$ and $0 < C_p < \infty$ such that for any partition $\pi : 0 = t_0 < t_1 < t_2 < \ldots < t_n = T$,*

$$\sum_{i=0}^{n-1} E\left[|f_{t_{i+1}} - f_{t_i}|^{2p}\right]^{1/p} \leq C_p.$$

Example 5.5.5. Let $f : \mathbb{R} \to \mathbb{R}$ be continuously differentiable with bounded first derivative. Then $f(B^{(H)}(s))$ is a bounded quadratic variation process. In fact, for any $p \geq 1$ and partition $\pi : 0 = t_0 < t_1 < t_2 < \ldots < t_n = T$,

$$\sum_{i=0}^{n-1} E\left[|f(B_{t_{i+1}}^{(H)}) - f(B_{t_i}^{(H)})|^{2p}\right]^{1/p}$$

$$= \sum_{i=0}^{n-1} E\left[\left(\left[\int_0^1 f'\left(B_{t_i}^{(H)} + \theta(B_{t_{i+1}}^{(H)} - B_{t_i}^{(H)})\right) d\theta\right] \cdot (B_{t_{i+1}}^{(H)} - B_{t_i}^{(H)})\right)^{2p}\right]^{1/p}$$

$$\leq C \sum_{i=0}^{n-1} E\left[|B_{t_{i+1}}^{(H)} - B_{t_i}^{(H)}|^{2p}\right]^{1/p}$$

$$\leq C \sum_{i=0}^{n-1} |t_{i+1} - t_i|^{2H} \leq CT.$$

Theorem 5.5.6. *Let $(f(t), 0 \leq t \leq T)$ be a bounded quadratic variation process. Let $(\pi_n)_{n \in \mathbb{N}}$ be a sequence of partitions of $[0,T]$ such that $|\pi_n| \to 0$ as $n \to \infty$ and*

$$\sum_{i=0}^{n-1} f(t_i^{(n)})[B^{(H)}(t_{i+1}^{(n)}) - B^{(H)}(t_i^{(n)})], \ n \in \mathbb{N},$$

converges to a random variable G in $L^2(\mathbb{P}^H)$, where $\pi_n = \left\{t_0^{(n)}, \ldots, t_n^{(n)}\right\}$. Then

$$\sum_{i=0}^{n-1} f(t_{i+1}^{(n)})[B^{(H)}(t_{i+1}^{(n)}) - B^{(H)}(t_i^{(n)})], \ n \in \mathbb{N},$$

also converges to G in $L^2(\mathbb{P}^H)$.

Proof. It suffices to show that $\sum_{i=0}^{n-1}(f_{t_{i+1}} - f_{t_i})[B_{t_{i+1}}^{(H)} - B_{t_i}^{(H)}]$ converges to 0 in $L^2(\mathbb{P}^H)$. Let p be a number as indicated in the definition of bounded quadratic variation for $(f_t, 0 \leq t \leq T)$,

$$E\left[\left(\sum_{i=0}^{n-1}(f_{t_{i+1}} - f_{t_i})(B_{t_{i+1}}^{(H)} - B_{t_i}^{(H)})\right)^2\right]^{1/2}$$

$$\leq \sum_{i=0}^{n-1} \left(E\left[(f_{t_{i+1}} - f_{t_i})^2 (B_{t_{i+1}}^{(H)} - B_{t_i}^{(H)})^2\right]\right)^{1/2}$$

$$\leq \sum_{i=0}^{n-1} E\left[(f_{t_{i+1}} - f_{t_i})^{2p}\right]^{1/(2p)} E\left[(B_{t_{i+1}}^{(H)} - B_{t_i}^{(H)})^{2q}\right]^{1/(2q)}$$

$$\leq \left\{\sum_{i=0}^{n-1} E\left[(f_{t_{i+1}} - f_{t_i})^{2p}\right]^{1/p}\right\}^{1/2} \left\{\sum_{i=0}^{n-1} E\left[|B_{t_{i+1}}^{(H)} - B_{t_i}^{(H)}|^{2q}\right]^{1/q}\right\}^{1/2}$$

$$\leq C \left(\sum_{i=0}^{n-1} |t_{i+1} - t_i|^{2H}\right)^{1/2}$$

$$\leq C \max_{0 \leq i \leq n-1}(t_{i+1} - t_i)^{H-1/2} \left(\sum_{i=0}^{n-1} |t_{i+1} - t_i|\right)^{1/2}$$

$$\leq C\sqrt{T} \max_{0 \leq i \leq n-1}(t_{i+1} - t_i)^{H-1/2}$$

where $1/p + 1/q = 1$. The proposition follows since $\max_{0 \leq i \leq n-1}(t_{i+1} - t_i)^{H-1/2} \to 0$ as $|\pi| \to 0$. □

It can also be shown with a slightly more lengthy argument that if $(f_s, s \geq 0)$ is a process with bounded quadratic variation and ξ_i is any point in $[t_i, t_{i+1}]$, then the limit of the Riemann sums $\sum_{i=0}^{n-1} f_{\xi_i}[B_{t_{i+1}}^{(H)} - B_{t_i}^{(H)}]$ converge in $L^2(\mathbb{P}^H)$ to $\int_0^T f_s \, d^{\circ} B^{(H)}(s)$ if it is true for any particular choice of such a ξ_i.

5.6 Relation with the WIS integral

We now clarify the relation between the forward integral and the WIS integral for every $H \in (0,1)$. Hence, in this section the integral $\int \psi \, dB^{(H)}$ must then be interpreted in the sense introduced in Chapter 4 and *our reference probability space is here* $(\Omega, \mathcal{F}, \mathbb{P})$, *where $\mathbb{P}$ is the probability measure induced by the standard Brownian motion* (see Definition A.1.1). At this purpose we first reformulate Definition 5.1.2 under the probability $\mathbb{P}$ and then extend it to $(\mathcal{S})^*$ following the approach of [33].

Definition 5.6.1.

1. The (classical) forward integral *of a realvalued measurable process Y with integrable trajectories is defined by*

$$\int_0^T Y(t) \, d^- B^{(H)}(t) = \lim_{\epsilon \to 0} \int_0^T Y(t) \frac{B^{(H)}(t+\epsilon) - B^{(H)}(t)}{\epsilon} \, dt,$$

provided that the limit exists in probability under $\mathbb{P}$.

2. The (generalized) forward integral *of a realvalued measurable process Y with integrable trajectories is defined by*

$$\int_0^T Y(t) \, d^- B^{(H)}(t) = \lim_{\epsilon \to 0} \int_0^T Y(t) \frac{B^{(H)}(t+\epsilon) - B^{(H)}(t)}{\epsilon} \, dt,$$

provided that the limit exists in $(\mathcal{S})^*$.

Note that in the generalized definition of forward integral, the limit is required to exist in the *Hida space of stochastic distributions* $(\mathcal{S})^*$ introduced in Definition A.1.4. Convergence in $(\mathcal{S})^*$ is also explained in Appendix A.

Corollary 5.6.2. *Let $\psi(t) = \psi(t,\omega)$ be a measurable forward integrable process and assume that ψ is càglàd. The forward integral of ψ with respect to the fBm $B^{(H)}$ coincides with*

$$\int_0^T \psi(t) \, d^- B^{(H)}(t) = \lim_{|\Delta| \to 0} \sum_{j=1}^N \psi(t_j) \Delta B_{t_j}^{(H)} \tag{5.10}$$

whenever the left-hand limit exists in probability, where $\pi : 0 = t_0 < t_1 < \ldots < t_N = T$ is a partition of $[0,T]$ with mesh size $|\Delta| = \sup_{j=0,\ldots,N-1} |t_{j+1} - t_j|$ and $\Delta B_{t_j}^{(H)} = B_{t_{j+1}}^{(H)} - B_{t_j}^{(H)}$.

Proof. Let ψ be a càglàd forward integrable process and
$$\psi^{(\Delta)}(t) = \sum_k \psi(t_k) I_{(t_k, t_{k+1}]}(t)$$
be a càglàd step function approximation to ψ. Then $\psi^{(\Delta)}(t)$ converges boundedly almost surely to $\psi(t)$ as $|\Delta| \longrightarrow 0$. The forward integral of $\psi^{(\Delta)}(t)$ is then given by

$$\int_0^T \psi^{(\Delta)}(t) \, d^- B^{(H)}(t) = \lim_{\epsilon \to 0} \int_0^T \psi^{(\Delta)}(s) \frac{B^{(H)}(s+\epsilon) - B^{(H)}(s)}{\epsilon} \, ds$$
$$= \lim_{\epsilon \to 0} \sum_k \psi(t_k) \int_{t_k}^{t_{k+1}} \frac{1}{\epsilon} \int_s^{s+\epsilon} dB^{(H)}(u) \, ds$$
$$= \lim_{\epsilon \to 0} \sum_k \psi(t_k) \int_{t_k}^{t_{k+1}} \frac{1}{\epsilon} \int_{u-\epsilon}^u ds \, dB^{(H)}(u)$$
$$= \sum_k \psi(t_k) \Delta B_{t_k}^{(H)}, \qquad (5.11)$$

where $\Delta B_{t_k}^{(H)} = B_{t_{k+1}}^{(H)} - B_{t_k}^{(H)}$. Hence (5.10) follows by the dominated convergence theorem and by (5.11). $\square$

For the sequel we will use the same notation as in Appendix A. We recall that $\mathcal{J}$ is the set of all multi-indices $\alpha = (\alpha_1, \alpha_2, \ldots)$ of finite length $l(\alpha) = \max\{i : \alpha_i \neq 0\}$ with $\alpha_i \in \mathbb{N}_0 = \{0, 1, 2, \ldots\}$ for all i and that ξ_n are the Hermite functions defined in (A.6).

Definition 5.6.3. *The space $\mathbb{L}_{1,2}^{(H)}$ consists of all càglàd processes $\psi(t) = \sum_{\alpha \in \mathcal{J}} c_\alpha(t) \mathcal{H}_\alpha(\omega) \in (\mathcal{S})^*$ for every $t \in [0,T]$ such that*

$$\|\psi\|_{\mathbb{L}_{1,2}^{(H)}}^2 := \sum_{\alpha \in \mathcal{J}} \sum_{i=1}^\infty \alpha_i \alpha! \|c_\alpha\|_{L^2([0,T])}^2 < \infty.$$

Note that if $\psi(t) \in (\mathcal{S})^*$ for every $t \in [0,T]$, then $D_s \psi(t)$ exists in $(\mathcal{S})^*$ by Theorem A.3.5. We recall that we denote by D the Malliavin derivative with respect to a standard Brownian motion.

We recall a preliminary lemma needed in the following.

Lemma 5.6.4. *Let $(\Gamma, \mathcal{G}, m)$ be a measure space. Let $f_\epsilon : \Gamma \to B$, $\epsilon \in \mathbb{R}$, be measurable functions with values in a Banach space $(B, \|\cdot\|_B)$. If $f_\epsilon(\gamma) \to f_0(\gamma)$ as $\epsilon \to 0$ for almost every $\gamma \in \Gamma$ and there exists $K < \infty$ such that*

$$\int_\Gamma \|f_\epsilon(\gamma)\|_B^2 \, dm(\gamma) < K$$

for all $\epsilon \in \mathbb{R}$, then

$$\int_\Gamma f_\epsilon(\gamma)\,dm(\gamma) \to \int_\Gamma f_0(\gamma)\,dm(\gamma)$$

in $\|\cdot\|_B$.

Proof. The proof is analogous to the one of Theorem II.21.2 of [196]. □

Lemma 5.6.5. *Suppose that $\psi \in \mathbb{L}_{1,2}^{(H)}$. Then*

$$M_{t+}D_{t+}\psi(t) := \lim_{\epsilon \to 0} \frac{1}{\epsilon} \int_t^{t+\epsilon} M_s D_s \psi(t)\,ds$$

exists in $L^2(\mathbb{P})$ for all t. Moreover,

$$\int_0^T M_{t+}D_{t+}\psi(t)\,dt = \lim_{\epsilon \to 0} \int_0^T \left[\frac{1}{\epsilon}\int_t^{t+\epsilon} M_s D_s \psi(t)\,ds\right] dt \quad (5.12)$$

in $L^2(\mathbb{P})$ and

$$E\left[\left(\int_0^T M_{s+}D_{s+}\psi(s)\,ds\right)^2\right] < \infty. \quad (5.13)$$

Proof. Suppose that $\psi(t)$ has the expansion

$$\psi(t) = \sum_{\alpha \in \mathcal{J}} c_\alpha(t) \mathcal{H}_\alpha(\omega).$$

In the sequel we drop ω in $\mathcal{H}_\alpha(\omega)$ for the sake of simplicity. Then, in the same notation as in Appendix A, we have

$$D_s\psi(t) = \sum_{\alpha \in \mathcal{J}} \sum_{i=1}^\infty c_\alpha(t)\alpha_i \mathcal{H}_{\alpha-\epsilon^{(i)}} \xi_i(s)$$

and

$$M_s D_s \psi(t) = \sum_{\alpha \in \mathcal{J}} \sum_{i=1}^\infty c_\alpha(t)\alpha_i \mathcal{H}_{\alpha-\epsilon^{(i)}} \eta_i(s),$$

where $\eta_i(s) = M\xi_i(s)$. Hence,

$$\frac{1}{\epsilon}\int_t^{t+\epsilon} M_s D_s \psi(t)\,ds = \sum_{\alpha \in \mathcal{J}} \sum_{i=1}^\infty \left[c_\alpha(t)\frac{1}{\epsilon}\int_t^{t+\epsilon} \eta_i(s)\,ds\right] \alpha_i \mathcal{H}_{\alpha-\epsilon^{(i)}}.$$

Since $\eta_i(s) = M\xi(s)$ is a continuous function, we have that

$$\frac{1}{\epsilon}\int_t^{t+\epsilon} \eta_i(s)\,ds \to \eta_i(t)$$

as $\epsilon \to 0$.

We apply now Lemma 5.6.4 with $\gamma = (\alpha, i)$, $dm(\gamma) = \sum_{\alpha \in \mathcal{J}} \sum_{i=1}^{\infty} \delta_{(\alpha,i)}$, where δ_x denotes the point mass at x, $B = L^2(\mathbb{P})$, and functions $f_\epsilon = [c_\alpha(t) \frac{1}{\epsilon} \int_t^{t+\epsilon} \eta_i(s)\, ds] \alpha_i \mathcal{H}_{\alpha - \epsilon^{(i)}}$. We obtain

$$\int_\Gamma \|f_\epsilon(\gamma)\|_B^2\, dm(\gamma) = \sum_{\alpha \in \mathcal{J}} \sum_{i=1}^{\infty} \|f_\epsilon(\gamma)\|_{L^2(\mathbb{P})}^2$$

$$= \sum_{\alpha \in \mathcal{J}} \sum_{i=1}^{\infty} \left[c_\alpha(t) \frac{1}{\epsilon} \int_t^{t+\epsilon} \eta_i(s)\, ds \right]^2 \alpha_i \alpha!$$

$$\leq \left[\frac{(t+\epsilon)^{2H} - t^{2H}}{\epsilon} \right]^2 \sum_{\alpha \in \mathcal{J}} \sum_{i=1}^{\infty} c_\alpha(t)^2 \alpha_i \alpha!$$

since

$$\frac{1}{\epsilon} \int_t^{t+\epsilon} \eta_i(s)\, ds = \langle M\xi_i, \frac{1}{\epsilon} I_{[t,t+\epsilon]} \rangle_{L^2(\mathbb{R})} = \langle M^2 e_i, \frac{1}{\epsilon} I_{[t,t+\epsilon]} \rangle_{L^2(\mathbb{R})}$$

$$= \langle e_i, \frac{1}{\epsilon} I_{[t,t+\epsilon]} \rangle_H \leq \|e_i\|_H \frac{1}{\epsilon} \|I_{[t,t+\epsilon]}\|_H = \frac{(t+\epsilon)^{2H} - t^{2H}}{\epsilon},$$

where we have used that the fact that $\|e_i\|_H = 1$ and (4.6). Since we have $\sum_{\alpha \in \mathcal{J}} \sum_{i=1} c_\alpha(t)^2 \alpha_i \alpha! < \infty$ for almost every t, by Lemma 5.6.4 it follows that $\sum_{\alpha \in \mathcal{J}} \sum_{i=1}^{\infty} [c_\alpha(t) \frac{1}{\epsilon} \int_t^{t+\epsilon} \eta_i(s)\, ds] \alpha_i \mathcal{H}_{\alpha - \epsilon^{(i)}}$ converges to

$$\sum_{\alpha \in \mathcal{J}} \sum_{i=1}^{\infty} c_\alpha(t) \eta_i(t) \alpha_i \mathcal{H}_{\alpha - \epsilon^{(i)}}$$

in $L^2(\mathbb{P})$.

We now prove (5.12). Consider

$$\int_0^T \frac{1}{\epsilon} \int_t^{t+\epsilon} M_s D_s \psi(t)\, ds\, dt = \sum_{\alpha \in \mathcal{J}} \sum_{i=1}^{\infty} \int_0^T \left[c_\alpha(t) \frac{1}{\epsilon} \int_t^{t+\epsilon} \eta_i(s)\, ds \right] dt\, \alpha_i \mathcal{H}_{\alpha - \epsilon^{(i)}}.$$

Now assuming $f_\epsilon = \int_0^T \left[c_\alpha(t) \frac{1}{\epsilon} \int_t^{t+\epsilon} \eta_i(s)\, ds \right] dt\, \alpha_i \mathcal{H}_{\alpha - \epsilon^{(i)}}$ and as before $\gamma = (\alpha, i)$, $B = L^2(\mathbb{P})$, $dm(\gamma) = \sum_{\alpha \in \mathcal{J}} \sum_{i=1}^{\infty} \delta_{\alpha,i}$, where δ_x denotes the point mass at x, we use again Lemma 5.6.4. We obtain

$$\int_\Gamma \|f_\epsilon(\gamma)\|_B^2\, dm(\gamma) = \sum_{\alpha \in \mathcal{J}} \sum_{i=1}^{\infty} \|f_\epsilon(\gamma)\|_{L^2(\mathbb{P})}^2$$

$$= \sum_{\alpha \in \mathcal{J}} \sum_{i=1}^{\infty} \left[\int_0^T c_\alpha(t) \frac{1}{\epsilon} \int_t^{t+\epsilon} \eta_i(s)\, ds\, dt \right]^2 \alpha_i \alpha!$$

$$\leq \sum_{\alpha \in \mathcal{J}} \sum_{i=1}^{\infty} \left[\int_0^T c_\alpha(t) \left(\frac{(t+\epsilon)^{2H} - t^{2H}}{\epsilon} \right) dt \right]^2 \alpha_i \alpha!$$

5.6 Relation with the WIS integral

$$\leq \sum_{\alpha \in \mathcal{J}} \sum_{i=1} \left[\int_0^T c_\alpha(t)^2 \, dt \right]$$
$$\cdot \left[\int_0^T \left(\frac{(t+\epsilon)^{2H} - t^{2H}}{\epsilon} \right)^2 dt \right] \alpha_i \alpha!. \quad (5.14)$$

Since $\psi \in \mathbb{L}_{1,2}^{(H)}$ by Lemma 5.6.4, we can conclude that the limit (5.12) exists in $L^2(\mathbb{P})$ and also that (5.13) holds. □

Lemma 5.6.6. *Suppose that* $\psi \in \mathbb{L}_{1,2}^{(H)}$ *and let*

$$\psi^{(\Delta)}(s) = \sum_k \psi(t_k) I_{(t_k, t_{k+1}]}(s)$$

be a càglàd step function approximation to ψ*, where* $\Delta = \max_i |\Delta t_i|$ *is the maximal length of the subinterval in the partition* $0 = t_0 < \ldots < t_n = T$ *of* $[0,T]$*. Then* $\psi^{(\Delta)} \in \mathbb{L}_{1,2}^{(H)}$ *for all* Δ *and*

$$\int_0^T M_{s+} D_{s+} \psi^{(\Delta)}(s) \, ds \longrightarrow \int_0^T M_{s+} D_{s+} \psi(s) \, ds \quad \text{in } L^2(\mathbb{P}) \quad (5.15)$$

as $|\Delta| \longrightarrow 0$.

Proof. Since $\psi^{(\Delta)}(s) = \sum_{\alpha \in \mathcal{J}} c_\alpha^{(\Delta)}(s) \mathcal{H}_\alpha(\omega)$ with

$$c_\alpha^{(\Delta)}(s) = \sum_k c_\alpha(t_k) I_{(t_k, t_{k+1}]}(s)$$

and

$$\|c_\alpha^{(\Delta)}\|_{L^2([0,T])} \leq \text{const.} \|c_\alpha\|_{L^2([0,T])} \quad \forall \alpha,$$

it follows that $\psi^{(\Delta)} \in \mathbb{L}_{1,2}^{(H)}$. We have

$$\frac{1}{\epsilon} \int_t^{t+\epsilon} M_s D_s \psi^{(\Delta)}(t) \, ds$$
$$= \sum_{\alpha \in \mathcal{J}} \sum_{i=1}^\infty \left\{ \int_0^T \left[c_\alpha^{(\Delta)}(t) \frac{1}{\epsilon} \int_t^{t+\epsilon} \eta_i(s) \, ds \right] dt \right\} \alpha_i \mathcal{H}_{\alpha - \epsilon^{(i)}}.$$

If we assume $\gamma = (\alpha, i)$, $B = L^2(\mathbb{P})$, $m(d\gamma) = \sum_{\alpha \in \mathcal{J}} \sum_{i=1}^\infty \delta_{(\alpha, i)}$, where δ_x denotes the point mass at x, and $f_\Delta = \left\{ \int_0^T \left[c_\alpha^{(\Delta)}(t) 1/\epsilon \int_t^{t+\epsilon} \eta_i(s) \, ds \right] dt \right\} \alpha_i \mathcal{H}_{\alpha - \epsilon^{(i)}}$, with the same argument as in (5.14) by Lemma 5.6.4, we obtain that

$$\int_0^T \left[\frac{1}{\epsilon} \int_t^{t+\epsilon} M_s D_s \psi(t) \, ds \right] dt$$
$$= \lim_{|\Delta| \to 0} \int_0^T \left[\frac{1}{\epsilon} \int_t^{t+\epsilon} M_s D_s \psi^{(\Delta)}(t) \, ds \right] dt \quad (5.16)$$

in $L^2(\mathbb{P})$ for almost every s, since $c_\alpha^{(\Delta)}$ converges by dominated convergence to c_α in $L^2(\mathbb{P})$ and $\psi^{(\Delta)} \in \mathbb{L}_{1,2}^{(H)}$. Using (5.16) and Lemma 5.6.5 we conclude that (5.15) holds. □

We now investigate the relation among forward integrals and WIS integrals.

Theorem 5.6.7. *Let $H \in (0,1)$. Suppose $\psi \in \mathbb{L}_{1,2}^{(H)}$ and that one of the following conditions holds:*

i) ψ is Wick-Itô-Skorohod integrable (Definition 4.2.2);
ii) ψ is forward integrable in $(\mathcal{S})^*$ (Definition 5.6.1).

Then

$$\int_0^T \psi(t)\, d^- B^{(H)}(t) = \int_0^T \psi(t)\, dB^{(H)}(t) + \int_0^T M_{t+} D_{t+} \psi(t)\, dt, \qquad (5.17)$$

holds as an identity in $(\mathcal{S})^*$, where here $\int_0^T \psi(t) dB^{(H)}(t)$ is the WIS integral of Definition 4.2.2.

Proof. We prove (5.17) assuming that hypothesis $i)$ is in force. The argument works symmetrically under hypothesis $ii)$. Let $\psi \in \mathbb{L}_{1,2}^{(H)}$. Since ψ is càglàd, we can approximate it as

$$\psi(t) = \lim_{|\Delta t| \to 0} \sum_j \psi(t_j) I_{(t_j, t_{j+1}]}(t) \qquad \text{a.e.,}$$

where for any partition $0 = t_0 < t_1 < \ldots < t_N = T$ of $[0,T]$, with $\Delta t_j = t_{j+1} - t_j$, we have put $|\Delta t| = \sup_{j=0,\ldots,N-1} \Delta t_j$.

As before we put $\psi^{(\Delta)}(t) = \sum_{j=0}^{N-1} \psi(t_k) I_{(t_k, t_{k+1}]}(t)$ and evaluate

$$\int_0^T \psi^{(\Delta)}(t)\, d^- B^{(H)}(t)$$

$$= \lim_{\epsilon \to 0} \int_0^T \psi^{(\Delta)}(t,\omega) \frac{B^{(H)}(t+\epsilon) - B^{(H)}(t)}{\epsilon}\, dt$$

$$= \lim_{\epsilon \to 0} \int_0^T \left[\sum_j \psi(t_j) I_{(t_j,t_{j+1}]}(t) \right] \frac{1}{\epsilon} \int_t^{t+\epsilon} dB^{(H)}(u)\, dt$$

$$= \lim_{\epsilon \to 0} \int_0^T \left[\sum_j \psi(t_j) I_{(t_j,t_{j+1}]}(t) \right] \diamond \frac{1}{\epsilon} \int_t^{t+\epsilon} dB^{(H)}(u)\, dt$$

$$+ \lim_{\epsilon \to 0} \sum_j \int_0^T I_{(t_j,t_{j+1}]}(t) \frac{1}{\epsilon} \int_{\mathbb{R}} I_{[t,t+\epsilon]}(u) M_u^2 D_u^{(H)} \psi(t_j)\, du\, dt.$$

The first limit is equal to

$$\lim_{\epsilon \to 0} \int_0^T \left[\sum_j \psi(t_j) I_{(t_j, t_{j+1}]}(t) \right] \diamond \frac{1}{\epsilon} \int_t^{t+\epsilon} dB^{(H)}(u) \, dt$$

$$= \lim_{\epsilon \to 0} \int_0^T \left[\sum_j \psi(t_j) I_{(t_j, t_{j+1}]}(t) \right] \diamond \frac{1}{\epsilon} \int_t^{t+\epsilon} W^{(H)}(u) \, du \, dt$$

$$= \lim_{\epsilon \to 0} \int_0^T \frac{1}{\epsilon} \left[\int_{u-\epsilon}^u \sum_j \psi(t_j) I_{(t_j, t_{j+1}]}(t) \right] \diamond W^{(H)}(u) \, du$$

$$= \int_0^T \psi^{(\Delta)}(u) \diamond W^{(H)}(u) \, du,$$

which converges in $(\mathcal{S})^*$ to $\int_0^T \psi(u) \diamond W^{(H)}(u) \, du = \int_0^T \psi(u) \, dB^{(H)}(u)$. For the second limit we get

$$\lim_{\epsilon \to 0} \frac{1}{\epsilon} \sum_j \int_0^T I_{(t_j, t_{j+1}]}(t) \int_t^{t+\epsilon} M_u^2 D_u^{(H)} \psi(t_j) \, du \, dt$$

$$= \lim_{\epsilon \to 0} \int_0^T \frac{1}{\epsilon} \int_t^{t+\epsilon} M_u^2 D_u^{(H)} \psi^{(\Delta)}(t) \, du \, dt$$

$$= \lim_{\epsilon \to 0} \int_0^T \frac{1}{\epsilon} \int_t^{t+\epsilon} M_u D_u \psi^{(\Delta)}(t) \, du \, dt.$$

By Lemmas 5.6.5 and 5.6.6 the last limit converges to

$$\int_0^T M_{u+} D_{u+} \psi(u) \, du$$

in $L^2(\mathbb{P})$. □

An analogous relation to the one of Theorem 5.6.7 between Stratonovich integrals and WIS integrals for *fBm* is proved under different conditions in [164].

An Itô formula for forward integrals with respect to classical Brownian motion was obtained by [201] and then extended to the *fBm* case in [101]. Here we prove the following Itô formula for forward integrals with respect to *fBm* as a consequence of Lemma 5.6.8.

Lemma 5.6.8. *Let* $G \in (\mathcal{S})^*$ *and suppose that* ψ *is forward integrable. Then*

$$G(\omega) \int_0^T \psi(t) \, d^- B^{(H)}(t) = \int_0^T G(\omega) \psi(t) \, d^- B^{(H)}(t). \quad (5.18)$$

Proof. This is immediate by Definition 5.6.1. □

Definition 5.6.9. *Let ψ be a forward integrable process and let $\alpha(s)$ be a measurable process such that $\int_0^t |\alpha(s)|\,ds < \infty$ almost surely for all $t \geq 0$. Then the process*

$$X(t) := x + \int_0^t \alpha(s)\,ds + \int_0^t \psi(s)\,d^- B^{(H)}(s), \quad t \geq 0 \qquad (5.19)$$

is called a fractional forward process. *As a shorthand notation for (5.19), we write*

$$d^- X(t) := \alpha(t)\,dt + \psi(t)\,d^- B^{(H)}(t), \quad X(0) = x.$$

Theorem 5.6.10. *Let*

$$d^- X(t) = \alpha(t)\,dt + \psi(t)\,d^- B^{(H)}(t), \quad X(0) = x$$

be a fractional forward process. Suppose $f \in C^2(\mathbb{R}^2)$ and put $Y(t) = f(t, X(t))$. Then if $1/2 < H < 1$, we have

$$d^- Y(t) = \frac{\partial f}{\partial t}(t, X(t))\,dt + \frac{\partial f}{\partial x}(t, X(t))\,d^- X(t).$$

Proof. Let $0 = t_0 < t_1 < \ldots < t_N = t$ be a partition of $[0, t]$. By using Taylor expansion and by equation (5.18), we get

$$Y(t) - Y(0)$$

$$= \sum_j Y(t_{j+1}) - Y(t_j)$$

$$= \sum_j f(t_{j+1}, X(t_{j+1})) - f(t_j, X(t_j))$$

$$= \sum_j \frac{\partial f}{\partial t}(t_j, X(t_j))\Delta t_j + \sum_j \frac{\partial f}{\partial x}(t_j, X(t_j))\Delta X(t_j)$$

$$+ \frac{1}{2}\sum_j \frac{\partial^2 f}{\partial x^2}(t_j, X(t_j))(\Delta X(t_j))^2 + \sum_j o((\Delta t_j)^2) + o((\Delta X(t_j))^2)$$

$$= \sum_j \frac{\partial f}{\partial t}(t_j, X(t_j))\Delta t_j + \sum_j \int_{t_j}^{t_{j+1}} \frac{\partial f}{\partial x}(t_j, X(t_j))\,d^- X_t$$

$$+ \frac{1}{2}\sum_j \frac{\partial^2 f}{\partial x^2}(t_j, X(t_j))(\Delta X(t_j))^2 + \sum_j o((\Delta t_j)^2) + o((\Delta X(t_j))^2),$$

where $\Delta X(t_j) = X(t_{j+1}) - X(t_j)$. Since $1/2 < H < 1$, the quadratic variation of the *fBm* is zero, and we are left with the first terms of the sum above, which converges to

$$\int_0^t \frac{\partial f}{\partial s}(s, X(s)) \, ds + \int_0^t \frac{\partial f}{\partial x}(s, X(s)) \, d^- X(s).$$

□

6
A useful summary

In Chapters 2 to 5 we have presented several ways of introducing a stochastic calculus with respect to the *fBm*. We have already underlined the relations among these different approaches, but in our opinion it is convenient to provide here a comprehensive summary, including a further investigation of their analogies and differences.

Moreover, in this chapter we present a general overview of the Itô formulas for the different definitions of stochastic integral for *fBm* together with an investigation of their relations.

6.1 Integrals with respect to *fBm*

We recall shortly the main approaches to define a stochastic integral with respect to *fBm*, and we refer the reader to the related chapters for further details.

6.1.1 Wiener integrals

In order to define the Wiener integrals with respect to $B^{(H)}$, we fix an interval $[0, T]$ and introduce the so-called *reproducing kernel Hilbert space*, denoted by $\mathcal{H}$. Recall that

$$R_H(t,s) := \frac{1}{2}(s^{2H} + t^{2H} - |t-s|^{2H}), \quad s,t > 0,$$

and let $K_H(t,s)$, $H \in (0,1)$, be a deterministic kernel such that

$$R_H(t,s) = \int_0^{t \wedge s} K_H(t,u) K_H(s,u) \, du.$$

For explicit expressions for $K_H(t,s)$, we refer to Chapter 2.

Definition 6.1.1. *The reproducing kernel Hilbert space (RKHS), denoted by $\mathcal{H}$, associated to $B^{(H)}$ for every $H \in (0,1)$, is defined as the closure of the vector space spanned by the set of functions $\{R_H(t, \cdot), t \in [0,T]\}$ with respect to the scalar product:*

$$\langle R_H(t, \cdot), R_H(s, \cdot) \rangle = R_H(t, s) \quad \forall t, s \in [0, T].$$

Definition 6.1.2. *For any $H \in (0,1)$, the (abstract) Wiener integral with respect to the fBm is defined as the linear extension from $\mathcal{H}$ in $L^2(\mathbb{P}^H)$ of the isometric map*

$$\mathcal{H} \longrightarrow L^2(\mathbb{P}^H),$$
$$R_H(t, \cdot) \longmapsto B^{(H)}(t).$$

We replace $\mathcal{H}$ by an isometrically isomorphic Hilbert space.

Definition 6.1.3. *By a representation of $\mathcal{H}$ we mean a pair $(\mathfrak{F}, i)$ composed of a functional space $\mathfrak{F}$ and a bijective isometry i between $\mathfrak{F}$ and $\mathcal{H}$.*

First we study the case $H > 1/2$ and consider the following representation for $\mathcal{H}$.

Theorem 6.1.4. *For any $H > 1/2$, consider $L^2([0,T])$ equipped with the twisted scalar product:*

$$\langle f, g \rangle_H = H(2H-1) \int_0^T \int_0^T f(s)g(t)|s-t|^{2H-2}\, ds\, dt. \tag{6.1}$$

Define the linear map i_2 on the space $\mathcal{E}$ of step functions on $[0,T]$ by

$$i_2 : (L^2([0,T]), <,>_H) \longrightarrow \mathcal{H},$$
$$I_{[0,t]} \longmapsto R_H(t, \cdot).$$

Then the extension of this map to the closure of $(L^2([0,T]), <,>_H)$ with respect to the scalar product defined in (6.1) is a representation of $\mathcal{H}$.

From now on we identify the RKHS $\mathcal{H}$ with $\overline{(L^2([0,T]), <,>_H)}$ through the representation map i_2. Note that the map i_2 induces an isometry, that associates $I_{[0,t]}$ to $B^{(H)}(t)$, between $\mathcal{H}$ and the chaos of first order associated with $B^{(H)}$, i.e., the closed subspace of $L^2(\Omega, \mathbb{P})$ generated by $B^{(H)}$. We denote the image of this isometry by $B^{(H)}(\phi)$ for $\phi \in \mathcal{H}$ and call it the *Wiener integral (of second type)* of $\phi \in \mathcal{H}$. In the sequel we use only the name Wiener integral dropping the additional specification "of second type".

In order to characterize this kind of integral, we introduce the linear operator K_H^* defined on $\phi \in \mathcal{E}$ as follows:

$$(K_H^* \phi)(s) := \int_s^T \phi(t) \frac{\partial K_H}{\partial t}(t, s)\, dt.$$

6.1 Integrals with respect to fBm

Then $(K_H^* I_{[0,t]})(s) = K_H(t,s) I_{[0,t]}(s)$ and the process $B(t)$ associated by the representation i_2 to $(K_H^*)^{-1}(I_{[0,t]})$, i.e.,

$$B(t) := B^{(H)}((K_H^*)^{-1} I_{[0,t]}),$$

is a Brownian motion (for the proof of this, see Chapter 2). Analogously, the stochastic process associated to

$$K_H^* I_{[0,t]} = K_H(t,s) I_{[0,t]}(s)$$

by the isometry induced by $B(t)$ on $L^2([0,T])$ is a fBm $B^{(H)}(t)$ with integral representation

$$B^{(H)}(t) = \int_0^T K_H^* I_{[0,t]} \, dB(s) = \int_0^t K_H(t,s) \, dB(s).$$

We obtain an expression of the Wiener integral with respect to $B^{(H)}$ *in terms of an integral with respect to the Brownian motion* B.

Proposition 6.1.5. *Let $H > 1/2$. If $\phi \in \mathcal{H}$, then*

$$B^{(H)}(\phi) = \int_0^T (K_H^* \phi)(s) \, dB(s).$$

As we have seen, Wiener integrals are introduced for deterministic integrands. In order to extend the definition of the Wiener integral to the general case of *stochastic integrands*, we follow the approach of [72] and use Theorem 2.1.7.

Definition 6.1.6. *Consider $H > 1/2$. Let u be a stochastic process $u.(\omega)$: $[0,T] \longrightarrow \mathcal{H}$ such that $K_H^* u$ is Skorohod integrable (Definition A.2.1) with respect to the standard Brownian motion $B(t)$. Then we define the stochastic integral of u with respect to the fBm $B^{(H)}$ as*

$$\int_0^T u(s) \, dB^{(H)}(s) := \int_0^T (K_H^* u)(s) \, \delta B(s),$$

where the integral on the right-hand side is a Skorohod integral with respect to $B(t)$ (Definition A.2.1).

We now focus on the case $H < 1/2$. As before we consider the following representation for the RKHS $\mathcal{H}$. Consider the space $\mathcal{E}$ of step functions on $[0,T]$ endowed with the inner product

$$\langle I_{[0,t]}, I_{[0,s]} \rangle_H := R_H(t,s), \quad 0 \le t, s \le T,$$

and the linear map i_2 on $\mathcal{E}$ given by

$$i_2 : (\mathcal{E}, <, >_H) \longrightarrow \mathcal{H},$$

$$I_{[0,t]} \longmapsto R_H(t,\cdot).$$

Then the extension of this map to the closure of $(\mathcal{E}, <,>_H)$ with respect to the scalar product defined in (2.7) is a representation of $\mathcal{H}$. From now on we identify $\mathcal{H} = \overline{(\mathcal{E}, <,>_H)}$ and we define the Wiener integral for $H < 1/2$ as the extension to $\psi \in \mathcal{H}$ of the isometry

$$B^{(H)} : \overline{(\mathcal{E}, <,>_H)} \longrightarrow L^2(\mathbb{P}^H),$$
$$I_{[0,t]}(\cdot) \longmapsto B^{(H)}(t),$$

induced by the representation i_2. Consider the linear operator K_H^* from the space $\mathcal{E}$ of step functions on $[0,T]$ to $L^2([0,T])$ defined by

$$(K_H^*\phi)(s) := K_H(T,s)\phi(s) + \int_s^T [\phi(t) - \phi(s)]\frac{\partial K_H}{\partial t}(t,s)\, dt. \tag{6.2}$$

Then (6.2) evaluated for $\phi = I_{[0,t]}$ gives

$$(K_H^* I_{[0,t]})(s) = K_H(t,s) I_{[0,t]}(s),$$

and for $H < 1/2$ we have

$$\mathcal{H} = (K_H^*)^{-1}(L^2([0,T])) = I_{T-}^{1/2-H}(L^2([0,T])).$$

This representation of $\mathcal{H}$ guarantees in addition that the inner product space $\mathcal{H}$ is complete when endowed with the inner product

$$\langle f, g \rangle = \int_0^T K_H^* f(s) K_H^* g(s)\, ds,$$

as shown in Lemma 5.6 of [188].

Hence we can conclude with the following:

Proposition 6.1.7. *For $H < 1/2$ the Wiener-type integral $B^{(H)}(\phi)$ with respect to fBm can be defined for $\phi \in \mathcal{H} = I_{T-}^{1/2-H}(L^2([0,T]))$ and the following holds:*

$$B^{(H)}(\phi) = \int_0^T (K_H^* \phi)(t)\, dB(t).$$

6.1.2 Divergence-type integrals

The *divergence operator* δ_H is the adjoint of the derivative operator. We say that a random variable $u \in L^2(\Omega; \mathcal{H})$ belongs to the domain $\operatorname{dom} \delta_H$ of the divergence operator if

$$\left| E\left[\langle D^{(H)} F, u \rangle_H \right] \right| \leq c_u \|F\|_{L^2(\Omega)}$$

for any $F \in S_H$, where S_H is the set of smooth cylindrical random variables of the form $F = f(B^{(H)}(\psi_1), \ldots, B^{(H)}(\psi_n))$, where $n \geq 1$, $f \in C_b^\infty(\mathbb{R}^n)$, and $\psi_i \in \mathcal{H}$.

Definition 6.1.8. *Let $H > 1/2$ and $u \in \mathrm{dom}\,\delta_H$. Then $\delta_H(u)$ is the element in $L^2(\Omega; \mathcal{H})$ defined by the duality relationship*

$$E[F\delta_H(u)] = E\left[\langle D^{(H)}F, u\rangle_H\right]$$

for any $F \in \mathbb{D}_H^{1,2}$.

For $H < 1/2$, the definition of a divergence-type integral is more delicate. Namely, by Proposition 2.2.6 it follows that processes of the form

$$B^{(H)}(t)I_{(a,b]}(t), \quad t \in \mathbb{R},$$

cannot be in $\mathrm{dom}\,\delta_H$ if $H \leq 1/4$. We consider then an extended domain for the divergence operator. Let $S_{\mathcal{K}}$ the space of smooth cylindrical variables of the form

$$F = f(B^{(H)}(\psi_1), \ldots, B^{(H)}(\psi_n)),$$

where $n \geq 1$, $f \in C_b^\infty(\mathbb{R}^n)$, i.e., f is bounded with smooth bounded partial derivatives, and $\psi_i \in \mathcal{K}$, where

$$\mathcal{K} := (K_H^*)^{-1}(K_H^{*,a})^{-1}(L^2([0,T]))$$

and $K_H^{*,a}$ is the adjoint of the operator K_H^* in $L^2([0,T])$.

Definition 6.1.9. *Let $u(t), t \in [0,T]$ be a measurable process such that $E\left[\int_0^T u^2(t)\,dt\right] < \infty$. We say that $u \in \mathrm{dom}^*\delta_H$ if there exists a random variable $\delta_H(u) \in L^2(\mathbb{P}^H)$ such that for all $F \in S_{\mathcal{K}}$, we have*

$$\int_0^T E\left[u(t)K_H^{*,a}K_H^* D_t^{(H)} F\right] dt = E[\delta_H(u)F].$$

Note that if $u \in \mathrm{dom}^*\delta_H$, then $\delta_H(u)$ is unique and the mapping

$$\delta_H : \mathrm{dom}^*\delta_H \longrightarrow \cup_{p>1} L^p(\Omega)$$

is linear.

6.1.3 fWIS integrals

Consider $H > 1/2$ and

$$\phi(s,t) = \phi_H(s,t) = H(2H-1)|s-t|^{2H-2}, \quad s,t \in \mathbb{R}.$$

Let $\mathcal{L}_\phi(0,T)$ be the family of stochastic processes F on $[0,T]$ with the following properties: $F \in \mathcal{L}_\phi(0,T)$ if and only if $E\left[\|F\|_H^2\right] < \infty$, F is ϕ-differentiable, the trace of $D_s^\phi F_t$, $0 \leq s,t \leq T$ exists, and $E\left[\int_0^T \int_0^T |D_s^\phi F_t|^2\,ds\,dt\right] < \infty$ and for each sequence of partitions $(\pi_n, n \in \mathbb{N})$ such that $|\pi_n| \to 0$ as $n \to \infty$,

152   6 A useful summary

$$\sum_{i,j=0}^{n-1} E\left[\int_{t_i^{(n)}}^{t_{i+1}^{(n)}} \int_{t_j^{(n)}}^{t_{j+1}^{(n)}} \left| D_s^\phi F_{t_i^{(n)}}^\pi D_t^\phi F_{t_j^{(n)}}^\pi - D_s^\phi F_t D_t^\phi F_s \right| ds\, dt \right]$$

and

$$E\left[\|F^\pi - F\|_H^2\right]$$

tend to 0 as $n \to \infty$, where $\pi_n : 0 = t_0^{(n)} < t_1^{(n)} < \ldots < t_{n-1}^{(n)} < t_n^{(n)} = T$.

Theorem 6.1.10. *Let $(F_t, t \in [0,T])$ be a stochastic process such that $F \in \mathcal{L}_\phi(0,T)$. The limit*

$$\int_0^T F_s\, dB_s^{(H)} := \int_0^T F_s\, dB_s^{(H)} = \lim_{|\pi| \to 0} \sum_{i=0}^{n-1} F_{t_i}^\pi \diamond (B_{t_{i+1}}^{(H)} - B_{t_i}^{(H)}) \quad (6.3)$$

exists in $L^2(\mathbb{P}^H)$, where $\mathbb{P}^H$ is the measure induced by $B^{(H)}$. Moreover, this integral satisfies

$$E\left[\int_0^T F_s\, dB_s^{(H)}\right] = 0$$

and

$$\left\|\int_0^T F_s\, dB_s^{(H)}\right\|_{\mathcal{L}_\phi(0,T)} := E\left[\left|\int_0^T F_s\, dB_s^{(H)}\right|^2\right]$$

$$= E\left[\int_0^T \int_0^T D_s^\phi F_t D_t^\phi F_s\, ds\, dt + \|1_{[0,T]} F\|_H^2\right].$$

This definition can be extended in the following way.

Definition 6.1.11. *Suppose $Y : \mathbb{R} \to (\mathcal{S})_H^*$ is a given function such that $Y(t) \diamond W^{(H)}(t)$ is dt-integrable in $(\mathcal{S})_H^*$ (Definition 3.1.11). Then we define its fWIS integral, $\int_\mathbb{R} Y(t) dB_t^{(H)}$, by*

$$\int_\mathbb{R} Y(t)\, dB_t^{(H)} := \int_\mathbb{R} Y(t) \diamond W^{(H)}(t)\, dt, \quad (6.4)$$

where the Wick product $\diamond$ is introduced in (3.22) and $W^{(H)}(t)$ is the fractional white noise defined in (3.19).

Note that in Chapter 3 we have followed the opposite direction, i.e., we have introduced the fWIS integral in the more general case (6.4) and then its L^2 version (6.3).

6.1.4 WIS integrals

Let M be the operator introduced in Definition 4.1.2. We define the WIS integral in the following way. We recall that the spaces $(\mathcal{S})$ and $(\mathcal{S})^*$ are introduced in Definition A.1.4.

Definition 6.1.12. *Suppose that $Z : \mathbb{R} \longrightarrow (\mathcal{S})^*$ is a given function with the property that*

$$\langle Z(t), \psi \rangle \in L^1(\mathbb{R}, dt) \quad \forall \psi \in (\mathcal{S}). \tag{6.5}$$

Then $\int_{\mathbb{R}} Z(t)dt$ is defined to be the unique element of $(\mathcal{S})^*$ such that

$$\langle \int_{\mathbb{R}} Z(t)\,dt, \psi \rangle = \int_{\mathbb{R}} \langle Z(t), \psi \rangle dt \quad \forall \psi \in (\mathcal{S}). \tag{6.6}$$

Just as in [109, Proposition 8.1], one can show that (6.6) defines $\int_{\mathbb{R}} Z(t)\,dt$ as an element of $(\mathcal{S})^*$. If (6.5) holds, we say that $Z(t)$ is *dt-integrable* in $(\mathcal{S})^*$.

Definition 6.1.13. *Let $H \in (0,1)$. Let $Y : \mathbb{R} \to (\mathcal{S})^*$ be such that $Y(t) \diamond W^{(H)}(t)$ is dt-integrable in $(\mathcal{S})^*$ (Definition 6.1.12). Then we say that Y is WIS integrable and we define the WIS integral of $Y(t) = Y(t,\omega)$ with respect to $B^{(H)}(t)$ by*

$$\int_{\mathbb{R}} Y(t,\omega)\,dB^{(H)}(t) := \int_{\mathbb{R}} Y(t) \diamond W^{(H)}(t)\,dt,$$

where the Wick product $\diamond$ is introduced in Definition A.1.7 and $W^{(H)}(t)$ is the fractional white noise defined in (4.16).

In particular, we define the stochastic integral over a finite interval $[0,T]$ as

$$\int_0^T Y(t,\omega)\,dB^{(H)}(t) := \int_{\mathbb{R}} Y(t)I_{[0,T]} \diamond W^{(H)}(t)\,dt.$$

Consider now the space $\mathcal{M}$ of the processes $Y(t) = \sum_{\alpha \in \mathcal{J}} c_\alpha(t)\mathcal{H}_\alpha(\omega) \in (\mathcal{S})^*$ such that $c_\alpha(\cdot)$ belongs to $L^2_H(\mathbb{R}) = \{f : \mathbb{R} \to \mathbb{R}; Mf(x) \in L^2(\mathbb{R})\}$ and $\sum_{\alpha \in \mathcal{J}} Mc_\alpha(t)\mathcal{H}_\alpha(\omega)$ converges in $(\mathcal{S})^*$ for all t. Then the following fundamental relation holds.

Proposition 6.1.14. *Let $H \in (0,1)$. If $Y : \mathbb{R} \to (\mathcal{S})^*$ is WIS integrable and $Y \in \mathcal{M}$, then*

$$\int_{\mathbb{R}} Y(t)\,dB^{(H)}(t) = \int_{\mathbb{R}} MY(t)\,\delta B(t),$$

where the integral on the right-hand side is a Skorohod integral with respect to $B(t)$ (Definition A.2.1).

We remark that

$$\int_0^T Y(t)\,dB^{(H)}(t) = \int_{\mathbb{R}} Y(t) I_{[0,T]}(t)\,dB^{(H)}(t)$$

$$= \int_{\mathbb{R}} M(Y I_{[0,T]})\,\delta B(t) \neq \int_0^T MY(t)\,\delta B(t).$$

6.1.5 Pathwise integrals

Consider $H \in (0,1)$.

Definition 6.1.15. *Let $(u_t)_{t \in [0,T]}$ be a process with integrable trajectories. The* symmetric integral *of u with respect to $B^{(H)}$ is defined as*

$$\int_0^T u(s)\,d^\circ B^{(H)}(s) = \lim_{\epsilon \to 0} \frac{1}{2\epsilon} \int_0^T u(s)[B^{(H)}(s+\epsilon) - B^{(H)}(s-\epsilon)]\,ds,$$

if the limit exists in probability.
The forward integral *of u with respect to $B^{(H)}$ is defined as*

$$\int_0^T u(s)\,d^- B^{(H)}(s) = \lim_{\epsilon \to 0} \frac{1}{\epsilon} \int_0^T u(s) \frac{B^{(H)}(s+\epsilon) - B^{(H)}(s)}{\epsilon}\,ds,$$

if the limit exists in probability. On the other hand, the backward integral *is defined as*

$$\int_0^T u(s)\,d^+ B^{(H)}(s) = \lim_{\epsilon \to 0} \frac{1}{\epsilon} \int_0^T u(s) \frac{B^{(H)}(s-\epsilon) - B^{(H)}(s)}{\epsilon}\,ds,$$

if the limit exists in probability.

The definition of a pathwise integral for *fBm* with Hurst index $H < 1/2$ is more delicate. For example, the forward integral $\int_0^T B^{(H)}(s)\,d^- B^{(H)}(s)$ does not exist in the L^2 sense, as it is shown in Chapter 5. However, sufficient conditions can be established to guarantee the existence of the symmetric integral for $H < 1/2$.

Let $K_H(t,s)$ be the kernel defined in (2.3). Consider, on the space $\mathcal{E}$ of step functions on $[0,T]$, the seminorm

$$\|\phi\|^2_{K_H} = \int_0^T \phi^2(s) K_H(T,s)^2\,ds$$

$$+ \int_0^T \left(\int_0^T |\phi(t) - \phi(s)|(t-s)^{H-\frac{3}{2}}\,dt \right)^2 ds,$$

and denote by $\mathcal{H}_{K_H}$ the completion of $\mathcal{E}$ with respect to $\|\cdot\|^2_{K_H}$.

Proposition 6.1.16. *Let $H < 1/2$ and $(u_t)_{t \in [0,T]}$ be a stochastic process in the space $\mathbb{D}^{1,2}(\mathcal{H}_{K_H})$. Suppose that the trace defined as*

$$\operatorname{Tr} D^{(H)} u := \lim_{\epsilon \to 0} \frac{1}{2\epsilon} \int_0^T \langle D^{(H)} u_s, I_{[s-\epsilon, s+\epsilon]} \rangle_H \, ds$$

exists as a limit in probability and that

$$E\left[\int_0^T u^2(s)(s^{2H-1} + (T-s)^{2H-1}) \, ds \right] < \infty,$$

$$E\left[\int_0^T \int_0^T (D_r^{(H)} u_s)^2 (s^{2H-1} + (T-s)^{2H-1}) \, ds \, dr \right] < \infty.$$

Then the symmetric integral $\int_0^T u(s) \, d^\circ B^{(H)}(s)$ of u with respect to the fBm defined as the limit in probability

$$\lim_{\epsilon \to 0} \int_0^T u_s \frac{B^{(H)}(s+\epsilon) - B^{(H)}(s-\epsilon)}{2\epsilon} \, ds$$

exists.

Moreover, the symmetric integral $\int_0^T f(B^{(H)}(t)) \, d^\circ B^{(H)}(t)$ of a general smooth function f exists in $L^2(\mathbb{P}^H)$ if and only if $H > 1/6$ (see [54]).

We consider the following extended definition of the forward integral, seen as an element of $(\mathcal{S})^*$.

Definition 6.1.17. *The (generalized) forward integral of a measurable process Y with integrable trajectories is defined by*

$$\int_0^T Y(t) \, d^- B^{(H)}(t) = \lim_{\epsilon \to 0} \int_0^T Y(t) \frac{B^{(H)}(t+\epsilon) - B^{(H)}(t)}{\epsilon} \, dt,$$

provided that the limit exists in $(\mathcal{S})^*$.

6.2 Relations among the different definitions of stochastic integral

We investigate here the relations among the different definitions of stochastic integrals for fBms summarized in Section 6.1. For the proofs of the following results, we refer to the related chapter.

6.2.1 Relation between Wiener integrals and the divergence

For $H > 1/2$ the divergence operator coincides with the Wiener integral introduced in Definition 6.1.2. In fact, it is sufficient to note that

$$\delta_H(\sum_{i=1}^n a_i I_{[t_i, t_{i+1}]}) = \sum_{i=1}^n a_i(B^{(H)}(t_{i+1}) - B^{(H)}(t_i))$$

for $0 \leq t_1 \leq t_2 \leq \ldots \leq t_{n+1} \leq T$. By a limiting procedure it follows that for any $\psi \in \mathcal{H}$, we have

$$\delta_H(\psi) = B^{(H)}(\psi).$$

This relation still holds for the extended Wiener integral for stochastic integrands. In fact, if $F \in \mathbb{D}_H^{1,2}$, we have

$$E\left[\langle u, D^{(H)} F \rangle_H \right] = E\left[\langle K_H^* u, DF \rangle_{L^2([0,T])} \right]$$

for any $u \in \mathcal{H}$, and the equality $K_H^* D^{(H)} F = DF$ holds. This implies that

$$\operatorname{dom} \delta_H = (K_H^*)^{-1}(\operatorname{dom} \delta),$$

where $\delta = \delta_{1/2}$ denotes the divergence operator with respect to the standard Brownian motion B. Hence, for any $\mathcal{H}$-valued random variable $u \in \operatorname{dom} \delta_H$, it holds that

$$\delta_H(u) = \delta(K_H^* u) = \int_0^T K_H^* u(s) \, \delta B(s),$$

where the integral on the right-hand side is a Skorohod integral with respect to the standard Brownian motion (Definition A.2.1). Hence, we have proved the following:

Proposition 6.2.1. *Let $u \in \operatorname{dom} \delta_H$. Then $\delta_H(u)$ coincides with the extended Wiener integral of u (Definition 6.1.6), i.e.,*

$$\delta_H(u) = B^{(H)}(u).$$

6.2.2 Relation between the divergence and the fWIS integral

Here we investigate the relation between the stochastic integral introduced in Chapter 3 and the divergence-type integral for $H > 1/2$. By Remark 3.1.1 we have $L_\phi^2(\mathbb{R}) \supseteq L_\phi^2([0,T]) = \mathcal{H} = \overline{(L^2([0,T]), <,>_H)}$ if we identify $\psi \in L_\phi^2([0,T])$ with $\psi I_{[0,T]}$. Hence, for every deterministic $\psi \in \mathcal{H}$, $B^{(H)}(\psi)$ coincides with the fWIS integral of ψ. Moreover, for any $\psi \in \mathcal{H}$ and any random variable $F \in L^2(\mathbb{P}^H)$, the Wick product $F \diamond B^{(H)}(\psi)$ exists in $L^2(\mathbb{P}^H)$ if and only if ψF belongs to $\operatorname{dom} \delta_H$, and we have that

$$F \diamond B^{(H)}(\psi) = \delta_H(F\psi).$$

6.2 Relations among the different definitions of stochastic integral

Proposition 6.2.2. *Consider $H > 1/2$. Let u be a process that is λ-Hölder continuous in the norm $\mathbb{D}^{1,2}(\mathcal{H})$ with $\lambda > H - 1/2$. Then*

$$\lim_{|\pi| \to 0^+} \sum_{i=1}^{n} u_{t_i} \diamond (B^{(H)}(t_{i+1}) - B^{(H)}(t_i)) = \delta_H(u),$$

where $\pi = \{0 = s_0 < s_1 < \cdots < s_{n+1} = T\}$ is a partition of $[0,T]$ with mesh size $|\pi| = \max\limits_{i=1,\ldots,n} |t_{i+1} - t_i|$ and the convergence holds in $L^2(\mathbb{P}^H)$.

Hence we can conclude that under the hypotheses of Proposition 3.12.1 the fWIS integral coincides with the divergence operator when $H > 1/2$.

6.2.3 Relation between the fWIS and the WIS integrals

Here we study the relation between the fWIS integral and the WIS integral for $H > 1/2$ as defined in Chapters 3 and 4, respectively. The WIS integral exists in $(\mathcal{S})^*$, and the fWIS integral admits a generalized definition as an element of $(\mathcal{S})^*_H$. Here we compare the generalized Definition 6.1.10 for the fWIS integral with Definition 6.1.13 for the case $H > 1/2$. We obtain the relations (6.10) and (6.11) that also justify why we use the same symbol to denote both the fWIS and the WIS integrals. To distinguish between these two types of integrals, we adopt in the sequel the symbols $\mathfrak{I}_{fWIS}$ and $\mathfrak{I}_{WIS}$ to denote, respectively, the fWIS and the WIS integral.

In Chapter 4 we have seen that the operator M induces an isometry between $L^2_H(\mathbb{R})$ and $L^2(\mathbb{R})$. Moreover, we have

$$\int_{\mathbb{R}} [Mf(t)]^2 \, dt = H(2H-1) \int \int_{\mathbb{R}^2} f(s)f(t) |s-t|^{2H-2} \, ds \, dt,$$

by (4.10). Hence, we can conclude that for $H > 1/2$, $L^2_H(\mathbb{R})$ coincides with the subspace generated by the deterministic functions in $L^2_\phi(\mathbb{R})$. Let $\{e_n, \, n = 1, 2, \ldots\}$ be an orthonormal basis of $L^2_\phi(\mathbb{R})$. Then

$$\{\xi_n, \, n = 1, 2, \ldots\} = \{Me_n, \, n = 1, 2, \ldots\} \tag{6.7}$$

is an orthonormal basis of $L^2(\mathbb{R})$. Consider an element $F(\omega) = \sum_{\alpha \in \mathcal{J}} c_\alpha \widetilde{\mathcal{H}}_\alpha(\omega) \in (\mathcal{S})^*_H$, where if $\alpha = (\alpha_1, \ldots, \alpha_m) \in \mathcal{J}$, we put

$$\widetilde{\mathcal{H}}_\alpha(\omega) := h_{\alpha_1}(\langle \omega, e_1 \rangle) \cdots h_{\alpha_m}(\langle \omega, e_m \rangle).$$

The operator M induces the function

$$\widetilde{M} : (\mathcal{S})^*_H \longrightarrow (\mathcal{S})^* \tag{6.8}$$

such that

$$\widetilde{M}\left(\sum_{\alpha\in\mathcal{J}}c_\alpha\widetilde{\mathcal{H}}_\alpha(\omega)\right) = \widetilde{M}\left(\sum_{\alpha\in\mathcal{J}}c_\alpha h_{\alpha_1}(\langle\omega,e_1\rangle)\cdots h_{\alpha_m}(\langle\omega,e_m\rangle)\right)$$

$$:= \sum_{\alpha\in\mathcal{J}}c_\alpha h_{\alpha_1}(\langle\omega,Me_1\rangle)\cdots h_{\alpha_m}(\langle\omega,Me_m\rangle)$$

$$= \sum_{\alpha\in\mathcal{J}}c_\alpha h_{\alpha_1}(\langle\omega,\xi_1\rangle)\cdots h_{\alpha_m}(\langle\omega,\xi_m\rangle)$$

$$= \sum_{\alpha\in\mathcal{J}}c_\alpha \mathcal{H}_\alpha(\omega). \tag{6.9}$$

Then the following relations between the WIS integral and the fWIS integral hold. If $Y : \mathbb{R} \to (\mathcal{S})^*$ is a given function such that $Y(t) \diamond W^{(H)}(t)$ is integrable in $(\mathcal{S})^*$, then $\widetilde{M}^{-1}Y : \mathbb{R} \to (\mathcal{S})_H^*$ is such that $\widetilde{M}^{-1}Y(t) \diamond W^{(H)}(t)$ is integrable in $(\mathcal{S})_H^*$ and

$$\mathcal{I}_{WIS}(Y) = \int_\mathbb{R} Y(t) \diamond W^{(H)}(t)\,dt$$
$$= \widetilde{M}\left(\int_\mathbb{R}\widetilde{M}^{-1}Y(t) \diamond W^{(H)}(t)\,dt\right) = \widetilde{M}\left(\mathcal{I}_{fWIS}(\widetilde{M}^{-1}Y)\right). \tag{6.10}$$

Analogously, suppose $Y : \mathbb{R} \to (\mathcal{S})_H^*$ is a given function such that $Y(t) \diamond W^{(H)}(t)$ is integrable in $(\mathcal{S})_H^*$. Then $\widetilde{M}Y : \mathbb{R} \to (\mathcal{S})^*$ is such that $\widetilde{M}Y(t) \diamond W^{(H)}(t)$ is integrable in $(\mathcal{S})^*$ and

$$\mathcal{I}_{fWIS}(Y) = \int_\mathbb{R} Y(t) \diamond W^{(H)}(t)\,dt$$
$$= \widetilde{M}^{-1}\left(\int_\mathbb{R}\widetilde{M}Y(t) \diamond W^{(H)}(t)\,dt\right) = \widetilde{M}^{-1}\left(\mathcal{I}_{WIS}(\widetilde{M}Y)\right). \tag{6.11}$$

For further details, we also refer to [116].

6.2.4 Relations with the pathwise integrals

We investigate here the relations among the stochastic integral of divergence, fWIS and WIS types and the symmetric (respectively, forward) integral.

In the divergence case, we need to distinguish between $H > 1/2$ and $H < 1/2$.

Proposition 6.2.3 (Relation between the symmetric integral and the divergence I). *Let $H > 1/2$. Suppose $(u_t)_{t\in[0,T]}$ is a stochastic process in $\mathbb{D}^{1,2}(|\mathcal{H}|)$ and that*

$$\int_0^T\int_0^T |D_s^{(H)}u(t)||t-s|^{2H-2}\,ds\,dt < \infty, \text{ a.s..}$$

Then the symmetric integral exists and the following relation holds:

6.2 Relations among the different definitions of stochastic integral

$$\int_0^T u(t)\, d^\circ B^{(H)}(t)$$
$$= \delta_H(u) + H(2H-1) \int_0^T \int_0^T D_s^{(H)} u(t) |t-s|^{2H-2}\, ds\, dt.$$

Note that under the assumptions of Proposition 6.2.3, the symmetric, backward and forward integrals coincide.

Proposition 6.2.4 (Relation between the symmetric integral and the divergence II). *Let $H < 1/2$ and assume that the hypotheses of Proposition 6.1.16 hold. Then the symmetric integral $\int_0^T u(s) d^\circ B^{(H)}(s)$ of u exists, and*

$$\int_0^T u(s)\, d^\circ B^{(H)}(s) = \delta_H(u) + \operatorname{Tr} D^{(H)} u.$$

In Proposition 5.5.1 we already investigated the relation between the fWIS integral and the symmetric integral.

Theorem 6.2.5 (Relation between the symmetric and fWIS integrals). *Let $H > 1/2$. If $F \in \mathcal{L}_\phi(0,T)$, then the symmetric integral $\int_0^T F_s\, d^\circ B^{(H)}(s)$ exists and the following equality is satisfied*

$$\int_0^T F_s\, d^\circ B^{(H)}(s) = \int_0^T F_s\, dB^{(H)}(s) + \int_0^T D_s^\phi F_s\, ds \quad a.s.,$$

where $\int_0^T F_s\, dB^{(H)}(s)$ is the fWIS integral.

We note that the result of Theorem 6.2.5 also follows by Propositions 6.2.2 and 6.2.3 and by (6.16), that we prove in the sequel.

Moreover, we recall the following relation between the generalized forward integrals and the WIS integral that holds for every $H \in (0,1)$. We recall that the space $\mathbb{L}_{1,2}^{(H)}$ consists of all càglàd processes $\psi(t) = \sum_{\alpha \in \mathcal{J}} c_\alpha \mathcal{H}_\alpha(\omega)$ such that

$$\|\psi\|_{\mathbb{L}_{1,2}^{(H)}}^2 := \sum_{\alpha \in \mathcal{J}} \sum_{i=1}^\infty \alpha_i \alpha! \|c_\alpha\|_{L^2([0,T])}^2 < \infty.$$

Lemma 6.2.6. *Suppose that $\psi \in \mathbb{L}_{1,2}^{(H)}$. Then*

$$M_{t+} D_{t+} \psi(t) := \lim_{\epsilon \to 0} \frac{1}{\epsilon} \int_t^{t+\epsilon} M_s D_s \psi(t)\, ds$$

exists in $L^2(\mathbb{P})$ for all t. Moreover,

$$\int_0^T M_{t+} D_{t+} \psi(t)\, dt = \lim_{\epsilon \to 0} \int_0^T \left[\frac{1}{\epsilon} \int_t^{t+\epsilon} M_s D_s \psi(t)\, ds \right] dt$$

in $L^2(\mathbb{P})$ and

$$E\left[\left(\int_0^T M_{s+} D_{s+} \psi(s)\, ds\right)^2\right] < \infty.$$

Theorem 6.2.7 (Relation between the (generalized) forward and WIS integral). *Let $H \in (0,1)$. Suppose $\psi \in \mathbb{L}_{1,2}^{(H)}$ and that one of the following conditions holds:*

1. *ψ is WIS integrable.*
2. *ψ is forward integrable in $(\mathcal{S})^*$.*

Then

$$\int_0^T \psi(t)\, d^-B^{(H)}(t) = \int_0^T \psi(t)\, dB^{(H)}(t) + \int_0^T M_{t+}^2 D_{t+}^{(H)} \psi(t)\, dt,$$

holds as an identity in $(\mathcal{S})^*$, where $\int_0^T \psi(t)\, dB^{(H)}(t)$ is the WIS integral.

Remark 6.2.8. We show now that if $(\psi_t)_{t \in [0,T]}$ satisfies the hypotheses of Proposition 6.2.3, then

$$\int_0^T M_{t+}^2 D_{t+}^{(H)} \psi(t)\, dt = H(2H-1) \int_0^T \int_0^T D_s^{(H)} \psi(t) |t-s|^{2H-2}\, ds\, dt \quad (6.12)$$

if we identify ψ_t with $\psi_t I_{[0,T]}$. We need only note that

$$\int_0^T \left[\frac{1}{\epsilon} \int_t^{t+\epsilon} M_s^2 D_s^{(H)} \psi(t)\, ds \right] dt$$
$$= \int_0^T \int_0^T \frac{1}{\epsilon} M^2[t, t+\epsilon](s) D_s^{(H)} \psi(t)\, ds\, dt, \quad (6.13)$$

where $M^2[t, t+\epsilon](s) = M_s^2(I_{[t,t+\epsilon]})$, and that by (4.3) and Lemma 3.1.2,

$$M^2[t, t+\epsilon](s) = k_H \frac{1}{\epsilon} \int_t^{t+\epsilon} \left(\int_{\mathbb{R}} |u-x|^{H-3/2} |u-s|^{H-3/2}\, du \right) dx$$
$$= H(2H-1) \frac{1}{\epsilon} \int_t^{t+\epsilon} |s-x|^{2H-2}\, dx,$$

where $k_H = (H(2H-1)\Gamma(3/2-H))/(2\Gamma(H-1/2)\Gamma(2-2H))$. By dominated convergence the limit of (6.13) exists almost surely as $\epsilon \to 0$ and (6.12) holds.

6.3 Itô formulas with respect to *fBm*

Several Itô formulas have been proposed in the literature according to the different definitions of stochastic integral for *fBm*. Here we present some of them and investigate their relations. We begin with the Itô formula for functionals of $B^{(H)}$.

6.3 Itô formulas with respect to fBm

Theorem 6.3.1. *Let $H \in (0,1)$. Assume that $f(s,x): \mathbb{R} \times \mathbb{R} \to \mathbb{R}$ belongs to $C^{1,2}(\mathbb{R} \times \mathbb{R})$, and assume that the random variables*

$$f(t, B^{(H)}(t)), \int_0^t \frac{\partial f}{\partial s}(s, B^{(H)}(s))\, ds, \text{ and } \int_0^t \frac{\partial^2 f}{\partial x^2}(s, B^{(H)}(s)) s^{2H-1}\, ds$$

are square integrable for every t. Then

$$f(t, B^{(H)}(t)) = f(0,0) + \int_0^t \frac{\partial f}{\partial s}(s, B^{(H)}(s))\, ds + \int_0^t \frac{\partial f}{\partial x}(s, B^{(H)}(s))\, dB^{(H)}(s)$$

$$+ H \int_0^t \frac{\partial^2 f}{\partial x^2}(s, B^{(H)}(s)) s^{2H-1}\, ds. \tag{6.14}$$

The Itô formula (6.14) is formulated in terms of WIS integrals and holds for every $H \in (0,1)$. The same formula is valid also for fWIS integrals and divergence-type ones for $H > 1/2$. To guarantee the existence of the divergence for $H < 1/2$, we need more restrictive hypotheses. We consider the following Theorem due to [54].

Theorem 6.3.2. *Let $H < 1/2$. Let F be a function of class $C^2(\mathbb{R})$ that satisfies the growth condition*

$$\max\left\{|F(x)|, |F'(x)|, |F''(x)|\right\} \leq c e^{\lambda x^2},$$

where $c, \lambda > 0$, and $\lambda < 1/(4T^{2H})$. Then for all $t \in [0,T]$, the process $F'(B^{(H)}(s)) I_{[0,t]}(s)$ belongs to $\mathrm{dom}^* \delta_H$, and we have

$$F(B_t^{(H)}) = F(0) + \delta_H(F'(B^{(H)}) I_{[0,t]}) + H \int_0^t F''(B^{(H)}(s)) s^{2H-1}\, ds$$

for all $t \in [0,T]$.

Proof. Here we sketch a proof by following [177]. For further details, we refer to Lemma 4.3 of [54]. We have that $F'(B^{(H)}(s)) I_{[0,t]}(s) \in L^2(\Omega \times [0,T])$ and

$$F(B_t^{(H)}) - F(0) - H \int_0^t F''(B^{(H)}(s)) s^{2H-1} ds \in L^2(\Omega).$$

Hence, it is sufficient to prove that for any $G \in \mathcal{S}_H$ we have

$$E\left[\langle F'(B^{(H)}(s)) I_{[0,t]}(s), D^{(H)} G \rangle_H \right]$$

$$= E\left[G(F(B_t^{(H)}) - F(0) - H \int_0^t F''(B^{(H)}(s)) s^{2H-1} ds)\right].$$

This equality can be proved by choosing smooth cylindrical random variable of the form $G = h_n(B^{(H)}(\psi))$, where h_n denotes the nth Hermite polynomial (see Appendix A), and applying the integration by parts formula. □

We now present Itô formulas for *functionals of integrals* for the different definitions of stochastic integral for *fBm* and clarify the relations among them. We start with the one for fWIS integral. In Theorem 3.7.2 we have proved the following result.

Theorem 6.3.3 (Itô formula for the fWIS integral). *Let $H > 1/2$. Let $\eta_t = \int_0^t F_u \, dB_u^{(H)}$, where $(F_u, 0 \leq u \leq T)$ is a stochastic process in $\mathcal{L}_\phi(0,T)$. Assume that there is an $\alpha > 1 - H$ such that*

$$E\left[|F_u - F_v|^2\right] \leq C|u-v|^{2\alpha},$$

where $|u-v| \leq \delta$ for some $\delta > 0$ and

$$\lim_{0 \leq u,v \leq t, |u-v| \to 0} E\left[|D_u^\phi(F_u - F_v)|^2\right] = 0.$$

Let $f : \mathbb{R}_+ \times \mathbb{R} \to \mathbb{R}$ be a function having the first continuous derivative in its first variable and the second continuous derivative in its second variable. Assume that these derivatives are bounded. Moreover, it is assumed that $E\left[\int_0^T |F_s D_s^\phi \eta_s| \, ds\right] < \infty$ and $(f'(s, \eta_s) F_s, s \in [0, T])$ is in $\mathcal{L}_\phi(0, T)$. Then for $0 \leq t \leq T$,

$$f(t, \eta_t) = f(0,0) + \int_0^t \frac{\partial f}{\partial s}(s, \eta_s) \, ds + \int_0^t \frac{\partial f}{\partial x}(s, \eta_s) F_s \, dB_s^{(H)}$$
$$+ \int_0^t \frac{\partial^2 f}{\partial x^2}(s, \eta_s) F_s D_s^\phi \eta_s \, ds \quad a.s. \tag{6.15}$$

By using the result of Proposition 6.2.2, Theorem 6.3.3 can be restated in terms of divergence-type integrals for $H > 1/2$. We only need to note that the stochastic gradient introduced in Definition 3.3.1 coincides with the derivative operator $D^{(H)}$ of (2.31) [with respect to the particular representation $\mathcal{H} = L_\phi^2([0,T])$] and that we have

$$D_t^\phi F = \int_0^T D_v^{(H)} F \phi(t,v) \, dv = H(2H-1) \int_0^T |t-v|^{2H-2} D_v^{(H)} F \, dv. \tag{6.16}$$

Here we recall an Itô formula for the divergence integral following the approach of [8, Theorem 8].

Theorem 6.3.4 (Itô formula for the divergence integral). *Let $H > 1/2$. Let ψ be a function of class $C^2(\mathbb{R})$. Assume that the process $(u_t)_{t \in [0,T]}$ belongs to $\mathbb{D}_{loc}^{2,2}(|\mathcal{H}|)$ and that the integral $X_t = \int_0^t u(s) \, dB^{(H)}(s)$ is almost surely continuous. Assume that $E\left[|u|^2\right]^{1/2}$ belongs to $\mathcal{H}$. Then for each $t \in [0, T]$ the following holds:*

$$\psi(X_t) = \psi(0) + \delta_H(\psi'(X))$$

$$+ \alpha_H \int_0^t \psi''(X_s)u(s)\left(\int_0^T |s-v|^{2H-2}\delta_H(D^{(H)}u\ I_{[0,s]})\right)ds$$

$$+ \alpha_H \int_0^t \psi''(X_s)u(s)\left(\int_0^s u(v)|s-v|^{2H-2}\,dv\right)ds. \tag{6.17}$$

Remark 6.3.5. Since $[(2H-1)/s^{2H-1}](s-v)^{2H-2}I_{[0,s]}(v)$ tends to the identity as H goes to $1/2$, we can formally recover the Itô formula for the Skorohod integral with respect to the standard Brownian motion proved in [179] by taking the limit as H converges to $1/2$ of equation (6.17).

Moreover, we remark that conditions under which the integral process admits a continuous modification are proved in [6] and [8].

Finally, we provide an Itô formula for the WIS integral proved in [33].

Theorem 6.3.6 (Itô formula for the WIS integral). *Suppose $1/2 < H < 1$. Let $\gamma(s)$ be a measurable process such that $\int_0^t |\gamma(s)|\,ds < \infty$ almost surely for all $t \geq 0$; let $\psi(t) = \sum_{\alpha \in \mathcal{J}} c_\alpha(t)\mathcal{H}_\alpha(\omega)$ be càglàd, WIS integrable and such that*

$$\sum_{\alpha \in \mathcal{J}} \sum_{i=1}^\infty \sum_{k=1}^\infty \|c_\alpha\|_{L^2([0,T])}\alpha_i(\alpha_k+1)\alpha! < \infty.$$

Suppose that $M_t D_t \psi(s)$ is also WIS integrable for almost all $t \in [0,T]$. Consider

$$X(t) = x + \int_0^t \gamma(s)\,ds + \int_0^t \psi(s)\,dB^{(H)}(s), \quad t \in [0,T],$$

or in shorthand notation,

$$dX(t) = \gamma(t)\,dt + \psi(t)\,dB^{(H)}(t), \quad X(0) = x.$$

Suppose X_t has a càdlàg version (Remark 6.3.5). Let $f \in C^2(\mathbb{R}^2)$ and put $Y(t) = f(t, X(t))$. Then on $[0,T]$,

$$\begin{aligned}dY(t) &= \frac{\partial f}{\partial t}(t,X(t))\,dt + \frac{\partial f}{\partial x}(t,X(t))\,dX(t) \\ &\quad + \frac{\partial^2 f}{\partial x^2}(t,X(t))\psi(t)M_{t+}D_{t+}X(t)\,dt,\end{aligned} \tag{6.18}$$

and equivalently,

$$\begin{aligned}dY(t) &= \frac{\partial f}{\partial t}(t,X(t))\,dt + \frac{\partial f}{\partial x}(t,X(t))\,dX(t) \\ &\quad + \frac{\partial^2 f}{\partial x^2}(t,X(t))\psi(t)M_t^2(\psi I_{[0,t]})_t\,dt \\ &\quad + \left[\frac{\partial^2 f}{\partial x^2}(t,X(t))\psi(t)\int_0^t M_t^2 D_t^{(H)}\psi(u)dB^{(H)}(u)\right]dt,\end{aligned}$$

where $M^2(\psi\chi_{[0,t]})_t = M^2(\psi\chi_{[0,t]})(t)$.

164   6 A useful summary

We now show that if $H > 1/2$, then (6.14) is a particular case of (6.18) when $X(t) = B^{(H)}(t)$.

Proposition 6.3.7. *For every $H \in (0,1)$ we have*
$$M_{t+}D_{t+}B^{(H)}(t) = Ht^{2H-1}, \quad t \geq 0.$$

Proof. Let $t \geq 0$. We recall that $D_t^{(H)}B^{(H)}(u) = I_{[0,u)}(t)$. Hence, we need to prove that
$$M_{t+}D_{t+}B^{(H)}(t) = \lim_{s \to t^+} \frac{1}{\epsilon} \int_t^{t+\epsilon} M_s^2 D_s^{(H)} B^{(H)}(t)\, ds$$
$$= [M_t^2 I_{[0,u)}(t)]_{u=t} = Ht^{2H-1}$$

We consider $\psi(u) = \int_{\mathbb{R}} (M_t I_{[0,u)}(t))^2\, dt$. Since, by [89], we have that $\psi(u) = u^{2H}$, we only need to show that $\psi'(u) = 2[M_t^2 I_{[0,u)}(t)]_{t=u}$. We rewrite $\psi(u)$ as follows:
$$\psi(u) = \int_{\mathbb{R}} (M_t I_{[0,u)}(t))^2\, dt = \int_{\mathbb{R}} I_{[0,u)}(t) M_t^2 I_{[0,u)}(t)\, dt = \int_0^u M_t^2 I_{[0,u)}(t)\, dt$$

by using the properties of the operator M. We compute
$$\frac{\psi(u+\epsilon) - \psi(u)}{\epsilon} = \frac{1}{\epsilon}\left(\int_0^{u+\epsilon} M_t^2 I_{[0,u+\epsilon]}(t)\, dt - \int_0^u M_t^2 I_{[0,u)}(t)\, dt\right)$$
$$= \frac{1}{\epsilon}\left\{\int_u^{u+\epsilon} M_t^2 I_{[0,u+\epsilon]}(t)\, dt \right.$$
$$\left. + \int_0^u [M_t^2 I_{[0,u+\epsilon]}(t) - M_t^2 I_{[0,u)}(t)]\, dt\right\}$$

by adding and subtracting $\int_0^u M_t^2 I_{[0,u+\epsilon]}(t)\, dt$. Since the operator M transforms $I_{[0,u)}(t)$ into a continuous function, we obtain

1. $\int_u^{u+\epsilon} M_t^2 I_{[0,u+\epsilon]}(t)\, dt = [M_t^2 I_{[0,u+\epsilon]}(t)]_{t=\xi_\epsilon}\epsilon$, where $u < \xi_\epsilon < u + \epsilon$. By writing
$$[M_t^2 I_{[0,u+\epsilon]}(t)]_{t=\xi_\epsilon} = [M_t^2 (I_{[0,u+\epsilon]} - I_{[0,u)})(t)]_{t=\xi_\epsilon} + [M_t^2 I_{[0,u)}(t)]_{t=\xi_\epsilon}$$
we obtain that when taking the limit as $\epsilon \longrightarrow 0$, the first term goes to zero while the second term converges to $[M_t^2 I_{[0,u)}(t)]_{t=u}$ since $\xi_\epsilon \longrightarrow u$ when $\epsilon \longrightarrow 0$.
2. We have that
$$\frac{1}{\epsilon}\int_0^u [M_t^2 I_{[0,u+\epsilon]}(t)\, dt - M_t^2 I_{[0,u)}(t)]\, dt = \frac{1}{\epsilon}\int_0^u M_t^2 [I_{(u,u+\epsilon]}(t)]\, dt$$
$$= \frac{1}{\epsilon}\int_0^T I_{[0,u)}(t)(M_t^2 [I_{(u,u+\epsilon]}(t)]\, dt$$

$$= \frac{1}{\epsilon} \int_u^{u+\epsilon} M_t^2[I_{[0,u)}(t)]\, dt$$

converges to $[M_t^2 I_{[0,u)}(t)]_{t=u}$ as $\epsilon \longrightarrow 0$.

Hence,

$$\psi'(u) = \lim_{\epsilon \to 0} \frac{\psi(u+\epsilon) - \psi(u)}{\epsilon} = 2[M_t^2 I_{[0,u)}(t)]_{t=u},$$

i.e., the equality $[M_t^2 I_{[0,u)}(t)]_{t=u} = H u^{2H-1}$ holds for every $H \in (0,1)$.

$\square$

We now show that Theorem 6.3.3 is a special case of Theorem 6.3.6 for integrals seen as elements of $L^2(\mathbb{P}^H)$. At this purpose, we need only to prove the following result that in addition justifies the use of the same notation for the Malliavin derivative introduced, respectively, in Definitions 3.3.3 and 4.4.7. To distinguish between the two, we denote in the sequel by $\widetilde{D}^{(H)}$ the derivative in $(\mathcal{S})_H^*$ (Definition 3.3.3) and by $D^{(H)}$ the one in $(\mathcal{S})^*$ (Definition 4.4.7).

As before, let $(e_i)_{i \in \mathbb{N}}$ be the orthonormal basis of $L^2_\phi(\mathbb{R})$ defined in (3.10) and $(\xi_i)_{i \in \mathbb{N}}$ be an orthonormal basis of $L^2(\mathbb{R})$ as in (6.7).

Proposition 6.3.8. *Consider the stochastic derivatives* $\widetilde{D}_t^{(H)}, D_t^\phi, D_t^{(H)}$ *defined, respectively, in Definitions 3.3.3, 3.5.1, and 4.4.4. Let* $F \in (\mathcal{S})_H^*$. *Then* $\widetilde{M} F \in (\mathcal{S})^*$ *and*

$$\widetilde{M} \widetilde{D}_t^{(H)} F = D_t^{(H)}(\widetilde{M} F). \tag{6.19}$$

Moreover, for every differentiable $F \in L^2(\mathbb{P}^H)$, *we obtain*

$$D_t^\phi F = M^2[\widetilde{M}^{-1} D_t^{(H)}(\widetilde{M} F)].$$

Proof. Let $F \in (\mathcal{S})_H^*$. Then

$$F(\omega) = \sum_{\alpha \in \mathcal{J}} c_\alpha \widetilde{\mathcal{H}}_\alpha(\omega) = \sum_{\alpha \in \mathcal{J}} c_\alpha h_{\alpha_1}(\langle \omega, e_1 \rangle) \cdots h_{\alpha_m}(\langle \omega, e_m \rangle)$$

and by (6.9)

$$\widetilde{M} F(\omega) = \sum_{\alpha \in \mathcal{J}} c_\alpha h_{\alpha_1}(\langle \omega, M e_1 \rangle) \cdots h_{\alpha_m}(\langle \omega, M e_m \rangle)$$

$$= \sum_{\alpha \in \mathcal{J}} c_\alpha h_{\alpha_1}(\langle \omega, \xi_1 \rangle) \cdots h_{\alpha_m}(\langle \omega, \xi_m \rangle).$$

By Definition 4.4.7 we have that

$$D_t^{(H)} \widetilde{M} F(\omega) = \sum_{\gamma \in \mathcal{J}} \left[\sum_{i=1}^{l(\gamma)} c_{\gamma + \varepsilon^{(i)}} (\gamma_i + 1) e_i(t) \right] \mathcal{H}_\gamma(\omega) \tag{6.20}$$

in $(\mathcal{S})^*$. Comparing (6.20) with Definition 3.10.6, by (6.8) we obtain immediately the relation (6.19).

Consider now $F \in L^2(\mathbb{P}^H)$. The second relation follows by (6.19), (3.31), and the fact that

$$\int_0^T M_u^2 D_u^{(H)} \widetilde{M} F \, du = \langle M_u^2 D_u^{(H)} \widetilde{M} F, I_{[0,T]} \rangle_{L^2(\mathbb{R})}$$

$$= \langle D_u^{(H)} \widetilde{M} F, I_{[0,T]} \rangle_H = \int_0^T \int_{\mathbb{R}} D_u^{(H)} \widetilde{M} F \phi(t,u) \, dt \, du.$$

□

To finish our overview of Itô formulas for *fBm*, we conclude with some remarks concerning other possible approaches. In [92] a basic stochastic calculus and an Itô formula are developed for finite cubic variation processes, which also apply to $B^{(H)}$ for $H \geq 1/3$. This extends a result from [42], where an Itô formula is established for $H > 1/6$ but with respect to $f \in C_b^5$. A more general Itô formula for $g \in C_b^3(\mathbb{R})$ valid for every $H \in (0,1)$ for cylindrical integrands is stated in [73].

Part III

Applications of stochastic calculus

7
Fractional Brownian motion in finance

As we all know, fBm cannot be used in finance, because it produces arbitrage. Therefore, fBm in finance is forbidden. But, as we also know, boys like to do forbidden things.

Esko Valkeila, in a talk given at the workshop Applications of Partial Differential Equations, Institut Mittag-Leffler, November 2007.

In view of the success – and at the same time limitations – of the classical Black Scholes market based on Brownian motion ($H = 1/2$), it is natural to ask if an extension to *fBm* ($0 < H < 1$) could give interesting financial models also. For example, if $H > 1/2$ then *fBm* has a certain memory (or *persistence*) feature (see Chapter 1), and this has been used in the modeling of weather derivatives (see, e.g., [38]). In addition if $H < 1/2$ then *fBm* has a certain turbulence (or *anti-persistence*) feature, which seems to be shared by electricity prices in the liberated Nordic electricity market (see [210]).

However, it is more controversial to let *fBm* simply replace classical Brownian motion in the classical Black Scholes market. Basically the problems are the following:

1. If we define, as in Chapter 5, the corresponding integration with respect to *fBm* in the *pathwise* (forward) way (which is a natural form from a modeling point of view and which makes mathematical sense for $H > 1/2$), then the corresponding financial market has *arbitrage*.
2. If we define the corresponding integration with respect to *fBm* in the *WIS* sense (see Chapter 4), then the corresponding market is free from (strong) arbitrage, but this integration is hard to justify from a modeling point of view.

We now discuss these two cases separately in more detail:

1. The pathwise (forward) integration (see Chapter 5).

2. The WIS integration (see Chapter 4).

7.1 The pathwise integration model $(1/2 < H < 1)$

For simplicity we concentrate on the simplest nontrivial type of market, namely, on the *fBm* version of the classical Black Scholes market, as follows. Suppose there are two investment possibilities:

1. A *safe* or *risk-free* investment, with price dynamics

$$dS_0(t) = rS_0(t)\,dt; \qquad S_0(0) = 1. \tag{7.1}$$

2. A *risky* investment, with price dynamics

$$d^-S_1(t) = \mu S_1(t)\,dt + \sigma S_1(t)\,d^-B^{(H)}(t); \qquad S_1(0) = x > 0. \tag{7.2}$$

Here $r, \mu, \sigma \neq 0$ and $x > 0$ are constants. By Theorem 5.6.10 we know that the solution of this equation is

$$S_1(t) = x\exp(\sigma B^{(H)}(t) + \mu t), \qquad t \geq 0.$$

Let $\{\mathcal{F}_t^{(H)}\}_{t\geq 0}$ be the filtration of $B^{(H)}(\cdot)$, i.e., $\mathcal{F}_t^{(H)}$ is the σ-algebra generated by the random variables $B^{(H)}(s)$, $s \leq t$.

A *portfolio* in this market is a two-dimensional $\mathcal{F}_t^{(H)}$-adapted stochastic process $\theta(t) = (\theta_0(t), \theta_1(t))$, where $\theta_i(t)$ gives the number of units of investment number i held at time t, $i = 0, 1$. The corresponding *wealth process* $V^\theta(t)$ is defined by

$$V^\theta(t) = \theta(t) \cdot S(t) = \theta_0(s)S_0(t) + \theta_1(t)S_1(t),$$

where

$$S(t) = (S_0(t), S_0(t)).$$

We say that θ is *pathwise self-financing* if

$$d^-V^\theta(t) = \theta(t) \cdot d^-S(t),$$

i.e.,

$$V^\theta(t) = V^\theta(0) + \int_0^t \theta_0(s)\,dS_0(s) + \int_0^t \theta_1(s)\,d^-S_1(s).$$

If, in addition, $V^\theta(t)$ is *lower bounded*, then we call the portfolio θ (pathwise) *admissible*.

Definition 7.1.1. *A pathwise admissible portfolio θ is called an* arbitrage *if the corresponding wealth process $V^\theta(t)$ satisfies the following three conditions:*

$$V^\theta(0) = 0, \tag{7.3}$$

$$V^\theta(T) \geq 0 \quad a.s., \tag{7.4}$$

$$\mathbb{P}(V^\theta(T) > 0) > 0. \tag{7.5}$$

7.1 The pathwise integration model ($1/2 < H < 1$)

Remark 7.1.2. The nonexistence of arbitrage in a market is a basic equilibrium condition. It is not possible to make a sensible mathematical theory for a market with arbitrage. Therefore, one of the first things to check in a mathematical finance model is whether arbitrage exists. In the above pathwise *fBm* market the existence of arbitrage was proved by [195] in 1997. Subsequently several simple examples of arbitrage were found. See, e.g., [70], [204], [208]. Note, however, that the existence of arbitrage in this pathwise model is already a direct consequence of Theorem 7.2 in [77]: There it is proved in general that if there is no arbitrage using simple portfolios (with pathwise products), then the price process is a semimartingale. Hence, since the process $S_1(t)$ given by (7.2) is not a semimartingale, an arbitrage must exist.

Here is a simple arbitrage example, due to [70], [204], [208]. For simplicity assume that

$$\mu = r \quad \text{and} \quad \sigma = x = 1. \tag{7.6}$$

Define

$$\theta_0(t) = 1 - \exp(2B^{(H)}(t)), \quad \theta_1(t) = 2[\exp(B^{(H)}(t)) - 1]. \tag{7.7}$$

Then the corresponding wealth process is

$$\begin{aligned} V^\theta(t) &= \theta_0(t) S_0(t) + \theta_1(t) S_1(t) \\ &= [1 - \exp(2B^{(H)}(t))] \exp(rt) \\ &\quad + 2[\exp(B^{(H)}(t)) - 1] \exp(B^{(H)}(t) + rt) \\ &= \exp(rt)[\exp(B^{(H)}(t)) - 1]^2 > 0 \quad \text{for a.a. } (t, \omega). \end{aligned} \tag{7.8}$$

This portfolio is self-financing, since

$$\begin{aligned} \theta_0(t)\, dS_0(t) + \theta_1(t)\, d^- S_1(t) &= [1 - \exp(2B^{(H)}(t))] r \exp(rt)\, dt \\ &\quad + 2[\exp(B^{(H)}(t)) - 1] S_1(t)[rt + d^- B^{(H)}(t)] \\ &= r \exp(rt)[\exp(B^{(H)}(t)) - 1]^2\, dt \\ &\quad + 2 \exp(rt)(\exp(B^{(H)}(t)) - 1) \exp(B^{(H)}(t))\, d^- B^{(H)}(t) \\ &= d(\exp(rt)[\exp(B^{(H)}(t)) - 1]^2) = d^- V^\theta(t). \end{aligned}$$

We have proved the following:

Theorem 7.1.3. *The portfolio $\theta(t) = (\theta_0(t), \theta_1(t))$ given by (7.7) is a (pathwise) arbitrage in the (pathwise) fractional Black Scholes market given by (7.1), (7.2), and (7.6).*

In view of this result the pathwise *fBm* model is not suitable in finance, at least not in this simple form (but possibly in combination with classical Brownian motion or other stochastic processes; see, e.g., Section 9.6).

7.2 The WIS integration model ($0 < H < 1$)

We now consider the WIS integration version of the market (7.1), (7.2). Mathematically the model below is an extension to $H \in (0,1)$ of the model introduced in [121] for $H \in (1/2, 1)$. (Subsequently a related model, also valid for all $H \in (0,1)$, was presented in [89]). However, compared to [121] we give a different *interpretation* of the mathematical concepts involved. This presentation is based on [183]. We also remark that for $H > 1/2$ we can repeat formally the same model construction by using the *fWIS integration* studied in Chapter 3 instead of the WIS calculus.

Assume that the values $S_0(t), S_1(t)$ of the risk-free (e.g., bond) and risky asset (e.g., stock), respectively, are given by

$$dS_0(t) = rS_0(t)\,dt, \qquad S_0(0) = 1 \qquad \text{(bond)} \qquad (7.9)$$

and

$$dS_1(t) = \mu S_1(t)\,dt + \sigma S_1(t)\,dB^{(H)}(t), \qquad S_1(0) = x > 0 \qquad \text{(stock)} \qquad (7.10)$$

where $r, \mu, \sigma \neq 0$ and $x > 0$ are constants and the integral appearing in (7.10) is the one introduced in Chapter 4.

By Theorem of 4.2.6 the solution of equation (7.10) is

$$S_1(t) = x\exp(\sigma B^{(H)}(t) + \mu t - \frac{1}{2}\sigma^2 t^{2H}), \qquad t \geq 0.$$

In this WIS model $S_1(t)$ does not represent the observed stock price at time t, but we give it a different interpretation. We assume that $S_1(t)$ represents in a broad sense the total value of the company and that it is *not* observed directly. Instead we adopt a quantum mechanical point of view, regarding $S_1(t, \omega)$ as a stochastic distribution in ω [represented mathematically as an element of $(\mathcal{S})^*$] and regarding the actual *observed stock price* $\bar{S}_1(t)$ as the result of applying $S_1(t, \cdot) \in (\mathcal{S})^*$ to a stochastic test function $\psi(\cdot) \in (\mathcal{S})$ (see Definition A.1.4). We call $S_1(t)$ the *generalized stock price*.
Hence, we have

$$\bar{S}_1(t) := \langle S_1(t, \cdot), \psi(\cdot) \rangle = \langle S_1(t), \psi \rangle,$$

where in general $\langle F, \psi \rangle$ denotes the action of a stochastic distribution $F \in (\mathcal{S})^*$ to a stochastic test function $\psi \in (\mathcal{S})$.

We call such stochastic test functions ψ *market observers*. We will assume that they have the form

$$\psi(\omega) = \exp^{\diamond}\left(\int_{\mathbb{R}} h(t)\,dB^{(H)}(t)\right)$$

$$= \exp\left(\int_{\mathbb{R}} h(t)\,dB^{(H)}(t) - \frac{1}{2}\|h\|_H^2\right) \qquad \text{for some } h \in L_H^2(\mathbb{R}). \qquad (7.11)$$

The set of all linear combinations of such ψ is dense in both $(\mathcal{S})$ and $(\mathcal{S})^*$. Moreover, these ψ are normalized, in the sense that

$$E\left[\exp^{\diamond}\left(\int_{\mathbb{R}} h(t)\,dB^{(H)}(t)\right)\right] = 1 \qquad \forall h \in L_H^2(\mathbb{R}).$$

We let $\mathcal{D}$ denote the set of all market observers of the form (7.11).

Similarly, a *generalized portfolio* is an adapted process

$$\theta(t) = \theta(t,\omega) = (\theta_0(t,\omega), \theta_1(t,\omega)), \qquad (t,\omega) \in [0,T] \times \Omega, \qquad (7.12)$$

such that $\theta(t,\omega)$ is measurable with respect to $\mathcal{B}[0,T] \otimes \mathcal{F}^{(H)}$, where $\mathcal{B}[0,T]$ is the Borel σ-algebra on $[0,T]$ and $\mathcal{F}^{(H)}$ is the σ-algebra generated by $\{B^{(H)}(s)\}_{s \geq 0}$, representing a general strategy for choosing the number of units of investment number i at time t, $i = 0, 1$.

For example, $\theta_1(t)$ could be the usual "buy and hold" strategy, consisting of buying a certain number of stocks at a stopping time $\tau_1(\omega)$ and holding them until another stopping time $\tau_2(\omega) > \tau_1(\omega)$, or $\theta_1(t)$ could be the strategy to hold a fixed fraction of the current wealth in stocks. If the actual observed price at time t is $\bar{S}_1(t) = \langle S_1(t,\cdot), \psi(\cdot) \rangle$, the actual number of stocks held is

$$\bar{\theta}_1(t) := \langle \theta_1(t,\cdot), \psi(\cdot) \rangle.$$

Thus the actual *observed wealth* $\bar{V}_1(t)$ held in the risky asset corresponding to this portfolio is

$$\bar{V}_1(t) = \langle \theta_1(t), \psi \rangle \cdot \langle S_1(t), \psi \rangle.$$

By Lemma 7.2.1 below this can be written

$$\bar{V}_1(t) = \langle \theta_1(t) \diamond S_1(t), \psi \rangle, \qquad (7.13)$$

where $\diamond$ denotes the Wick product. In fact, $F := \theta_1(t) \diamond S_1(t)$ is the *unique* $F \in (\mathcal{S})^*$ such that

$$\langle F, \psi \rangle = \langle \theta_1(t), \psi \rangle \cdot \langle S_1(t), \psi \rangle \qquad \forall \psi \in \mathcal{D}.$$

In view of this it is natural to define the *generalized total wealth process* $V(t,\omega)$ associated to $\theta(t,\omega)$ by the Wick product

$$V(t,\cdot) = \theta(t,\cdot) \diamond S(t,\cdot) = \theta_0(t)S_0(t) + \theta_1(t) \diamond S_1(t). \qquad (7.14)$$

Similarly, if we consider a discrete time market model and keep the generalized portfolio process
$$\theta(t) = \theta(t_k, \omega), \qquad t_k \leq t < t_{k+1},$$

constant from $t = t_k$ to $t = t_{k+1}$, the corresponding change in the generalized wealth process is

$$\Delta V(t_k) = \theta(t_k) \diamond \Delta S(t_k),$$

where

174   7 Fractional Brownian motion in finance

$$\Delta V(t_k) = V(t_{k+1}) - V(t_k), \quad \Delta S(t_k) = S(t_{k+1}) - S(t_k).$$

If we sum this over k and take the limit as $\Delta t_k = t_{k+1} - t_k$ goes to 0, we end up with the following *generalized wealth process formula*:

$$V(T) = V(0) + \int_0^T \theta(t) \diamond dS(t) = V(0) + \int_0^T \theta(t)\, dS(t),$$

where $dS(t)$ means that the integral is interpreted in the WIS sense.

Therefore, by (7.9) and (7.10) this argument leads to the following *(WIS) self-financing property*:

$$V(T) = V(0) + \int_0^T r\theta_0(t)S_0(t)\, dt + \int_0^T \mu\theta_1(t) \diamond S_1(t)\, dt \qquad (7.15)$$
$$+ \int_0^T \sigma\theta_1(t) \diamond S_1(t)\, dB^{(H)}(t),$$

where the integral to the right of (7.15) is a WIS integral. We now prove the fundamental result that explains why the Wick product suddenly appears in (7.13) above:

Lemma 7.2.1. 1. *Let $F, G \in (\mathcal{S})^*$. Then*

$$\langle F \diamond G, \psi \rangle = \langle F, \phi \rangle \cdot \langle G, \psi \rangle \qquad \forall \psi \in \mathcal{D}.$$

2. *Moreover, if $Z \in (\mathcal{S})^*$ is such that*

$$\langle Z, \psi \rangle = \langle F, \psi \rangle \cdot \langle G, \psi \rangle \qquad \forall \psi \in \mathcal{D},$$

then

$$Z = F \diamond G.$$

Proof. 1. Choose $\psi = \exp^\diamond \left(\int_{\mathbb{R}} h(t)dB^{(H)}(t) \right) \in \mathcal{D}$. We may assume that

$$F = \exp^\diamond \left(\int_{\mathbb{R}} f(t)\, dB^{(H)}(t) \right) \quad \text{and} \quad G = \exp^\diamond \left(\int_{\mathbb{R}} g(t)\, dB^{(H)}(t) \right)$$

for some $f, g \in L_H^2(\mathbb{R})$, because the set of all linear combinations of such Wick exponentials is dense in $(\mathcal{S})^*$. For such F, G, ψ we have

$$\langle F, \psi \rangle = E[F \cdot \psi] \quad \text{and} \quad \langle G, \psi \rangle = E[G \cdot \psi].$$

Therefore

$$\langle F \diamond G, \psi \rangle = E\left[\exp^\diamond \left(\int_{\mathbb{R}} (f+g)\, dB^{(H)} \right) \cdot \exp^\diamond \left(\int_{\mathbb{R}} h\, dB^{(H)} \right) \right]$$
$$= E\left[\exp \left(\int_{\mathbb{R}} (f+g)\, dB^{(H)} - \frac{1}{2}\|f+g\|_H^2 \right) \right.$$

$$\cdot \exp\left(\int_{\mathbb{R}} h\, dB^{(H)} - \frac{1}{2}\|h\|_H^2\right)\Bigg]$$
$$= E\Bigg[\exp\left(\int_{\mathbb{R}}(f+g+h)\, dB^{(H)} - \frac{1}{2}\|f\|_H^2 - \frac{1}{2}\|g\|_H^2 \right.$$
$$\left. - \frac{1}{2}\|h\|_H^2 - \langle f, g\rangle_H\right)\Bigg]$$
$$= E\Bigg[\exp\left(\int_{\mathbb{R}}(f+g+h)\, dB^{(H)} - \frac{1}{2}\|f+g+h\|_H^2 + \langle f,h\rangle_H \right.$$
$$\left. + \langle g,h\rangle_H\right)\Bigg]$$
$$= E\Bigg[\exp^{\diamond}\left(\int_{\mathbb{R}}(f+g+h)\, dB^{(H)}\right) \cdot \exp\langle f+g, h\rangle_H\Bigg]$$
$$= \exp\langle f+g, h\rangle_H. \tag{7.16}$$

On the other hand, a similar computation gives

$$\langle F, \psi\rangle \cdot \langle G, \psi\rangle = E\Bigg[\exp^{\diamond}\left(\int_{\mathbb{R}} f\, dB^{(H)}\right) \cdot \exp^{\diamond}\left(\int_{\mathbb{R}} h\, dB^{(H)}\right)\Bigg]$$
$$\cdot E\Bigg[\exp^{\diamond}\left(\int_{\mathbb{R}} g\, dB^{(H)}\right) \cdot \exp^{\diamond}\left(\int_{\mathbb{R}} h\, dB^{(H)}\right)\Bigg]$$
$$= \exp\langle f, h\rangle_H \cdot \exp\langle g, h\rangle_H = \exp\langle f+g, h\rangle_H. \tag{7.17}$$

Comparing (7.16) and (7.17) we get 1.
2. This follows from the fact that the set of linear combinations of elements of $\mathcal{D}$ is dense in $(\mathcal{S})$ and $(\mathcal{S})^*$ is the dual of $(\mathcal{S})$. □

Remark 7.2.2. We emphasize that this model for fBm in finance does not a priori assume that the Wick product models the growth of wealth. In fact, the Wick product comes as a mathematical consequence of the basic assumption that the observed value is the result of applying a test function to a distribution process describing in a broad sense the value of a company. This way of thinking stems from microcosmos (quantum mechanics), but it has been argued that it is often a good description of *macrocosmos* situations as well. Here is an example.

An agent from an opinion poll firm stops a man on the street and asks him what political party he would vote for if there was an election today. Often this man on the street does not really have a firm opinion about this beforehand (he is in a diffuse state of mind politically), but the contact with the agent forces him to produce an answer. In a similar sense the general state of a company does not really have a noted stock price a priori, but it brings out a number (price) when confronted with a market observer (the stock market).

In view of the above we now make the following definitions:

Definition 7.2.3. *1. The* total wealth process $V^\theta(t)$ *corresponding to a generalized portfolio* $\theta(t)$ *defined in (7.12) in the WIS model is defined by*

$$V^\theta(t) = \theta(t) \diamond S(t). \tag{7.18}$$

2. *A generalized portfolio $\theta(t)$ is called* **WIS self-financing** *if*

$$dV^\theta(t) = \theta(t)\, dS(t),$$

i.e.,

$$V^\theta(t) = V^\theta(0) + \int_0^t \theta_0(s)\, dS_0(s) + \int_0^t \theta_1(s)\, dS_1(s), \tag{7.19}$$

where the integral to the right of (7.19) is a WIS integral. In particular, we assume that the two integrals in (7.19) exist.

By the Girsanov theorem for *fBm* for every $H \in (0,1)$, as described in Section 4.3, there exists a probability measure $\hat{\mathbb{P}}$ on $(\Omega, \mathcal{F})$ such that $\hat{\mathbb{P}}$ is equivalent to $\mathbb{P}$ (i.e., $\hat{\mathbb{P}}$ has the same null sets as $\mathbb{P}$) with Radon–Nikodym density

$$\frac{d\hat{\mathbb{P}}}{d\mathbb{P}} = \exp\left(\int_\mathbb{R} \psi(s)\, dB^{(H)}(s) - \frac{1}{2}\|\psi\|_{L^2(\mathbb{R})}^2\right), \tag{7.20}$$

where

$$\psi(t) = \frac{(\mu - r)[(T-t)^{1/2-H} + t^{1/2-H}]}{2\sigma \Gamma(3/2 - H)\cos[\pi/2(1/2 - H)]},$$

and such that

$$\hat{B}^{(H)}(t) := \frac{\mu - r}{\sigma} t + B^{(H)}(t)$$

is a *fBm* with respect to $\hat{\mathbb{P}}$. Replacing $B^{(H)}(t)$ by $\hat{B}^{(H)}(t)$ in (7.19), we get

$$e^{-rt}V^\theta(t) = V^\theta(0) + \int_0^t e^{-rs}\sigma\theta_1(s) \diamond S_1(s)\, d\hat{B}^{(H)}(s). \tag{7.21}$$

Definition 7.2.4. *We call a generalized portfolio $\theta(t)$* WIS admissible *if it is WIS self-financing and $\theta_1(s) \diamond S_1(s)$ is Skorohod integrable with respect to $\hat{B}^{(H)}(s)$.*

Definition 7.2.5. *A WIS admissible portfolio $\theta(t)$ is called a* strong arbitrage *if the corresponding total wealth process $V^\theta(t)$ satisfies*

$$V^\theta(0) = 0, \tag{7.22}$$
$$V^\theta(T) \in L^2(\hat{\mathbb{P}}) \quad \text{and} \quad V^\theta(T) \geq 0 \quad a.s., \tag{7.23}$$
$$\mathbb{P}(V^\theta(T) > 0) > 0. \tag{7.24}$$

The following result was first proved by [121] for the case $1/2 < H < 1$ and then extended to arbitrary $H \in (0,1)$ by [89] (in a related model):

Theorem 7.2.6. *There is no strong arbitrage in the WIS fractional Black Scholes market (7.9) and (7.10).*

Proof. If we take the expectation with respect to $\hat{\mathbb{P}}$ of both sides of (7.21) with $t = T$, we get, by (7.10),

$$e^{-rT}\hat{E}[V^\theta(T)] = V^\theta(0).$$

From this we see that (7.22) to (7.24) cannot hold. □

Remark 7.2.7. Note that the nonexistence of a strong arbitrage in this market [where the value process $S_1(t)$ is *not* a semimartingale] is not in conflict with the result of [77] mentioned in Remark 7.1.2, because in this market the underlying products are Wick products, not ordinary pathwise products.

We proceed to discuss *completeness* in this market.

Definition 7.2.8. *The market is called* (WIS) complete *if for every $\mathcal{F}_T^{(H)}$-measurable random variable $F \in L^2(\hat{\mathbb{P}})$ there exists an admissible portfolio $\theta(t) = (\theta_0(t), \theta_1(t))$ such that*

$$F = V^\theta(T) \quad a.s.$$

By (7.21) we see that this is equivalent to requiring that there exists g such that

$$e^{-rT}F(\omega) = e^{-rT}\hat{E}[F] + \int_0^T g(s,\omega)\, d\hat{B}^{(H)}(s),$$

where

$$g(s) = e^{-rs}\sigma\, \theta_1(s) \diamond S_1(s).$$

If such a g can be found, then we put

$$\theta_1(s) = \sigma^{-1} e^{rs} S_1(s)^{\diamond(-1)} \diamond g(s).$$

It was proved by [121] (for $1/2 < H < 1$) and subsequently by [89] in a related market [for arbitrary $H \in (0,1)$] that this market is complete. In fact, we have

Theorem 7.2.9. *Let $F \in L^2(\hat{\mathbb{P}})$ be $\mathcal{F}_T^{(H)}$-measurable. Then $F = V^\theta(T)$ almost surely for $\theta(t) = (\theta_0(t), \theta_1(t))$ with*

$$\theta_1(t) = \sigma^{-1} e^{-\rho(T-t)} S_1(t)^{\diamond(-1)} \diamond \tilde{E}_{\hat{\mathbb{P}}}[\hat{D}_t^{(H)} F \mid \mathcal{F}_t^{(H)}],$$

where $\tilde{E}_{\hat{\mathbb{P}}}[\cdot|\cdot]$ denotes the quasi-conditional expectation *and $\hat{D}_t^{(H)}$ is the fractional Hida–Malliavin derivative with respect to $\hat{B}^{(H)}(\cdot)$. The other component, $\theta_0(t)$, is then uniquely determined by the self-financing condition (7.19).*

In the *Markovian* case, i.e., when

$$F(\omega) = f(B^{(H)}(T))$$

for some integrable Borel function $f: \mathbb{R} \to \mathbb{R}$, we can give a more explicit expression for the replicating portfolio $\theta(t)$. This is achieved by using the following representation theorem, due to [14]. It has the same form as in the well-known classical case ($H = 1/2$):

Theorem 7.2.10. *Let* $f : \mathbb{R} \to \mathbb{R}$ *be a Borel function such that*

$$E[f^2(B^{(H)}(T))] < \infty.$$

Then

$$f(B^{(H)}(T)) = E[f(B(T))] + \int_0^T g(t,\omega) \, dB^{(H)}(t),$$

where

$$g(t,\omega) = \left\{ \frac{\partial}{\partial x} E[f(x + B^{(H)}(T-t))] \right\}_{x=B^{(H)}(t)}.$$

In view of the interpretation of the observed wealth $\bar{V}(t)$ as the result of applying a test function $\psi \in \mathcal{D}$ to the general wealth process $V(t)$, i.e.,

$$\bar{V}(t) = \langle V(t), \psi \rangle,$$

the following alternative definition of an arbitrage is natural (compare with Definition 7.2.5).

Definition 7.2.11. *A WIS admissible portfolio* $\theta(t)$ *is called a* **weak arbitrage** *if the corresponding total wealth process* $V^\theta(t)$ *satisfies*

$$V^\theta(0) = 0, \tag{7.25}$$

$$\langle V^\theta(T), \psi \rangle \geq 0 \quad \forall \psi \in \mathcal{D}, \tag{7.26}$$

$$\langle V^\theta(T), \psi \rangle > 0 \quad \text{for some } \psi \in \mathcal{D}. \tag{7.27}$$

Do weak arbitrages exist? The answer is yes. Here is an example, found in [17].

Example 7.2.12 (A weak arbitrage). For $\varepsilon > 0$ define

$$K_\varepsilon(x) = \begin{cases} -1 & \text{if } |x| \leq \varepsilon, \\ 1 & \text{if } |x| > \varepsilon. \end{cases}$$

Then there exists $\varepsilon_0 > 0$ such that

$$\int_\mathbb{R} K_{\varepsilon_0}(x) \exp\left(-\frac{1}{2}x^2\right) dx = 0. \tag{7.28}$$

By a variant of Lemma 2.6 in [15] we have

$$E\left[K(\langle \omega, f \rangle) \exp(\langle \omega, g \rangle - \frac{1}{2}\|g\|_H^2)\right]$$

$$= (2\pi)^{-1/2} \|f\|_H \int_\mathbb{R} K(u) \exp\left(-\frac{(u - \langle f, g \rangle_H)^2}{2\|f\|_H^2}\right) du \tag{7.29}$$

for all bounded measurable $K : \mathbb{R} \to \mathbb{R}$, $f, g \in L_H^2(\mathbb{R})$.

Applying (7.29) to $f = I_{[0,1]}$ and $\langle \omega, f \rangle = B^{(H)}(1)$, we get

$$E[K_{\varepsilon_0}(B^{(H)}(1))] = 0 \tag{7.30}$$

$$E\big[K_{\varepsilon_0}(B^{(H)}(1))\exp(\langle\omega,g\rangle - \frac{1}{2}\|g\|_H^2)\big] \geq 0 \quad \forall g \in L_H^2(\mathbb{R})$$

$$E\big[K_{\varepsilon_0}(B^{(H)}(1))\exp(\langle\omega,I_{[0,1]}\rangle - \frac{1}{2}\|I_{[0,1]}\|_H^2)\big] > 0. \tag{7.31}$$

Now consider the Skorohod fractional market (7.9) and (7.10) with $r = \mu = 0$, $\sigma = T = 1$. Then $S_0(t) = 1$ and $S_1(t) = x\exp(B^{(H)}(t) - 1/2t^{2H})$. Moreover, $\hat{B}^{(H)}(t) = B^{(H)}(t)$ and $\mathbb{P} = \hat{\mathbb{P}}$. Hence, by Theorem 7.2.9 and (7.28) there exists a Skorohod self-financing portfolio $\theta(t) = (\theta_0(t), \theta_1(t))$ such that

$$K_{\varepsilon_0}(B^{(H)}(1)) = V^\theta(1) = \int_0^T \theta_1(s)\, dS(s) \quad \text{a.s.}$$

Then $V^\theta(0) = 0$ and by (7.11), (7.30) and (7.31) we see that (7.26) and (7.27) hold. Hence, $\theta(t)$ is a weak arbitrage.

7.3 A connection between the pathwise and the WIS model

In spite of the fundamental differences in the features of the pathwise model and the WIS model, it turns out that there is a close relation between them. Assume $H \in (1/2, 1)$.

Fix $\psi \in \mathcal{D}$ and define the function $b_H : [0, T] \to \mathbb{R}$ by

$$b_H(t) = \langle B^{(H)}(t), \psi \rangle = E[B^{(H)}(t) \cdot \psi].$$

Then for $p > 1$ and any partition $\mathcal{P} : 0 = t_0 < t_1 < \cdots < t_N = T$ of $[0, T]$, we have

$$\sum_{j=0}^{N-1} |b_H(t_{j+1}) - b_H(t_j)|^p = \sum_{j=0}^{N-1} |E[(B^{(H)}(t_{j+1}) - B^{(H)}(t_j)) \cdot \psi]|^p$$

$$\leq \sum_{j=0}^{N-1} (E[|B^{(H)}(t_{j+1}) - B^{(H)}(t_j)|^p]^{1/p} \cdot E[\psi^q]^{1/q})^p$$

$$\leq C \sum_{j=0}^{N-1} E[|B^{(H)}(t_{j+1}) - B^{(H)}(t_j)|^p],$$

where $1/p + 1/q = 1$. Hence, by a known property of fBm,

$$\sup_{\mathcal{P}} \sum_{j=0}^{N-1} |b_H(t_{j+1}) - b_H(t_j)|^p < \infty \quad \text{if and only if } p \geq \frac{1}{H}.$$

In this sense the continuous function $b_H(t)$ is at least as regular as a generic path of a fBm $B^{(H)}(t,\omega)$. Therefore, we can define integration with respect to $b_H(t)$ just as we define pathwise integration with respect to $B^{(H)}(t)$. Now suppose we start with the wealth generating formula in the WIS model

$$V^\theta(T) = V^\theta(0) + \int_0^T \phi(s,\omega)\,dB^{(H)}(s).$$

Suppose ϕ is càglàd and $\psi \in \mathcal{D}$. Then this gives

$$\bar{V}^\theta(T) = \langle V^\theta(T), \psi \rangle = V^\theta(0) + \left\langle \int_0^T \phi(s,\omega)\,dB^{(H)}(s), \psi \right\rangle$$

$$= V^\theta(0) + \lim_{\Delta t_j \to 0} \left\langle \sum_{j=0}^{N-1} \phi(t_j) \diamond (B^{(H)}(t_{j+1}) - B^{(H)}(t_j)), \psi \right\rangle$$

$$= V^\theta(0) + \lim_{\Delta t_j \to 0} \sum_{j=0}^{N-1} \langle \phi(t_j), \psi \rangle \langle B^{(H)}(t_{j+1}) - B^{(H)}(t_j), \psi \rangle$$

$$= V^\theta(0) + \lim_{\Delta t_j \to 0} \sum_{j=0}^{N-1} \bar{\phi}(t_j)(b_H(t_{j+1}) - b_H(t_j))$$

$$= V^\theta(0) + \int_0^T \bar{\phi}(t)\,db_H(t). \tag{7.32}$$

We can summarize this as follows:

Theorem 7.3.1. *If $H > 1/2$ the mapping $F \to \langle F, \psi \rangle$, $F \in L^2(\mathbb{P})$, transforms the WIS fBm model into the pathwise fBm model. If $H = 1/2$ this mapping transforms the Wick Itô Skorohod Brownian motion model into the classical Brownian motion model.*

7.4 Concluding remarks

At first glance there seems to be a disagreement between the existence of arbitrage in the (fractional) pathwise model (see Theorem 7.1.3) and the nonexistence of a (strong) arbitrage in the WIS model (Theorem 7.2.6). The above discussion, including, in particular, Theorem 7.3.1, serves to explain this apparent contradiction. The arbitrages in the pathwise model correspond to the *weak* arbitrages in the WIS model (see Example 7.2.12) and not to the (nonexistent) *strong* arbitrages. In addition in [104] and [105] it is shown that there is no arbitrage with the *fBm* pathwise integral financial model when transaction costs are taken into account.

In spite of the mathematical coherence of the WIS model, there is still a lot of controversy about its economic interpretation and features. We refer to the discussions in [35] and [215] for more details. For other models for financial markets with *fBm*, we refer also to [21].

8

Stochastic partial differential equations driven by fractional Brownian fields

This chapter is devoted to the study of stochastic Poisson and stochastic heat equations driven by fractional white noise. The equations are solved both in the setting of white noise analysis and the setting of L^2 space. The main references for this chapter are [126] and [184].

8.1 Fractional Brownian fields

We start by recalling the standard white noise construction of multiparameter *classical* Brownian motion $B(x), x \in \mathbb{R}^d$, because we need the construction of the multiparameter fractional Brownian field for the study of stochastic partial differential equations.

Let $\mathcal{S} = \mathcal{S}(\mathbb{R}^d)$ be the Schwartz space of rapidly decreasing smooth functions on $\mathbb{R}^d$, and let $\Omega := \mathcal{S}'(\mathbb{R}^d)$ be its dual, *the space of tempered distributions*. By the Bochner–Minlos theorem (see [144], [109]) there exists a probability measure $\mathbb{P}$ on the Borel σ-algebra $\mathcal{F} = \mathcal{B}(\Omega)$ such that

$$\int_\Omega e^{i\langle \omega, f \rangle}\, d\mathbb{P}(\omega) = e^{-1/2\|f\|^2_{L^2(\mathbb{R}^d)}}, \qquad f \in \mathcal{S}(\mathbb{R}^d) \tag{8.1}$$

where $\langle \omega, f \rangle = \omega(f)$ denotes the action of $\omega \in \Omega = \mathcal{S}'(\mathbb{R}^d)$ applied to $f \in \mathcal{S}(\mathbb{R}^d)$. From (8.1) one can deduce that

$$E[\langle \omega, f \rangle] = 0 \qquad \forall f \in \mathcal{S}(\mathbb{R}^d),$$

where E denotes the expectation with respect to $\mathbb{P}$. Moreover, we have the isometry

$$E[\langle \omega, f \rangle \langle \omega, g \rangle] = \langle f, g \rangle_{L^2(\mathbb{R}^d)}, \qquad f, g \in \mathcal{S}(\mathbb{R}^d). \tag{8.2}$$

Using this isometry, we can extend the definition of $\langle \omega, f \rangle \in L^2(\mathbb{P})$ from $\mathcal{S}(\mathbb{R}^d)$ to $L^2(\mathbb{R}^d)$ as follows:

$$\langle \omega, f \rangle = \lim_{n \to \infty} \langle \omega, f_n \rangle \qquad \text{[limit in } L^2(\mathbb{P})\text{]}$$

when $f_n \in \mathcal{S}(\mathbb{R}^d)$, $f_n \to f \in L^2(\mathbb{R}^d)$ [limit in $L^2(\mathbb{R}^d)$].
In particular, we can now define, for $x = (x_1, \ldots, x_d) \in \mathbb{R}^d$,

$$\widetilde{B}(x) = \widetilde{B}(x, \omega) = \langle \omega, I_{[0,x]}(\cdot) \rangle, \qquad \omega \in \Omega,$$

where

$$I_{[0,x]}(y) = \prod_{i=1}^{d} I_{[0,x_i]}(y_i) \qquad \text{for } y = (y_1, \ldots, y_d) \in \mathbb{R}^d \tag{8.3}$$

and

$$I_{[0,x_i]}(y_i) = \begin{cases} 1 & \text{if } 0 \le y_i \le x_i, \\ -1 & \text{if } x_i \le y_i \le 0 \text{ except } x_i = y_i = 0, \\ 0 & \text{otherwise}. \end{cases} \tag{8.4}$$

By Kolmogorov's continuity theorem the process $\{\widetilde{B}(x)\}$ has a continuous version, which we will denote by $\{B(x)\}$. By (8.1) to (8.2) it follows that $\{B(x)\}$ is a Gaussian process with mean

$$E[B(x)] = B(0) = 0$$

and covariance [using (8.2)]

$$E[B(x)B(y)] = \langle I_{[0,x]}, I_{[0,y]} \rangle_{L^2(\mathbb{R}^d)}$$

$$= \begin{cases} \prod_{i=1}^{d} x_i \wedge y_i & \text{if } x_i, y_i \ge 0 \text{ for all } i, \\ \prod_{i=1}^{d} (-x_i) \wedge (-y_i) & \text{if } x_i, y_i \le 0 \text{ for all } i, \\ 0 & \text{otherwise}. \end{cases}$$

Therefore $\{B(x)\}_{x \in \mathbb{R}^d}$ is a d-parameter Brownian motion.

We now use this Brownian motion in order to construct d-parameter fBm $B^{(H)}(x)$ for all Hurst parameters $H = (H_1, \ldots, H_d) \in (0,1)^d$ extending the approach of Chapter 4.

For $0 < H_j < 1$ put

$$K_j = \left[2\Gamma\left(H_j - \frac{1}{2}\right) \cos\left(\frac{\pi}{2}\left(H_j - \frac{1}{2}\right)\right) \right]^{-1} [\sin(\pi H_j) \Gamma(2H_j + 1)]^{1/2}$$

and if $g \in \mathcal{S}(\mathbb{R}^d)$, $x = (x_1, \ldots, x_d) \in \mathbb{R}^d$, define $m_j g(\cdot) : \mathbb{R}^d \to \mathbb{R}$ by

$$m_j g(x) = \begin{cases} K_j \int_{\mathbb{R}} \dfrac{g(x - t\varepsilon^{(j)}) - g(x)}{|t|^{3/2 - H_j}} \, dt & \text{if } 0 < H_j < 1/2, \\ g(x) & \text{if } H_j = 1/2, \\ K_j \int_{\mathbb{R}} \dfrac{g(x_1, \ldots, x_{j-1}, t, x_{j+1}, \ldots, x_d) \, dt}{|x_j - t|^{3/2 - H_j}} & \text{if } 1/2 < H_j < 1, \end{cases}$$

$$\tag{8.5}$$

where
$$\varepsilon^{(j)} = (0, 0, \ldots, 1, \ldots, 0), \quad \text{the } j\text{th unit vector.}$$
Then define
$$M_H f(x) := m_1(m_2(\cdots(m_{d-1}(m_d f))\cdots))(x), \quad f \in \mathcal{S}(\mathbb{R}^d).$$
Note that if $f(x) = f_1(x) \cdots f_d(x_d) =: (f_1 \otimes \cdots \otimes f_d)(x)$ is a tensor product, then
$$M_H f(x) = \prod_{j=1}^{d} (M_{H_j} f_j)(x_j), \tag{8.6}$$
where
$$M_{H_j} f_j(x_j) = \begin{cases} K_j \displaystyle\int_{\mathbb{R}} \dfrac{f_j(x_j - t) - f_j(x_j)}{|t|^{3/2 - H_j}} \, dt & \text{if } 0 < H_j < 1/2 \\ f_j(x_j) & \text{if } H_j = 1/2 \\ K_j \displaystyle\int_{\mathbb{R}} \dfrac{f_j(t) \, dt}{|t - x_j|^{3/2 - H_j}} & \text{if } 1/2 < H_j < 1 \end{cases}$$

Therefore, if
$$\mathcal{F}g(\xi) := \hat{g}(\xi) := \int_{\mathbb{R}^d} e^{-ix\cdot\xi} g(x) \, dx, \quad \xi = (\xi_1, \ldots, \xi_d) \in \mathbb{R}^d, \tag{8.7}$$
denotes the Fourier transform of g, by (8.6) we have
$$\widehat{M_H f}(\xi) = \prod_{j=1}^{d} \widehat{M_{H_j} f_j}(\xi_j) = \prod_{j=1}^{d} |\xi_j|^{1/2 - H_j} \hat{f}_j(\xi_j) \tag{8.8}$$
and
$$\widehat{M_H^{-1} f}(\xi) = \left(\prod_{j=1}^{d} |\xi_j|^{1/2 - H_j} \right)^{-1} \hat{f}(\xi).$$

We now construct d-parameter fBm $B^{(H)}(x)$ with Hurst parameter $H = (H_1, \ldots, H_d) \in (0, 1)^d$ as follows. First, define
$$\widetilde{B}^{(H)}(x) = \widetilde{B}^{(H)}(x, \omega) = \langle \omega, M_H(I_{[0,x]}(\cdot)) \rangle \tag{8.9}$$
with $I_{[0,x]}(\cdot)$ as in (8.3) and (8.4). Then $\widetilde{B}^{(H)}(x)$ is a Gaussian process with mean
$$E[\widetilde{B}^{(H)}(x)] = \widetilde{B}^{(H)}(0) = 0 \tag{8.10}$$
and covariance (using (8.6) and [89, (1.13)])

$$E[\widetilde{B}^{(H)}(x)\widetilde{B}^{(H)}(y)] = \int_{\mathbb{R}^d} M_H(I_{[0,x]}(z))M_H(I_{[0,y]}(z))\,dz$$

$$= \int_{\mathbb{R}^d} \prod_{i=1}^{d} M_{H_i} I_{[0,x_i]}(z_i) \cdot \prod_{j=1}^{d} M_{H_j} I_{[0,y_j]}(z_j)\,dz_1 \cdots dz_d$$

$$= \prod_{j=1}^{d} \int_{\mathbb{R}} M_{H_j} I_{[0,x_j]}(t) \cdot M_{H_j} I_{[0,y_j]}(t)\,dt$$

$$= (\tfrac{1}{2})^d \prod_{j=1}^{d} \left(|x_j|^{2H_j} + |y_j|^{2H_j} - |x_j - y_j|^{2H_j}\right), \quad x, y \in \mathbb{R}^d. \quad (8.11)$$

By Kolmogorov's continuity theorem we get that $\{\widetilde{B}^{(H)}(x)\}$ has a continuous version, which we denote by $\{B^{(H)}(x)\}$. From (8.10) and (8.11) we conclude that $B^{(H)}(x)$ is a d-parameter fBm with Hurst parameter $H = (H_1, \ldots, H_d) \in (0,1)^d$.

If f is a simple deterministic function of the form

$$f(x) = \sum_{j=1}^{N} a_j I_{[0,y^{(j)}]}(x), \quad x \in \mathbb{R}^d,$$

for some $a_j \in \mathbb{R}, y^{(j)} \in \mathbb{R}^d$, and $N \in \mathbb{N}$, then we define its integral with respect to $B^{(H)}$ by

$$\int_{\mathbb{R}^d} f(x)\,dB^{(H)}(x) = \sum_{j=1}^{N} a_j B^{(H)}(y^{(j)}).$$

Note that by (8.9) this coincides with $<\omega, M_H f>$, and we have the isometry

$$E[(\int_{\mathbb{R}^d} f(x)\,dB^{(H)}(x))^2] = E[<\omega, M_H f>^2] = \|M_H f\|_{L^2(\mathbb{R}^d)}^2.$$

We can extend the definition of this integral to all $g \in L_H^2(\mathbb{R}^d)$, where

$$L_H^2(\mathbb{R}^d) = \{g : \mathbb{R}^d \to \mathbb{R}; \|g\|_{L_H^2(\mathbb{R}^d)} := \|M_H g\|_{L^2(\mathbb{R}^d)} < \infty\},$$

by setting

$$\int_{\mathbb{R}^d} g(x)\,dB^{(H)}(x) := \langle \omega, M_H g \rangle \quad \forall g \in L_H^2(\mathbb{R}^d). \quad (8.12)$$

Moreover, if $f, g \in L_H^2(\mathbb{R}^d)$, then we have the isometry

$$E\Big[\Big(\int_{\mathbb{R}^d} f(x)\,dB^{(H)}(x)\Big)\Big(\int_{\mathbb{R}^d} g(x)\,dB^{(H)}(x)\Big)\Big]$$
$$= E[\langle \omega, M_H f\rangle\langle \omega, M_H g\rangle]$$
$$= \langle M_H f, M_H g\rangle_{L^2(\mathbb{R}^d)} = \langle f, g\rangle_{L_H^2(\mathbb{R}^d)}. \quad (8.13)$$

8.2 Multiparameter fractional white noise calculus

With the processes $B^{(H)}(x)$ constructed in Section 8.1 as a starting point, we proceed to develop a d-parameter white noise theory. Let

$$h_n(t) = (-1)^n e^{t^2/2} \frac{d^n}{dt^n}(e^{-t^2/2}), \qquad n = 0, 1, 2, \ldots, \ t \in \mathbb{R},$$

be the , *Hermite polynomials* and let

$$\xi_n(t) = \pi^{-1/4}\bigl[(n-1)!\bigr]^{-1/2} h_{n-1}(\sqrt{2}t) e^{-t^2/2}, \qquad n = 1, 2, \ldots, \ t \in \mathbb{R}$$

be the *Hermite functions*.

If $\alpha = (\alpha_1, \ldots, \alpha_d) \in \mathbb{N}^d$ (with $\mathbb{N} = \{1, 2, \ldots\}$) and $x = (x_1, \ldots, x_d) \in \mathbb{R}^d$, define

$$\eta_\alpha(x) = \xi_{\alpha_1}(x_1) \cdots \xi_{\alpha_d}(x_d) = (\xi_{\alpha_1} \otimes \cdots \otimes \xi_{\alpha_d})(x)$$

and

$$e_\alpha(x) = \bigl(M_{H_{\alpha_1}}^{-1} \xi_{\alpha_1}\bigr)(x_1) \cdots \bigl(M_{H_{\alpha_d}}^{-1} \xi_{\alpha_d}\bigr)(x_d) = (M_H^{-1} \eta_\alpha)(x). \qquad (8.14)$$

Let $\{\alpha^{(i)}\}_{i=1}^\infty$ be a fixed ordering of $\mathbb{N}^d$ with the property that, with $|\alpha^{(i)}| = \alpha_1^{(i)} + \cdots + \alpha_d^{(i)}$,

$$i < j \Rightarrow |\alpha^{(i)}| \leq |\alpha^{(j)}|.$$

This implies that there exists a constant $C < \infty$ such that

$$|\alpha^{(k)}| \leq Ck \qquad \forall k. \qquad (8.15)$$

With a slight abuse of notation let us write

$$\eta_n(x) := \eta_{\alpha^{(n)}}(x) = M_H e_n(x)$$

and

$$e_n(x) := e_{\alpha^{(n)}}(x) = M_H^{-1} \eta_n(x), \qquad n = 1, 2, \ldots$$

Now let $\mathcal{J} = (\mathbb{N}_0^\mathbb{N})_c$ denote the set of all finite sequences $\alpha = (\alpha_1, \ldots, \alpha_m)$ with $\alpha_j \in \mathbb{N}_0 = \mathbb{N} \cup \{0\}$, $m = 1, 2, \ldots$ Then if $\alpha = (\alpha_1, \ldots, \alpha_m) \in \mathcal{J}$, we define

$$\mathcal{H}_\alpha(\omega) = h_{\alpha_1}(\langle \omega, \eta_1 \rangle) \cdots h_{\alpha_m}(\langle \omega, \eta_m \rangle)$$

In particular, note that by (8.12) we have

$$\mathcal{H}_{\varepsilon^{(i)}}(\omega) = h_1(\langle \omega, \eta_i \rangle) = \langle \omega, \eta_i \rangle = \int_{\mathbb{R}^d} \eta_i(x)\, dB(x)$$

$$= \int_{\mathbb{R}^d} M_H e_i(x)\, dB(x) = \langle \omega, M_H e_i \rangle = \int_{\mathbb{R}^d} e_i(x)\, dB^{(H)}(x) \qquad (8.16)$$

for $i = 1, 2, \ldots$

Note that if $f \in \mathcal{S}(\mathbb{R}^d)$, then $M_H f \in L^2(\mathbb{R}^d)$. Moreover, if $f, g \in \mathcal{S}(\mathbb{R}^d)$, then
$$\langle g, M_H f \rangle_{L^2(\mathbb{R}^d)} = \langle \widehat{g}, \widehat{M_H f} \rangle_{L^2(\mathbb{R}^d)} = \langle M_H g, f \rangle_{L^2(\mathbb{R}^d)}.$$
Therefore, since the action of $\omega \in \Omega = \mathcal{S}'(\mathbb{R}^d)$ extends to $L^2(\mathbb{R}^d)$, we can extend the definition of the operator M_H from $\mathcal{S}(\mathbb{R}^d)$ to $\Omega = \mathcal{S}'(\mathbb{R}^d)$ by setting
$$\langle M_H \omega, f \rangle = \langle \omega, M_H f \rangle, \qquad f \in \mathcal{S}(\mathbb{R}), \ \omega \in \mathcal{S}'(\mathbb{R}).$$

Example 8.2.1. The chaos expansion of classical Brownian motion $B(x) \in L^2(\mathbb{P})$ is

$$B(x) = \langle \omega, I_{[0,x]} \rangle = \sum_{k=1}^{\infty} \langle I_{[0,x]}, \eta_k \rangle_{L^2(\mathbb{R}^d)} \langle \omega, \eta_k \rangle$$
$$= \sum_{k=1}^{\infty} \left(\int_0^x \eta_k(y) dy \right) \cdot \mathcal{H}_{\varepsilon^{(k)}}(\omega), \qquad (8.17)$$

where, in general, we put
$$\int_0^x g(y)\, dy = \int_0^{x_d} \cdots \int_0^{x_1} g(y)\, dy_1 \ldots dy_d, \qquad x = (x_1, \ldots, x_d) \in \mathbb{R}^d.$$

Hence, by (8.9) the chaos expansion of fBm $B^{(H)}(x)$ is

$$B^{(H)}(x) = \langle \omega, M_H I_{[0,x]} \rangle \qquad (8.18)$$
$$= \sum_{k=1}^{\infty} \langle M_H I_{[0,x]}, \eta_k \rangle_{L^2(\mathbb{R}^d)} \langle \omega, \eta_k \rangle = \sum_{k=1}^{\infty} \langle I_{[0,x]}, M_H \eta_k \rangle_{L^2(\mathbb{R}^d)} \mathcal{H}_{\varepsilon^{(k)}}(\omega)$$
$$= \sum_{k=1}^{\infty} \left[\int_0^x M_H \eta_k(y)\, dy \right] \mathcal{H}_{\varepsilon^{(k)}}(\omega). \qquad (8.19)$$

Similarly, if $f \in L_H^2(\mathbb{R}^d)$, then by (8.12)

$$\int_{\mathbb{R}} f(x)\, dB^{(H)}(x) = \langle \omega, M_H f \rangle = \sum_{k=1}^{\infty} \langle M_H \eta_k, f \rangle_{L^2(\mathbb{R}^d)} \mathcal{H}_{\varepsilon^{(k)}}(\omega).$$

Next, we define the *d-parameter Hida test function and distribution spaces* $(\mathcal{S})$ and $(\mathcal{S})^*$, respectively.

Definition 8.2.2. *1. For $k = 1, 2, \ldots$, let $(\mathcal{S})^{(k)}$ be the set of $G \in L^2(\mathbb{P})$ with expansion*
$$G(\omega) = \sum_{\alpha} c_\alpha \mathcal{H}_\alpha(\omega)$$
such that
$$\|G\|_k^2 := \sum_{\alpha} \alpha!\, c_\alpha^2 (2\mathbb{N})^{\alpha k} < \infty,$$

where
$$(2\mathbb{N})^\beta = (2 \cdot 1)^{\beta_1}(2 \cdot 2)^{\beta_2} \cdots (2m)^{\beta_m} \quad \text{if } \beta = (\beta_1, \ldots, \beta_m) \in \mathcal{J}.$$

The space of Hida test functions, $(\mathcal{S})$, is defined by
$$(\mathcal{S}) = \bigcap_{k=1}^{\infty} (\mathcal{S})^{(k)},$$
equipped with the projective topology.

2. For $q = 1, 2, \ldots$, let $(\mathcal{S})^{(-q)}$ be the set of all formal expansions
$$G = \sum_\alpha c_\alpha \mathcal{H}_\alpha(\omega)$$
such that
$$\|G\|_q^2 := \sum_\alpha \alpha! \, c_\alpha^2 (2\mathbb{N})^{-q\alpha} < \infty.$$

The space of Hida distributions, $(\mathcal{S})^*$, is defined by
$$(\mathcal{S})^* = \bigcup_{q=1}^{\infty} (\mathcal{S})^{(-q)},$$
equipped with the inductive topology.

Note that with this definition we have
$$(\mathcal{S}) \subset L^2(\mathbb{P}) \subset (\mathcal{S})^*.$$

Example 8.2.3. Define *fractional white noise*, $W^{(H)}(x)$, by
$$W^{(H)}(x) = \sum_{k=1}^{\infty} M_H \eta_k(x) \mathcal{H}_{\varepsilon^{(k)}}(\omega), \quad x \in \mathbb{R}^d. \tag{8.20}$$

Then $W^{(H)}(x) \in (\mathcal{S})^*$ because in this case, by (8.14) and (8.15),
$$\sum_\alpha \alpha! c_\alpha^2 (2\mathbb{N})^{-q\alpha} = \sum_{k=1}^{\infty} (M_H \eta_k)^2(x)(2\mathbb{N})^{-q\varepsilon^{(k)}}$$
$$= \sum_{k=1}^{\infty} (M_{H_{\alpha_1^{(k)}}} \xi_{\alpha_1^{(k)}})^2(x_1) \cdots (M_{H_{\alpha_d^{(k)}}} \xi_{\alpha_d^{(k)}})^2(x_d)(2k)^{-q}$$
$$\leq \sum_{k=1}^{\infty} C_1 \Big(\prod_{j=1}^{d} (\alpha_j^{(k)})^{2/3 - H_{\alpha_j^{(k)}}/2} \Big)^2 (2k)^{-q}$$
$$\leq C_1 \sum_{k=1}^{\infty} (2k)^{4d/3 - q} < \infty$$

for $q > 4d/3 + 1$ (C_1 is a constant). Here we have used the estimate

$$|M_{H_j}\xi_n(t)| \leq C_2 n^{2/3-H_j/2} \quad \forall t, (C_2 \text{ constant}), \tag{8.21}$$

from Section 3 of [89].

Note that from (8.20) and (8.19) we have that

$$\frac{\partial^d}{\partial x_1 \cdots \partial x_d} B^{(H)}(x) = W^{(H)}(x) \quad [\text{in } (\mathcal{S})^*] \text{ for all } x \in \mathbb{R}^d.$$

This justifies the name *fractional white noise* for the process $W^{(H)}(x)$.

Choose $g \in \mathcal{S}(\mathbb{R}^d)$ and let $m_j g$ be as in (8.5). We establish a useful formula for the $L^2(\mathbb{R})$ norm of $m_j g$.

Theorem 8.2.4. *Let f and g be elements in $\mathcal{S}(\mathbb{R})$ (the space of rapidly decreasing smooth functions). If $0 < H_j < 1/2$, then there is a constant κ such that*

$$\int_\mathbb{R} m_j f(x) m_j g(x)\, dx = \kappa \int_\mathbb{R} \int_\mathbb{R} |x-y|^{2H_j} f'(x) g'(y)\, dx\, dy.$$

Proof. From (8.7) and (8.8) we have

$$\mathcal{F}(m_j f)(\xi) = K_j |\xi|^{1/2 - H_j} \hat{f}(\xi).$$

Thus,

$$\int_\mathbb{R} m_j f(x) m_j g(x)\, dx = \frac{1}{2\pi} K_j^2 \int_\mathbb{R} |\xi|^{1-2H_j} \overline{\hat{f}(\xi)} \hat{g}(\xi)\, d\xi.$$

For $\alpha > 0$ define

$$I^\alpha \phi(x) = \gamma_\alpha \int_\mathbb{R} \frac{\phi(t)}{|t-x|^{1-\alpha}}\, dt,$$

where $\gamma_\alpha = 2\Gamma(\alpha) \cos(\alpha \pi /2)$. By [206], we have

$$\mathcal{F}(I^\alpha \phi)(\xi) = |\xi|^{-\alpha} \hat{\phi}(\xi).$$

Therefore,

$$\int_\mathbb{R} f'(x) I^\alpha g'(x)\, dx = \frac{1}{2\pi} \int_\mathbb{R} |\xi|^{2-\alpha} \overline{\hat{f}(\xi)} \hat{g}(\xi)\, d\xi.$$

Hence,

$$\int_\mathbb{R} m_j f(x) m_j g(x)\, dx = K_j^2 \int_\mathbb{R} f'(x) I^\alpha g'(x)\, dx$$

if $1 - 2H_j = 2 - \alpha$. That is,

$$\alpha = 1 + 2H_j.$$

When the above identity is true, we have

$$\int_{\mathbb{R}} f'(x) I^\alpha g'(x)\, dx = \kappa \int_{\mathbb{R}^2} |x-y|^{2H_j} f'(x) g'(y)\, dx\, dy,$$

where
$$\kappa = \gamma_\alpha K_j^2.$$

□

Remark 8.2.5. It is easy to extend the identity to more general functions.

8.3 The stochastic Poisson equation

We now illustrate the use of the theory above by solving the Poisson equation with fractional white noise heat source. Let $D \subset \mathbb{R}^d$ be a given bounded domain with smooth (C^∞) boundary. We want to find $U(\cdot): \bar{D} \to (\mathcal{S})^*$ such that

$$\begin{cases} \Delta U(x) = -W^{(H)}(x) & \text{for } x \in D, \\ U(x) = 0 & \text{for } x \in \partial D, \end{cases} \quad (8.22)$$

where $\Delta = 1/2 \sum_{i=1}^d \partial^2/\partial x_i^2$ is the Laplacian operator, and such that U is continuous on the closure $\bar{D}$ of D.

From classical potential theory we are led to the solution candidate

$$U(x) = \int_D G(x,y) W^{(H)}(y)\, dy = \int_D G(x,y)\, dB^{(H)}(y), \quad (8.23)$$

where G is the classical Green function for the Dirichlet Laplacian.

We first verify that $U(x) \in (\mathcal{S})^*$ for all x. To this end, consider the expansion of $U(x)$:

$$U(x) = \int_D G(x,y) \sum_{k=1}^\infty M_H \eta_k(y) \mathcal{H}_{\varepsilon(k)}(\omega)\, dy$$
$$= \sum_{k=1}^\infty a_k(x) \mathcal{H}_{\varepsilon(k)}(\omega),$$

where
$$a_k(x) = \int_D G(x,y) M_H \eta_k(y)\, dy.$$

By the estimate (8.21) we have

$$|a_k(x)| \leq C_3 k^{2d/3} \int_D G(x,y)\, dy \leq C_4 k^{2d/3}, \quad (8.24)$$

and therefore,

$$\sum_{k=1}^{\infty} a_k^2(x)(2\mathbb{N})^{-q\varepsilon_k} \leq C_4^2 \sum_{k=1}^{\infty} (2k)^{4d/3}(2k)^{-q} < \infty$$

for $q > 4d/3 + 1$. This proves that $U(x) \in (\mathcal{S})^*$, and the same estimate gives that $U : \bar{D} \to (\mathcal{S})^*$ is continuous.

To prove $\Delta U(x) = -W^{(H)}(x)$, introduce

$$U_n(x) = \sum_{k=1}^{n} a_k(x) \mathcal{H}_{\varepsilon(k)}(\omega).$$

We have

$$\Delta U_n(x) = \sum_{k=1}^{n} \Delta a_k(x) \mathcal{H}_{\varepsilon(k)}(\omega) = -\sum_{k=1}^{n} M_H \eta_k(x) \mathcal{H}_{\varepsilon(k)}(\omega).$$

By the estimate in (8.21), we see that $\Delta U_n(x)$ converges to $-W^{(H)}(x)$ in $(\mathcal{S})^*$ uniformly for $x \in \bar{D}$. On the other hand, (8.24) implies that $U_n(x)$ converges to $U(x)$ in $(\mathcal{S})^*$ uniformly for $x \in \bar{D}$. Therefore, we conclude $\Delta U(x) = -W^{(H)}(x)$. Thus we have the following:

Theorem 8.3.1. *Let $H = (H_1, \ldots, H_d) \in (0,1)^d$. The stochastic fractional Poisson equation (8.22) has a unique solution $U(x) \in (\mathcal{S})^*$ given by*

$$U(x) = \int_D G(x,y) \, dB^{(H)}(y), \tag{8.25}$$

where $G(x,y)$ is the classical Green function for the Laplacian.

Next we discuss when this solution $U(x)$ belongs to $L^2(\mathbb{P})$.

Theorem 8.3.2. *Let $H = (H_1, \ldots, H_d) \in (0,1)^d$ and $\mathbb{H} := \{i, \, H_i < 1/2\}$. Suppose that $\mathbb{H}$ contains at most 1 element and*

$$\sum_{i=1}^{d} H_i > d - 2. \tag{8.26}$$

Then the solution $U(x)$ given by (8.25) belongs to $L^2(\mathbb{P})$ for all x.

Proof. By (8.23) and (8.13) we have

$$E[U^2(x)] = \langle G(x,\cdot), G(x,\cdot) \rangle_{L^2_H(\mathbb{R}^d)}$$

$$= \langle M_H G(x,\cdot), M_H G(x,\cdot) \rangle_{L^2(\mathbb{R}^d)} = \int_{\mathbb{R}^d} [M_H G(x,y)]^2 \, dy, \tag{8.27}$$

where the operator M_H acts on y, and we have extended the function $G(x,\cdot)$, and $M_H G(x,\cdot)$, to $\mathbb{R}^d$ by defining it to be zero outside D. Without loss of generality we can assume that

$$H_1 < 1/2 \quad \text{and} \quad H_i > 1/2 \quad \text{for } i > 1.$$

Since ∂D is smooth, there exists a constant C such that

$$\left|\frac{\partial^k}{\partial y_1 \cdots \partial y_k} G(x,y)\right| \le C \frac{\prod_{i=1}^k |x_i - y_i|}{|x-y|^{d+2k-2}}.$$

Recall that if $1/2 < H_j < 1$,

$$\int_R m_j f(x)^2 \, dx = c_j \int_R \int_R f(x) |x-y|^{2H_j - 2} f(y) \, dx \, dy$$

Hence, by (8.27) and Theorem 8.3.1

$$E[U^2(x)] \le C \int_D \int_D \frac{|x_1 - y_1|}{|x-y|^d} \phi(y,z) \frac{|x_1 - z_1|}{|x-z|^d} \, dy \, dz, \tag{8.28}$$

where

$$\phi(y,z) = |y_1 - z_1|^{2H_1} \prod_{i=2}^d |y_i - z_i|^{2H_i - 2}.$$

Let $z' = x - z$ and $y' = x - y$. Since D is bounded, there exists a positive constant R such that, using (8.28),

$$E[U^2(x)] \le C \int_{-R}^R \cdots \int_{-R}^R \frac{|y_1'|}{|y'|^d} \phi(y', z') \frac{|z_1'|}{|z'|^d} \, dy' \, dz'$$

$$= C \int_{-R}^R \cdots \int_{-R}^R \frac{|y_1||z_1||y_1 - z_1|^{2H_1}}{|y|^d} \frac{\prod_{i=2}^d |y_i - z_i|^{2H_i - 2}}{|z|^d} \, dy \, dz. \tag{8.29}$$

Next, we are going to show that the integral above is finite.

For notational simplicity, we assume that $R = 1$. For any $a > 0, b > 0, d > 0$, we claim that there is a constant c, independent of a and b, such that

$$\int_{-1}^1 \int_{-1}^1 \frac{|y||z||y-z|^{2H_1}}{(y^2 + a)^{d/2}(z^2 + b)^{d/2}} \, dy \, dz \le c \frac{a^{H_1} + b^{H_1}}{a^{d/2 - 1} b^{d/2 - 1}}. \tag{8.30}$$

Indeed,

$$\int_{-1}^1 \int_{-1}^1 \frac{|y||z||y-z|^{2H_1}}{(y^2+a)^{d/2}(z^2+b)^{d/2}} \, dy \, dz$$

$$\le c \left\{ \int_{-1}^1 dy \int_{-1}^1 \frac{|y||z|(|y|^{2H_1} + |z|^{2H_1})}{(y^2+a)^{d/2}(z^2+b)^{d/2}} \, dz \right\}$$

$$\le c \left\{ \int_0^1 dy \, \frac{y^{2H_1 + 1}}{(y^2+a)^{d/2}} \int_0^1 \frac{z}{(z^2+b)^{d/2}} \, dz \right.$$

$$+ \int_0^1 dz \frac{z^{2H_1+1}}{(z^2+b)^{d/2}} \int_0^1 \frac{y}{(y^2+a)^{d/2}} dy \bigg\}$$

$$\leq c \int_0^1 \frac{y^{2H_1+1}}{(y^2+a)^{d/2}} [b^{1-d/2} - (1+b)^{1-d/2}] dy$$

$$+ c \int_0^1 \frac{z^{2H_1+1}}{(z^2+b)^{d/2}} [a^{1-d/2} - (1+a)^{1-d/2}] dz$$

$$\leq cb^{1-d/2} \int_0^1 \frac{y}{(y^2+a)^{d/2-H_1}} dy + ca^{1-d/2} \int_0^1 \frac{z}{(z^2+b)^{d/2-H_1}} dz$$

$$\leq cb^{1-d/2} a^{1-d/2+H_1} + a^{1-d/2} b^{1-d/2+H_1} = c \frac{a^{H_1} + b^{H_1}}{a^{d/2-1} b^{d/2-1}}.$$

Applying (8.30) for $a = \sum_{i=2}^d y_i^2, b = \sum_{i=2}^d z_i^2$, we have

$$\int_{-1}^1 \cdots \int_{-1}^1 \frac{|y_1||z_1||y_1-z_1|^{2H_1}}{|y|^d} \frac{\prod_{i=2}^d |y_i - z_i|^{2H_i-2}}{|z|^d} dy\, dz$$

$$= \int_{-1}^1 \cdots \int_{-1}^1 \prod_{i=2}^d |y_i - z_i|^{2H_i-2} dy_2 \cdots dy_d\, dz_2 \cdots dz_d$$

$$\cdot \int_{-1}^1 \int_{-1}^1 \frac{|y_1||z_1||y_1-z_1|^{2H_1}}{|y|^d |z|^d} dy_1\, dz_1$$

$$\leq c \int_{-1}^1 \cdots \int_{-1}^1 \frac{[(\sum_{i=2}^d y_i^2)^{H_1} + (\sum_{i=2}^d z_i^2)^{H_1}]}{(\sum_{i=2}^d y_i^2)^{d/2-1}(\sum_{i=2}^d z_i^2)^{d/2-1}}$$

$$\cdot \prod_{i=2}^d |y_i - z_i|^{2H_i-2} dy_2 \cdots dy_d\, dz_2 \cdots dz_d$$

$$= \int_{-1}^1 \cdots \int_{-1}^1 \frac{\prod_{i=2}^d |y_i - z_i|^{2H_i-2}}{(\sum_{i=2}^d y_i^2)^{d/2-1-H_1}(\sum_{i=2}^d z_i^2)^{d/2-1}} dy_2 \cdots dy_d\, dz_2 \cdots dz_d$$

$$+ \int_{-1}^1 \cdots \int_{-1}^1 \frac{\prod_{i=2}^d |y_i - z_i|^{2H_i-2}}{(\sum_{i=2}^d y_i^2)^{d/2-1}(\sum_{i=2}^d z_i^2)^{d/2-1-H_1}} dy_2 \cdots dy_d\, dz_2 \cdots dz_d$$

$$=: I + II.$$

We now prove that both I and II are finite. By symmetry, we only look at I. For any choice of positive numbers $\alpha_i > 0$ with $\sum_{i=2}^d \alpha_i = 1$ and positive numbers $\beta_i > 0$ with $\sum_{i=2}^d \beta_i = 1$, we have

$$I \leq \int_{-1}^1 \cdots \int_{-1}^1 \frac{\prod_{i=2}^d |y_i - z_i|^{2H_i-2}}{\prod_{i=2}^d |y_i|^{\alpha_i(d-2-2H_1)} |z_i|^{\beta_i(d-2)}} dy_2 \cdots dy_d\, dz_2 \cdots dz_d.$$

Therefore, I is finite if the following conditions are met:

8.3 The stochastic Poisson equation 193

$$\alpha_i(d-2-2H_1) < 1, \qquad i = 2, \cdots, d;$$
$$\beta_i(d-2) < 1, \qquad i = 2, \cdots, d;$$
$$\alpha_i(d-2-2H_1) + \beta_i(d-2) - 2H_i + 2 < 2, \qquad i = 2, \cdots, d. \quad (8.31)$$

Adding these inequalities in (8.31), we see that it is sufficient to have

$$2(d-2) - 2H_1 < \sum_{i=2}^{d} 2H_i,$$

namely,

$$\sum_{i=1}^{d} H_i > d - 2.$$

This completes the proof. □

Remark 8.3.3. It is natural to ask if condition (8.26) is also *necessary* to have $U(x) \in L^2(\mathbb{P})$. Now we give a discussion.

We need the following. For any $a > 0, b > 0, d > 0$ (a and b bounded), there is a constant c, independent of a and b, such that

$$\int_{-1}^{1} \int_{-1}^{1} \frac{|y||z||y-z|^{2H_1}}{(y^2+a)^{d/2}(z^2+b)^{d/2}} \, dy \, dz \geq c \frac{a^{H_1}}{a^{d/2-1}b^{d/2-1}}$$

and

$$\int_{-1}^{1} \int_{-1}^{1} \frac{|y||z||y-z|^{2H_1}}{(y^2+a)^{d/2}(z^2+b)^{d/2}} \, dy \, dz \geq c \frac{b^{H_1}}{a^{d/2-1}b^{d/2-1}} \quad (8.32)$$

In fact,

$$\int_{-1}^{1} \int_{-1}^{1} \frac{|y||z||y-z|^{2H_1}}{(y^2+a)^{d/2}(z^2+b)^{d/2}} \, dy \, dz$$

$$\geq \int_{0}^{1} \int_{-1}^{1} \frac{y|z||y-z|^{2H_1}}{(y^2+a)^{d/2}(z^2+b)^{d/2}} \, dy \, dz$$

$$\geq \int_{0}^{1} \int_{-1}^{0} \frac{y|z|^{2H_1+1}}{(y^2+a)^{d/2}(z^2+b)^{d/2}} \, dy \, dz$$

$$\geq c \int_{0}^{1} \frac{y}{(y^2+a)^{d/2}} \, dy \int_{0}^{1} \frac{z^{2H_1+1}}{(z^2+b)^{d/2}} \, dz$$

$$= c a^{-d/2+1} b^{-d/2+1+H_1} \int_{0}^{1/\sqrt{a}} \frac{u}{(u^2+1)^{d/2}} \, du \int_{0}^{1/\sqrt{b}} \frac{v^{2H_1+1}}{(v^2+1)^{d/2}} \, dv$$

$$\geq c \frac{b^{H_1}}{a^{d/2-1}b^{d/2-1}}.$$

Let us consider (8.29) in the case when $d = k = 2$, namely,

$$\int_{-1}^{1}\cdots\int_{-1}^{1}\frac{|y_1||z_1||y_1-z_1|^{2H_1}}{|y|^{d+2}}\frac{|y_2||z_2||y_2-z_2|^{2H_2}}{|z|^{d+2}}\,dy\,dz,$$

where $0 < H_1, H_2 < 1/2$. Applying (8.32) for $a = y_2^2, b = z_2^2$, we have

$$\int_{-1}^{1}\cdots\int_{-1}^{1}\frac{|y_1||z_1||y_1-z_1|^{2H_1}}{|y|^{d+2}}\frac{|y_2||z_2||y_2-z_2|^{2H_2}}{|z|^{d+2}}\,dy\,dz$$
$$\geq c\int_{-1}^{1}\int_{-1}^{1}\frac{z_2^{H_1}|y_2||z_2||y_2-z_2|^{2H_2}}{y_2^2 z_2^2}\,dy_2\,dz_2$$

which is divergent. Thus we conjecture that $U(t,x)$ is in L^2 only if at most one Hurst exponent is less than $1/2$.

8.4 The linear heat equation

In this section we consider the linear stochastic fractional heat equation

$$\begin{cases} \dfrac{\partial U}{\partial t}(t,x) = \dfrac{1}{2}\Delta U(t,x) + W^{(H)}(t,x), & t \in (0,\infty),\ x \in D \subset \mathbb{R}^d; \\ U(0,x) = 0, & x \in D; \\ U(t,x) = 0, & t \geq 0,\ x \in \partial D. \end{cases} \qquad (8.33)$$

Here $W^{(H)}(t,x)$ is the fractional white noise with Hurst parameter $H = (H_0, H_1, \ldots, H_d) \in (0,1)^{d+1}$, $\Delta = \sum_{i=1}^{d} \partial^2/\partial x_i^2$ is the Laplace operator, and $D \subset \mathbb{R}^d$ is a bounded open set with smooth boundary ∂D. We are looking for a solution $U : [0,\infty) \times \bar{D} \to (\mathcal{S})^*$ that is continuously differentiable in (t,x) and twice continuously differentiable in x, i.e., belongs to $C^{1,2}((0,\infty) \times D; (\mathcal{S})^*)$, and which satisfies (8.33) in the strong sense [as an $(\mathcal{S})^*$-valued function]. Based on the corresponding solution in the deterministic case [with $W^{(H)}(t,x)$ replaced by a bounded deterministic function], it is natural to guess that the solution will be

$$U(t,x) = \int_0^t \int_D W^{(H)}(s,y) G_{t-s}(x,y)\,dy\,ds \qquad (8.34)$$

where $G_{t-s}(x,y)$ is the Green function for the heat operator $\partial/\partial t - 1/2\Delta$. It is well known [70] that G is smooth in $(0,\infty) \times D$ and that

$$|G_u(x,y)| \sim u^{-d/2}\exp\left(-\frac{|x-y|^2}{\delta u}\right) \quad \text{in } (0,\infty) \times D,$$

and

$$\left|\frac{\partial G_u(x,y)}{\partial y_i}\right| \sim u^{-d/2-1}|x_i - y_i|\exp\left(-\frac{|x-y|^2}{\delta u}\right) \quad \text{in } (0,\infty) \times D,$$

where the notation $X \sim Y$ means that
$$\frac{1}{C}X \leq Y \leq CX \quad \text{in } (0,\infty) \times D$$
for some positive constant $C < \infty$ depending only on D. We use this to verify that $U(t,x) \in (\mathcal{S})^*$ for all $(t,x) \in [0,\infty) \times \bar{D}$. Using (8.20), we see that the expansion of $U(t,x)$ is
$$U(t,x) = \int_0^t \int_D G_{t-s}(x,y)) \sum_{k=1}^\infty M_H \eta_k(s,y) \mathcal{H}_{\varepsilon(k)}(\omega) \, dy \, ds$$
$$= \sum_{k=1}^\infty b_k(t,x) \mathcal{H}_{\varepsilon(k)}(\omega), \tag{8.35}$$
where
$$b_k(t,x) = b_{\varepsilon(k)}(t,x) = \int_0^t \int_D G_{t-s}(x,y) M_H \eta_k(s,y) \, dy \, ds.$$
In the following C denotes a generic constant, not necessarily the same from place to place. From (8.21) we obtain that
$$|b_k(t,x)| \leq Ck^{2(d+1)/3} \int_0^t \int_D G_{t-s}(x,y) \, dy \, ds$$
$$= Ck^{2(d+1)/3} t. \tag{8.36}$$
Therefore,
$$\sum_{k=1}^\infty b_k^2(t,x)(2\mathbb{N})^{-q\varepsilon^{(k)}} \leq C(t) \sum_{k=1}^\infty k^{4(d+1)/3}(2k)^{-q} < \infty$$
for $q > 4(d+1)/3 + 1$.

Hence $U(t,x) \in (\mathcal{S})^{-q^*}$ for all $q > 4(d+1)/3 + 1$, for all t,x. In fact, this estimate also shows that $U(t,x)$ is uniformly continuous as a function from $[0,T] \times \bar{D}$ into $(\mathcal{S})^*$ for any $T < \infty$. Moreover, by the properties of $G_{t-s}(x,y)$, we get from (8.34) that
$$\frac{\partial U}{\partial t}(t,x) - \Delta U(t,x)$$
$$= \int_0^t \int_D W^{(H)}(s,y) \left(\frac{\partial}{\partial t} - \Delta\right) G_{t-s}(x,y) \, dy \, ds + W^{(H)}(t,x)$$
$$= W^{(H)}(t,x). \tag{8.37}$$
So $U(t,x)$ satisfies (8.33).

Next we study the L^2-integrability of $U(t,x)$. In the *standard* white noise case ($H_i = 1/2$ for all i) the same solution formula (8.34) holds. In this case we see that the solution $U(t,x)$ belongs to $L^2(\mathbb{P})$ if and only if

$$E[U^2(t,x)] = \int_0^t \int_D G_{t-s}^2(x,y)\,dy\,ds < \infty.$$

Now, if $D \subset (-1/2R, 1/2R)^d$ and we put $F = [-R, R]^d$,

$$\int_0^t \int_D G_{t-s}^2(x,y)\,ds\,dy \sim \int_0^t \int_D s^{-d} \exp\left(-\frac{2y^2}{\delta s}\right) dy\,ds$$

$$\sim \int_0^t \left[\int_{F/\sqrt{s}} s^{-d/2} \exp\left(-\frac{2z^2}{\delta}\right) dz\right] ds.$$

Hence, if $H_i = 1/2$ for all $i = 0, 1, \ldots, d$, we have

$$E[U^2(t,x)] < \infty \iff d = 1. \tag{8.38}$$

Now, consider the *fractional* case. Assume $1/2 < H_0 < 1$, and because of (8.38) we may assume that at most one of the indices: $H_1, H_2, \ldots, H_d$ is less than $1/2$, say, $0 < H_1 < 1/2$. Then

$$E[U^2(t,x)] = \int (M_H G_{t-\cdot}(x,\cdot)(s,y))^2 ds\,dy \leq C \int_0^t \int_0^t \int_D \int_D \left|\frac{\partial G_{t-s}(x,y)}{\partial y_1}\right|$$
$$\cdot \left|\frac{\partial G_{t-r}(x,z)}{\partial z_1}\right| |r-s|^{2H_0-2} |y_1-z_1|^{2H_1}$$
$$\cdot \prod_{i=2}^d |y_i-z_i|^{2H_i-2} dy_1 \cdots dy_d\,dz_1 \cdots dz_d\,dr\,ds$$
$$\sim \int_0^t \int_0^t \int_D \int_D r^{-d/2-1} s^{-d/2-1} |x_1-y_1||x_1-z_1| \exp\left(-\frac{|x-y|^2}{\delta r}\right)$$
$$\cdot \exp\left(-\frac{|x-z|^2}{\delta s}\right) |r-s|^{2H_0-2} |y_1-z_1|^{2H_1}$$
$$\cdot \prod_{i=2}^d |y_i-z_i|^{2H_i-2} dy_1 \cdots dy_d\,dz_1 \cdots dz_d\,dr\,ds. \tag{8.39}$$

Note that

$$\int_{-1/2R}^{1/2R} \int_{-1/2R}^{1/2R} |x_1-y_1||x_1-z_1| \exp\left(-\frac{|x_1-y_1|^2}{\delta r} - \frac{|x_1-z_1|^2}{\delta s}\right)$$
$$\cdot |y_1-z_1|^{2H_1} dy_1\,dz_1$$
$$\leq Crs(r^{H_1} + s^{H_1}). \tag{8.40}$$

Using (see Inequality (2.1) of [161])

$$\int_R \int_R |f(x)||g(y)||x-y|^{2H-2} dx\,dy \leq C\|f\|_{L^{1/H}} \|g\|_{L^{1/H}}, \tag{8.41}$$

we have

$$\prod_{i=2}^{d} \int_{-1/2R}^{1/2R} \int_{-1/2R}^{1/2R} \exp\left(-\frac{|x_i - y_i|^2}{\delta r} - \frac{|x_i - z_i|^2}{\delta s}\right) |y_i - z_i|^{2H_i - 2} dy_i \, dz_i$$

$$\leq C \prod_{i=2}^{d} \left\{ \left[\int_{-1/2R}^{1/2R} \exp\left(-\frac{|x_i - y_i|^2}{H_i \delta r}\right) dy_i \right]^{H_i} \right.$$

$$\left. \cdot \left[\int_{-1/2R}^{1/2R} \exp\left(-\frac{|x_i - z_i|^2}{H_i \delta s}\right) dz_i \right]^{H_i} \right\}$$

$$\sim (rs)^{1/2 \sum_{i=2}^{d} H_i}. \tag{8.42}$$

Substituting (8.42) into (8.39), we have

$$E[U^2(t,x)] \leq C \int_0^t \int_0^t (rs)^{-d/2 + 1/2 \sum_{i=2}^{d} H_i} (r^{H_1} + s^{H_1})$$
$$\cdot |r - s|^{2H_0 - 2} \, dr \, ds < \infty$$

if $d/2 - 1/2 \sum_{i=2}^{d} H_i < 1$ and $2 - 2H_0 + 2(d/2 - \sum_{i=2}^{d} 1/2 H_i) - H_1 < 2$. From this we obtain that

$$E[U^2(t,x)] < \infty \qquad \text{if } [(2H_0 + H_1) \wedge 2] + \sum_{i=2}^{d} H_i > d.$$

Now let $1/2 < H_i < 1$ for all $1 \leq i \leq d$. Then

$$E[U^2(t,x)] = \int (M_H G_{t-\cdot}(x, \cdot)(s, y))^2 \, ds \, dy$$

$$\sim \int_0^t \int_0^t \int_D \int_D |G_{t-s}(x,y) G_{t-r}(x,z)| \cdot |r - s|^{2H_0 - 2}$$
$$\cdot \prod_{i=1}^{d} |y_i - z_i|^{2H_i - 2} \, dy_1 \cdots dy_d \, dz_1 \cdots dz_d \, dr \, ds$$

$$\sim \int_0^t \int_0^t \int_D \int_D r^{-d/2} s^{-d/2} \exp\left(-\frac{|x-y|^2}{\delta r}\right) \exp\left(-\frac{|x-z|^2}{\delta s}\right)$$
$$\cdot |r - s|^{2H_0 - 2}$$
$$\cdot \prod_{i=1}^{d} |y_i - z_i|^{2H_i - 2} \, dy_1 \cdots dy_d \, dz_1 \cdots dz_d \, dr \, ds. \tag{8.43}$$

By (8.41), we have

$$\prod_{i=1}^{d} \int_{-1/2R}^{1/2R} \int_{-1/2R}^{1/2R} \exp\left(-\frac{|x_i - y_i|^2}{\delta r} - \frac{|x_i - z_i|^2}{\delta s}\right) |y_i - z_i|^{2H_i - 2} dy_i \, dz_i$$

$$\leq C \prod_{i=1}^{d} \left\{ \left[\int_{-1/2R}^{1/2R} \exp\left(-\frac{|x_i - y_i|^2}{H_i \delta r}\right) dy_i \right]^{H_i} \right.$$
$$\left. \cdot \left[\int_{-1/2R}^{1/2R} \exp\left(-\frac{|x_i - z_i|^2}{H_i \delta s}\right) dz_i \right]^{H_i} \right\}$$
$$\sim (rs)^{1/2 \sum_{i=1}^{d} H_i}. \tag{8.44}$$

Substituting (8.44) into (8.43), we have

$$E[U^2(t,x)] \leq C \int_0^t \int_0^t (rs)^{-d/2 + 1/2 \sum_{i=1}^d H_i} |r-s|^{2H_0 - 2} \, dr \, ds < \infty \tag{8.45}$$

if $2H_0 + \sum_{i=1}^d H_i > d$. We summarize what we have proved:

Theorem 8.4.1. *1. For any space dimension d there is a unique strong solution $U(t,x) : [0,\infty) \times D \to (\mathcal{S})^*$ of the fractional heat equation (8.33). The solution is given by*

$$U(t,x) = \int_0^t \int_D W^{(H)}(s,y) G_{t-s}(x,y) \, dy \, ds.$$

It belongs to $C^{1,2}((0,\infty) \times D \to (S)^*) \cap C([0,\infty) \times \bar{D} \to (S)^*)$.
2. If $0 < H_1 < 1/2$, $1/2 < H_i < 1$ for $i = 0, 2, 3, \ldots, d$ and

$$[(2H_0 + H_1) \wedge 2] + \sum_{i=2}^{d} H_i > d,$$

then $U(t,x) \in L^2(\mathbb{P})$ for all $t \geq 0$, $x \in \bar{D}$.
3. If $1/2 < H_i < 1$ for $i = 0, 1, \ldots, d$ and

$$2H_0 + \sum_{i=1}^{d} H_i > d,$$

then $U(t,x) \in L^2(\mathbb{P})$ for all $t \geq 0$, $x \in \bar{D}$.

8.5 The quasi-linear stochastic fractional heat equation

Let $f : \mathbb{R} \to \mathbb{R}$ be a function satisfying

$$|f(x) - f(y)| \leq L|x - y| \quad \text{for all } x, y \in \mathbb{R} \tag{8.46}$$
$$|f(x)| \leq M(1 + |x|) \quad \text{for all } x \in \mathbb{R}, \tag{8.47}$$

8.5 The quasi-linear stochastic fractional heat equation

where L and M are positive constants.

In this section we consider the following quasi-linear equation:
$$\begin{cases} \dfrac{\partial U}{\partial t}(t,x) = \dfrac{1}{2}\Delta U(t,x) + f(U(t,x)) + W^{(H)}(t,x), & t > 0,\ x \in \mathbb{R}^n, \\ U(0,x) = U_0(x), & x \in \mathbb{R}^n, \end{cases} \quad (8.48)$$

where $U_0(x)$ is a given bounded deterministic function on $\mathbb{R}^n$. We say that $U(t,x)$ is a solution of (8.48) if

$$\int_{\mathbb{R}^n} U(t,x)\varphi(x)\,dx - \int_{\mathbb{R}^n} U_0(x)\varphi(x)\,dx$$
$$= \frac{1}{2}\int_0^t\int_{\mathbb{R}^n} U(s,x)\Delta\varphi(x)\,dx\,ds + \int_0^t\int_{\mathbb{R}^n} f(U(s,x))\varphi(x)\,dx\,ds$$
$$+ \int_0^t\int_{\mathbb{R}^n} \varphi(x)\,dB^{(H)}(s,x) \quad (8.49)$$

for all $\varphi \in C_0^\infty(\mathbb{R}^n)$. As in Walsh [234] we can show that $U(t,x)$ solves (8.49) if and only if it satisfies the following integral equation:

$$U(t,x) = \int_{\mathbb{R}^n} U_0(y)G_t(x,y)\,dy + \int_0^t\int_{\mathbb{R}^n} f(U(s,y))G_{t-s}(x,y)\,dy\,ds$$
$$+ \int_0^t\int_{\mathbb{R}^n} G_{t-s}(x,y)\,dB^{(H)}(s,y), \quad (8.50)$$

where

$$G_{t-s}(x,y) = (2\pi(t-s))^{-n/2}\exp\left(-\frac{|x-y|^2}{2(t-s)}\right), \quad s < t,\ x \in \mathbb{R}^n, \quad (8.51)$$

is the Green function for the heat operator $\partial/\partial t - 1/2\Delta$ in $(0,\infty) \times \mathbb{R}^n$.

For the proof of our main result, we need the following two lemmas. Let $0 < \alpha < 1$. Define, for $u > 0$,

$$g(u,y) = \int_{\mathbb{R}} |y-z|^{-\alpha}\frac{1}{\sqrt{u}}\exp\left(-\frac{z^2}{2u}\right)dz.$$

Lemma 8.5.1. *Assume $p > 1/(1-\alpha)$. Then $g(u,y) \le C(1 + u^{-1/2(1-1/p)})$, where C is a constant independent of y and u.*

Proof. In the proof, we will use C to denote a generic constant independent of y and u. First, we have

$$g(u,y) = \int_{|z-y|\le 1} |y-z|^{-\alpha}\frac{1}{\sqrt{u}}\exp\left(-\frac{z^2}{2u}\right)dz$$
$$+ \int_{|z-y|>1} |y-z|^{-\alpha}\frac{1}{\sqrt{u}}\exp\left(-\frac{z^2}{2u}\right)dz.$$

By the Hölder inequality,

$$g(u,y) \leq C\Big\{1 + \Big[\int_{|z-y|\leq 1} |y-z|^{-\alpha p/(p-1)} dz\Big]^{(p-1)/p}$$

$$\cdot \Big[\int_{|z-y|\leq 1} \frac{1}{u^{1/2p}} \exp\Big(-\frac{pz^2}{2u}\Big) dz\Big]^{1/p}\Big\}$$

$$\leq C(1 + u^{-1/2(1-1/p)}) \qquad (8.52)$$

□

Let $F(y_1, y_2, \ldots, y_n)$ denote a function on $\mathbb{R}^n$.

Lemma 8.5.2. *Let $h = (h_1, h_2, \ldots, h_n)$ with $h_i \geq 0$, $1 \leq i \leq n$. Assume that F and all its partial derivatives of first order are integrable with respect to the Lebesgue measure. Then*

$$\int_{\mathbb{R}^n} |F(y-h) - F(y)|\, dy \leq \sum_{i=1}^n \Big(\int_{\mathbb{R}^n} \Big|\frac{\partial F}{\partial y_i}(y_1, y_2, \ldots, y_n)\Big|\, dy\Big) h_i. \qquad (8.53)$$

Proof. Observe that

$$F(y-h) - F(y)$$

$$= \sum_{i=1}^n (F(y_1, \ldots, y_{i-1}, y_i - h_i, y_{i+1} - h_{i+1}, \ldots, y_n - h_n)$$

$$\quad - F(y_1, \ldots, y_{i-1}, y_i, y_{i+1} - h_{i+1}, \ldots, y_n - h_n))$$

$$= \sum_{i=1}^n \int_{y_i - h_i}^{y_i} -\frac{\partial F}{\partial y_i}(y_1, \ldots, y_{i-1}, z, y_{i+1} - h_{i+1}, \ldots, y_n - h_n)\, dz. \qquad (8.54)$$

Integrating equation (8.54), we get

$$\int_{\mathbb{R}^n} |F(y-h) - F(y)|\, dy$$

$$\leq \sum_{i=1}^n \int_{\mathbb{R}^{n-1}} dy_1 \cdots dy_{i-1}\, dy_{i+1} \cdots dy_n$$

$$\cdot \int_{\mathbb{R}} dy_i \int_{y_i - h_i}^{y_i} \Big|\frac{\partial F}{\partial y_i}\Big|(y_1, \cdots, y_{i-1}, z, y_{i+1} - h_{i+1}, \cdots, y_n - h_n)\, dz$$

$$= \sum_{i=1}^n \int_{\mathbb{R}^{n-1}} dy_1 \cdots dy_{i-1}\, dy_{i+1} \cdots dy_n$$

$$\cdot \int_{\mathbb{R}} dz \left| \frac{\partial F}{\partial y_i} \right| (y_1, \cdots, y_{i-1}, z, y_{i+1} - h_{i+1}, \cdots, y_n - h_n) \int_z^{z+h_i} dy_i$$

$$= \sum_{i=1}^n \left[\int_{\mathbb{R}^n} \left| \frac{\partial F}{\partial y_i} (y_1, y_2, \ldots, y_n) \right| dy \right] h_i.$$

□

Our main result is the following:

Theorem 8.5.3. Let $H = (H_0, H_1, \ldots, H_n) \in (1/2, 1)^{n+1}$ with

$$H_i > 1 - \frac{1}{n} \quad \text{for } i = 1, 2, \ldots, n.$$

Then there exists a unique $L^2(\mathbb{P})$-valued random field solution $U(t, x)$, $t \geq 0$, $x \in \mathbb{R}^n$ of (8.48). Moreover, the solution has a jointly continuous version in (t, x) if $H_0 > 3/4$.

Proof. Define

$$V(t, x) = \int_0^t \int_{\mathbb{R}^n} G_{t-s}(x, y) \, dB^{(H)}(s, y).$$

Dividing R into regions $\{z : |z - y| \leq 1\}$ and $\{z : |z - y| > 1\}$, we see that a slight modification of the arguments in Section 8.4 gives $E[V^2(t, x)] < \infty$; so $V(t, x)$ exists as an ordinary random field. The existence of the solution now follows by usual Picard iteration. Define

$$U_0(t, x) = U_0(x)$$

and iteratively

$$U_{j+1}(t, x) = \int_{\mathbb{R}^n} U_0(y) G_t(x, y) \, dy$$
$$+ \int_0^t \int_{\mathbb{R}^n} f(U_j(s, y)) G_{t-s}(x, y) \, dy \, ds + V(t, x)$$

for $j = 0, 1, 2, \ldots$ Then by (8.47) $U_j(t, x) \in L^2(\mathbb{P})$ for all j. We have

$$U_{j+1}(t, x) - U_j(t, x) = \int_0^t \int_{\mathbb{R}^n} [f(U_j(s, y)) - f(U_{j-1}(s, y))] G_{t-s}(x, y) \, dy \, ds,$$

and therefore by (8.46) if $t \in [0, T]$,

$$E\left[|U_{j+1}(t, x) - U_j(t, x)|^2\right]$$
$$\leq LE\left[\left(\int_0^t \int_{\mathbb{R}^n} |U_j(s, y) - U_{j-1}(s, y)| G_{t-s}(x, y) \, dy \, ds\right)^2\right]$$

$$\leq L\left(\int_0^t \int_{\mathbb{R}^n} G_{t-s}(x,y)\,dy\,ds\right)$$
$$\cdot E\left[\int_0^t \int_{\mathbb{R}^n} |U_j(s,y) - U_{j-1}(s,y)|^2 G_{t-s}(x,y)\,dy\,ds\right]$$
$$\leq C_T \int_0^t \sup_y E[|U_j(s,y) - U_{j-1}(s,y)|^2]\,ds$$
$$\leq C_T^j \int_0^t \int_0^{s_1} \cdots \int_0^{s_{j-1}} \sup_y E[|U_1(s,y) - U_0(s,y)|^2]\,ds\,ds_{j-1}\cdots ds_1$$
$$\leq A_T C_T^j \frac{T^j}{(j)!}$$

for some constants A_T, C_T. It follows that the sequence $\{U_j(t,x)\}_{j=1}^\infty$ of random fields converges in $L^2(\mathbb{P})$ to a random field $U(t,x)$. Letting $k \to \infty$ in (8.53), we see that $U(t,x)$ is a solution of (8.48). The uniqueness follows from the Gronwall's inequality. It is not difficult to see that both

$$\int_{\mathbb{R}^n} U_o(y) G_t(x,y)\,dy \quad \text{and} \quad \int_0^t \int_{\mathbb{R}^n} f(U(s,y)) G_{t-s}(x,y)\,dy\,ds$$

are jointly continuous in (t,x). So to finish the proof of the theorem, it suffices to prove that $V(t,x)$ has a jointly continuous version. To this end, consider for $h \in \mathbb{R}$,

$$V(t+h,x) - V(t,x)$$
$$= \int_t^{t+h} \int_{\mathbb{R}^d} G_{t+h-s}(x,y)\,dB^{(H)}(s,y) \tag{8.55}$$
$$+ \int_0^t \int_{\mathbb{R}^n} (G_{t+h-s}(x,y) - G_{t-s}(x,y))\,dB^{(H)}(s,y)$$

By the estimate in (8.45) it follows that

$$E\left[\left|\int_t^{t+h} \int_{\mathbb{R}^n} G_{t+h-s}(x,y)\,dB^{(H)}(s,y)\right|^2\right] \leq C\int_t^{t+h} (u-t)^{2H_0-2}\,du$$
$$\leq Ch^{2H_0-1}. \tag{8.56}$$

To estimate the second term on the right-hand side of (8.55), we proceed as follows:

$$E\left[\left|\int_0^t \int_{\mathbb{R}^n} (G_{t+h-s}(x,y) - G_{t-s}(x,y))\,dB^{(H)}(s,y)\right|^2\right]$$
$$\leq C \int_{\mathbb{R}} \int_{\mathbb{R}} I_{[0,t]}(r) I_{[0,t]}(s) |r-s|^{2H_0-2}$$
$$\cdot \left\{\int_{\mathbb{R}^n} \int_{\mathbb{R}^n} \left[(t+h-r)^{-n/2} \exp\left(-\frac{|x-z|^2}{2(t+h-r)}\right)\right.\right.$$

8.5 The quasi-linear stochastic fractional heat equation

$$-(t-r)^{-n/2}\exp\left(-\frac{|x-z|^2}{2(t-r)}\right)\Bigg]$$

$$\cdot\left[(t+h-s)^{-n/2}\exp\left(-\frac{|x-y|^2}{2(t+h-s)}\right)\right.$$

$$\left.-(t-s)^{-n/2}\exp\left(-\frac{|x-y|^2}{2(t-s)}\right)\right]$$

$$\cdot\prod_{i=1}^{n}|y_i-z_i|^{2H_i-2}\,dy\,dz\bigg\}\,dr\,ds \qquad (8.57)$$

$$\leq C\int_{\mathbb{R}}\int_{\mathbb{R}} I_{[0,t]}(r)I_{[0,t]}(s)|r-s|^{2H_0-2}$$

$$\cdot\left\{\int_{\mathbb{R}^n}\int_{\mathbb{R}^n}\left[(r+h)^{-n/2}\exp\left(-\frac{|z|^2}{2(r+h)}\right)-r^{-n/2}\exp\left(-\frac{|z|^2}{2r}\right)\right]\right.$$

$$\cdot\left[(s+h)^{-n/2}\exp\left(-\frac{|y|^2}{2(s+h)}\right)-s^{-n/2}\exp\left(-\frac{|y|^2}{2s}\right)\right]$$

$$\cdot\prod_{i=1}^{n}|y_i-z_i|^{2H_i-2}\,dy\,dz\bigg\}\,dr\,ds. \qquad (8.58)$$

From (8.57) to (8.58), we first perform the change of variables, $x-y=y'$, $x-z=z'$, $t-r=r'$, $t-s=s'$, and then we change the name of y',z',r',s' back to y,z,r,s again for simplicity. Inequality (8.58) is less than

$$C\int_0^t ds\int_0^s dr(s-r)^{2H_0-2}$$

$$\cdot\left\{\int_{\mathbb{R}^n}dy\int_s^{s+h}\left[-\frac{n}{2}v^{-n/2-1}\exp\left(-\frac{|y|^2}{2v}\right)\right.\right.$$

$$\left.+\frac{1}{2}v^{-n/2-2}|y|^2\exp\left(-\frac{|y|^2}{2v}\right)\right]dv$$

$$\cdot\int_{\mathbb{R}^n}\left[(r+h)^{-n/2}\exp\left(-\frac{|z|^2}{2(r+h)}\right)-r^{-n/2}\exp\left(-\frac{|z|^2}{2r}\right)\right]$$

$$\cdot\prod_{i=1}^{n}|y_i-z_i|^{2H_i-2}\,dz\bigg\}$$

$$\leq C\int_0^t ds\int_0^s dr(s-r)^{2H_0-2}$$

$$\cdot\left\{\int_s^{s+h}dv\int_{\mathbb{R}^n}\left[\frac{n}{2}v^{-n/2-1}\exp\left(-\frac{|y|^2}{2v}\right)\right.\right.$$

$$\left.+\frac{1}{2}v^{-n/2-2}|y|^2\exp\left(-\frac{|y|^2}{2v}\right)\right]dy$$

$$\cdot \int_{\mathbb{R}^n} \left[(r+h)^{-n/2} \exp\left(-\frac{|z|^2}{2(r+h)}\right) + r^{-n/2} \exp\left(-\frac{|z|^2}{2r}\right) \right]$$
$$\cdot \prod_{i=1}^{n} |y_i - z_i|^{2H_i - 2} \, dz \bigg\}. \tag{8.59}$$

Choose $p > 1$ such that

$$\frac{1}{2H_i - 1} < p < \frac{n}{n-2} \quad \text{for } i = 1, 2, \ldots, n.$$

This is possible since $H_i > 1 - 1/n$ for $i = 1, 2, \ldots, n$. Then

$$\frac{2p}{p-1} > d \quad \text{and} \quad \frac{p}{p-1}(2H_i - 2) > -1, \quad i = 1, 2, \ldots, n.$$

Now applying Lemma 8.5.1 repeatedly to this choice of p and to $\alpha = 2 - 2H_i$, we get

$$(8.59) \leq C \int_0^t ds \int_0^s dr (s-r)^{2H_0 - 2}$$
$$\cdot \int_s^{s+h} dv \frac{1}{v} \left(1 + C r^{-1/2(1-1/p)}\right)^n$$
$$\leq C \int_0^t ds \int_0^s dr$$
$$\cdot \int_s^{s+h} dv \frac{1}{v} \left(1 + r^{-n/2(1-1/p)}\right) (s-r)^{2H_0 - 2}. \tag{8.60}$$

Choose β such that $2 - 2H_0 < \beta < 1$. It follows that (8.60) is dominated by

$$C \int_0^t ds \int_0^s dr \frac{1}{s^{1-\beta}} \int_s^{s+h} dv \frac{1}{v^\beta} \left(1 + r^{-n/2(1-1/p)}\right) (s-r)^{2H_0 - 2}$$
$$\leq C h^{1-\beta} \int_0^t ds \int_0^s dr \frac{1}{s^{1-\beta}} r^{-n/2(1-1/p)} (s-r)^{2H_0 - 2}$$
$$= C h^{1-\beta} \int_0^t ds \frac{1}{s^{1-\beta}} \left[\int_0^{s/2} r^{-n/2(1-1/p)} (s-r)^{2H_0 - 2} \, dr \right.$$
$$\left. + \int_{s/2}^s r^{-n/2(1-1/p)} (s-r)^{2H_0 - 2} \, dr \right]$$
$$\leq C h^{1-\beta} \int_0^t \frac{1}{s^{1-\beta}} s^{1-n/2(1-1/p)-(2-2H_0)} \, ds \leq C h^{1-\beta}. \tag{8.61}$$

On the other hand, for $k \in \mathbb{R}^n$ we have

$$V(t, x+k) - V(t, x) = \int_0^t \int_{\mathbb{R}^n} [G_{t-s}(x+k, y) - G_{t-s}(x, y)] \, dB^{(H)}(s, y)$$

8.5 The quasi-linear stochastic fractional heat equation

Hence, by (8.51),

$$E[|V(t,x+k) - V(t,x)|^2]$$

$$\leq C \int_0^t \int_0^t |r-s|^{2H_0-2}$$

$$\cdot \int_{\mathbb{R}^n} \int_{\mathbb{R}^n} \left\{ (t-r)^{-n/2} \left[\exp\left(-\frac{|x+k-y|^2}{2(t-r)}\right) - \exp\left(-\frac{|x-y|^2}{2(t-r)}\right) \right] \right\}$$

$$\cdot \left\{ (t-s)^{-n/2} \left[\exp\left(-\frac{|x+k-z|^2}{2(t-s)}\right) - \exp\left(-\frac{|x-z|^2}{2(t-s)}\right) \right] \right\}$$

$$\cdot \prod_{i=1}^n |y_i - z_i|^{2H_i-2} \, dy \, dz \, dr \, ds$$

$$\leq C \int_0^t \int_0^t |r-s|^{2H_0-2}$$

$$\cdot \int_{\mathbb{R}^n} \int_{\mathbb{R}^n} \left\{ r^{-n/2} \left[\exp\left(-\frac{|y+k|^2}{2r}\right) - \exp\left(-\frac{|y|^2}{2r}\right) \right] \right\}$$

$$\cdot \left\{ s^{-n/2} \left[\exp\left(-\frac{|z+k|^2}{2s}\right) - \exp\left(-\frac{|z|^2}{2s}\right) \right] \right\}$$

$$\cdot \prod_{i=1}^n |y_i - z_i|^{2H_i-2} \, dy \, dz \, dr \, ds$$

$$\leq C \int_0^t ds \int_0^s dr (s-r)^{2H_0-2}$$

$$\cdot \int_{\mathbb{R}^d} dy \left| s^{-n/2} \left[\exp\left(-\frac{|y+k|^2}{2s}\right) - \exp\left(-\frac{|y|^2}{2s}\right) \right] \right|$$

$$\cdot \int_{\mathbb{R}^n} dz \left\{ r^{-n/2} \left[\exp\left(-\frac{|z+k|^2}{2r}\right) - \exp\left(-\frac{|z|^2}{2r}\right) \right] \right\}$$

$$\cdot \prod_{i=1}^n |y_i - z_i|^{2H_i-2}. \tag{8.62}$$

Applying Lemma 8.5.1 and Lemma 8.5.2 we get

$$(8.62) \leq C \int_0^t ds \int_0^s dr (s-r)^{2H_0-2} \left(1 + r^{-1/2(1-1/p)}\right)^n$$

$$\cdot \sum_{i=1}^n |k_i| \int_{\mathbb{R}^n} s^{-n/2-1} \exp\left(-\frac{|y|^2}{2s}\right) |y_i| \, dy$$

$$\leq C|k| \int_0^t ds \int_0^s dr \frac{1}{s^{1/2}} (s-r)^{2H_0-2} r^{-n/2(1-1/p)}$$

$$\leq C|k| \int_0^t ds \frac{1}{s^{1/2}} \Big[\int_0^{s/2} dr (s-r)^{2H_0-2} r^{-n/2(1-1/p)}$$
$$+ \int_{s/2}^s dr (s-r)^{2H_0-2} r^{-n/2(1-1/p)} \Big]$$
$$\leq C|k| \int_0^t s^{2H_0-n/2(1-1/p)-3/2} ds \leq C|k|, \qquad \text{if } H_0 > 3/4. \quad (8.63)$$

Combining the estimates (8.56), (8.61), and (8.63) we get, for some $\beta < 1$,

$$E[\,|V(t+h, x+k) - V(t,x)|^2] \leq C[h^{1-\beta} + |k|\,].$$

Since $V(t+h, x+k) - V(t,x)$ is a Gaussian random variable with mean zero, it follows that for any $m \geq 1$,

$$E[\,|V(t+h, x+k) - V(t,x)|^{2m}] \leq C_m E[\,|V(t+h, x+k) - V(t,x)|^2]^m$$
$$\leq C_m [h^{1-\beta} + |k|\,]^m \leq C_m [h^{1-\beta} + |k|]^m.$$

Hence, by Kolmogorov's theorem we conclude that $V(t,x)$ admits a jointly continuous version. □

9

Stochastic optimal control and applications

Stochastic control has many important applications and is a crucial branch of mathematics. Some textbooks contain fundamental theory and examples of applications of stochastic control theory for systems driven by standard Brownian motion (see, for example, [96], [97], [182], [231]). In this chapter we shall deal with the stochastic control problem where the controlled system is driven by a *fBm*.

Even in the stochastic optimal control of systems driven by Brownian motion case or even for deterministic optimal control the explicit solution is difficult to obtain except for linear systems with quadratic control. There are several approaches to the solution of classical stochastic control problem. One is the Pontryagin maximum principle, another one is the Bellman dynamic programming principle. For linear quadratic control one can use the technique of completing squares. There are also some other methods for specific problems. For example, a famous problem in finance is the optimal consumption and portfolio studied by Merton (see [162]), and one of the main methods to solve this problem is the martingale method combined with Lagrangian multipliers. See [135] and the reference therein.

The dynamic programming method seems difficult to extend to *fBm* since *fBm* – and solutions of stochastic differential equations driven by *fBm* – are not Markov processes. However, we shall extend the Pontryagin maximum principle to general stochastic optimal control problems for systems driven by *fBm*s. To do this we need to consider backward stochastic differential equations driven by *fBm*.

9.1 Fractional backward stochastic differential equations

Let $B_t^{(H)}, t \geq 0$, be a *fBm* with Hurst index $H > 1/2$ on the probability space $(\Omega, \mathcal{F}^{(H)}, \mathbb{P}^H)$ endowed with the natural filtration $\mathcal{F}_t^{(H)}$ of $B^{(H)}$ and $\mathcal{F}^{(H)} = \vee_{t \geq 0} \mathcal{F}_t^{(H)}$. Let $b : [0, T] \times \mathbb{R} \times \mathbb{R} \to \mathbb{R}$ be a given function and let

$F : \Omega \to \mathbb{R}$ be a given $\mathcal{F}_T^{(H)}$-measurable random variable, where $T > 0$ is a constant. Consider the problem of finding $\mathcal{F}^{(H)}$-*adapted processes* $p(t)$, $q(t)$ such that

$$dp(t) = b(t, p(t), q(t))\, dt + q(t)\, dB^{(H)}(t), \quad t \in [0, T], \tag{9.1}$$

$$P(T) = F \quad \text{a.s..} \tag{9.2}$$

This is a *fractional backward stochastic differential equation* in the two unknown processes $p(t)$ and $q(t)$ since the terminal condition instead of initial condition is given. This equation is the generalization of backward stochastic differential equation for Brownian motion case to *fBm* case. We will not discuss general theory for such equations. Instead we shall present a detailed study of linear variant of (9.1) and (9.2), namely,

$$dp(t) = [\alpha(t) + b_t p(t) + c_t q(t)]\, dt + q(t)\, dB^{(H)}(t), \quad t \in [0, T], \tag{9.3}$$

$$P(T) = F \quad \text{a.s.,} \tag{9.4}$$

where b_t and c_t are given continuous deterministic functions and $\alpha(t) = \alpha(t, \omega)$ is a given $\mathcal{F}^{(H)}$-adapted process such that $\int_0^T |\alpha(t, \omega)|\, dt < \infty$ almost surely.

To solve (9.3) and (9.4) we proceed as follows. By the fractional Girsanov theorem 3.2.4 we can rewrite (9.3) as

$$dp(t) = [\alpha(t) + b_t p(t)]\, dt + q(t)\, d\hat{B}^{(H)}(t), \quad t \in [0, T], \tag{9.5}$$

where

$$\hat{B}^{(H)}(t) = B^{(H)}(t) + \int_0^t c_s\, ds$$

is a *fBm* (with Hurst parameter H) under the new probability measure $\hat{\mathbb{P}}^H$ on $\mathcal{F}_T^{(H)}$ defined by

$$\frac{d\hat{\mathbb{P}}^H(\omega)}{d\mathbb{P}^H(\omega)} = \exp^\diamond(-\langle \omega, \hat{c}\rangle) = \exp\left(-\int_0^T \hat{c}(s)\, dB^{(H)}(s) - \tfrac{1}{2}\|\hat{c}\|_H^2\right),$$

where $\hat{c} = \hat{c}_t$ is the continuous function with supp$(\hat{c}) \subset [0, T]$ satisfying

$$\int_0^T \hat{c}_s \phi(s, t)\, ds = c_t, \quad 0 \leq t \leq T,$$

where ϕ is defined in (3.1) and here

$$\|\hat{c}\|_H^2 = \int_0^T \int_0^T \hat{c}(s)\hat{c}(t)\phi(s, t)\, ds\, dt.$$

If we multiply (9.5) by the integrating factor

$$\beta_t := \exp(-\int_0^t b_s\, ds),$$

we get
$$d(\beta_s p(s)) = \beta_s \alpha(s)\, ds + \beta_s q(s)\, d\hat{B}^{(H)}(s), \tag{9.6}$$

or, by integrating (9.6) from $s = t$ to $s = T$,

$$\beta_T F = \beta_t p(t) + \int_t^T \beta_s \alpha(s)\, ds + \int_t^T \beta_s q(s)\, d\hat{B}^{(H)}(s). \tag{9.7}$$

Assume from now on that

$$\|\alpha\|_{\hat{\mathcal{L}}_\phi(0,T)}^2 := E_{\hat{\mathbb{P}}^H}\left[\int_0^T \int_0^T \alpha(s)\alpha(t)\phi(s,t)\, ds\, dt\right]$$
$$+ E_{\hat{\mathbb{P}}^H}\left[\int_0^T \int_0^T \hat{D}_s^\phi \alpha(t) \hat{D}_t^\phi \alpha(s)\, ds\, dt\right] < \infty, \tag{9.8}$$

where $\hat{D}^\phi$ denotes the ϕ-derivative at s with respect to $\hat{B}^{(H)}(\cdot)$. By the fractional Itô isometry (3.41) applied to $\hat{B}^{(H)}$ and $\hat{\mathbb{P}}^H$ we then have

$$E_{\hat{\mathbb{P}}^H}\left[\left(\int_0^T \alpha(s)\, d\hat{B}^{(H)}(s)\right)^2\right] = \|\alpha\|_{\hat{\mathcal{L}}_\phi(0,T)}^2. \tag{9.9}$$

From now on let us also assume that

$$E_{\hat{\mathbb{P}}^H}\left[F^2\right] < \infty. \tag{9.10}$$

We now apply the quasi-conditional expectation operator (3.50)

$$\tilde{E}_{\hat{\mathbb{P}}^H}\left[\cdot | \mathcal{F}_t^{(H)}\right]$$

to both sides of (9.7) and get

$$\beta_T \tilde{E}_{\hat{\mathbb{P}}^H}\left[F | \mathcal{F}_t^{(H)}\right] = \beta_t p(t) + \int_t^T \beta_s \tilde{E}_{\hat{\mathbb{P}}^H}\left[\alpha(s) | \mathcal{F}_t^{(H)}\right] ds. \tag{9.11}$$

Here we have used that $p(t)$ is $\mathcal{F}_t^{(H)}$-measurable, that the filtration $\hat{\mathcal{F}}_t^{(H)}$ generated by $\hat{B}^{(H)}(s)\,; s \leq t$ is the same as $\mathcal{F}_t^{(H)}$, and that

$$\tilde{E}_{\hat{\mathbb{P}}^H}\left[\int_t^T f(s,\omega)\, d\hat{B}^{(H)}(s) | \hat{\mathcal{F}}_t^{(H)}\right] = 0, \quad \forall t \leq T,$$

for all $f \in \hat{\mathcal{L}}_\phi(0,T)$.

From (9.11) we get the solution

$$p(t) = \exp\left(-\int_t^T b_s\,ds\right)\tilde{E}_{\hat{\mathbb{P}}^H}\left[F|\mathcal{F}_t^{(H)}\right]$$
$$+\int_t^T \exp\left(-\int_t^s b_r\,dr\right)\tilde{E}_{\hat{\mathbb{P}}^H}\left[\alpha(s)|\mathcal{F}_t^{(H)}\right]ds, \quad t \le T. \tag{9.12}$$

In particular, choosing $t = 0$, we get

$$p(0) = \exp\left(-\int_0^T b_s\,ds\right) E_{\hat{\mathbb{P}}^H}[F]$$
$$+ \int_0^T \exp\left(-\int_0^s b_r dr\right) E_{\hat{\mathbb{P}}^H}[\alpha(s)]\,ds. \tag{9.13}$$

Note that $p(0)$ is $\mathcal{F}_0^{(H)}$-measurable and hence a constant. Choosing $t = 0$ in (9.7), we get

$$G = \int_0^T \beta_s q(s)\,d\hat{B}^{(H)}(s), \tag{9.14}$$

where

$$G = G(\omega) = \beta_T F(\omega) - \int_0^T \beta_s \alpha(s,\omega)\,ds - p(0),$$

with $p(0)$ given by (9.13).

By the fractional Clark Hausmann Ocone theorem (3.10.8) applied to $(\hat{B}^{(H)}, \hat{\mathbb{P}}^H)$ we have

$$G = E_{\hat{\mathbb{P}}^H}[G] + \int_0^T \tilde{E}_{\hat{\mathbb{P}}^H}\left[\hat{D}_s G|\mathcal{F}_s^{(H)}\right] d\hat{B}^{(H)}(s). \tag{9.15}$$

Comparing (9.14) and (9.15), we see that we can choose

$$q(t) = \exp\left(\int_0^t b_r\,dr\right) \tilde{E}_{\hat{\mathbb{P}}^H}\left[\hat{D}_t G|\mathcal{F}_t^{(H)}\right]. \tag{9.16}$$

We have proved the first part of the following result:

Theorem 9.1.1 ([31]).
Assume that (9.8) and (9.10) hold. Then a solution $(p(t),q(t))$ of (9.3) and (9.4) is given by (9.12) and (9.16). The solution is unique among all $\mathcal{F}_\cdot^{(H)}$-adapted processes $p(\cdot), q(\cdot) \in \hat{\mathcal{L}}_\phi(0,T)$.

Proof. It remains to prove uniqueness. The uniqueness of $p(\cdot)$ follows from the way we deduced formula (9.12) from (9.3) and (9.4). The uniqueness of q is deduced from (9.14) and (9.15) by the following argument: Substituting (9.15) from (9.14) and using that $E_{\hat{\mathbb{P}}^H}[G] = 0$, we get

$$0 = \int_0^T \left\{\beta_s q(s) - \tilde{E}_{\hat{\mathbb{P}}^H}\left[\hat{D}_s G|\mathcal{F}_s^{(H)}\right]\right\} d\hat{B}^{(H)}(s).$$

Hence by the fractional Itô isometry (9.9)

$$0 = E_{\hat{\mathbb{P}}^H}\left[\left(\int_0^T \left\{\beta_s q(s) - \tilde{E}_{\hat{\mathbb{P}}^H}\left[\hat{D}_s G | \hat{\mathcal{F}}_s^{(H)}\right]\right\} d\hat{B}^{(H)}(s)\right)^2\right]$$
$$= \|\beta_s q(s) - \tilde{E}_{\hat{\mathbb{P}}^H}\left[\hat{D}_s G | \hat{\mathcal{F}}_s^{(H)}\right]\|_{\hat{\mathcal{L}}_\phi(0,T)}^2,$$

from which it follows that

$$\beta_s q(s) - \tilde{E}_{\hat{\mathbb{P}}^H}\left[\hat{D}_s G | \hat{\mathcal{F}}_s^{(H)}\right] = 0 \quad \text{for} \quad a.a. \ (s,\omega) \in [0,T] \times \Omega.$$

□

For more information about backward stochastic differential equations driven by *fBm*, see [18].

9.2 A stochastic maximum principle

We now apply the theory in the previous section to prove a maximum principle for systems driven by *fBm*. See, e.g., [107], [186] and [231] and the references therein for more information about the maximum principle in the classical Brownian motion case.

Consider an m-dimensional *fBm* $B^{(H)}(t)$ with Hurst parameter $H = (H_1, H_2, \ldots, H_m)$, $1/2 < H_i < 1$, $i = 1, \ldots, m$, on the probability space $(\Omega, \mathcal{F}, \mathbb{P}^H)$ endowed with the natural filtration $\mathcal{F}_t^{(H)}$ generated by $B^{(H)}$, where $\mathbb{P}^H$ is the probability measure defined in (3.55) of Chapter 3. Suppose $X(t) = X^{(u)}(t)$ is a controlled system of the form

$$dX(t) = b(t, X(t), u(t))\, dt + \sigma(t, X(t), u(t))\, dB^{(H)}(t), \ X(0) = x \in \mathbb{R}^n, \quad (9.17)$$

where $b : [0,T] \times \mathbb{R}^n \times U \to \mathbb{R}^n$ and $\sigma : [0,T] \times \mathbb{R}^n \times U \to \mathbb{R}^{n \times m}$ are given C^1 functions. The control process $u(\cdot) : [0,T] \times \Omega \to U \subset \mathbb{R}^k$ is assumed to be $\mathcal{F}^{(H)}$-adapted, and U is a given closed convex set in $\mathbb{R}^k$.

Let $f : [0,T] \times \mathbb{R}^n \times U \to \mathbb{R}$, $g : \mathbb{R}^n \to \mathbb{R}$, and $G : \mathbb{R}^n \to \mathbb{R}^N$ be given lower bounded C^1 functions and define the *performance functional* $J(u)$ by

$$J(u) = E\left[\int_0^T f(t, X(t), u(t))\, dt + g(X(T))\right] \quad (9.18)$$

and the *terminal condition* by

$$E\left[G(X(T))\right] = 0.$$

Definition 9.2.1. *Let $\mathcal{A}$ denote the set of all $\mathcal{F}_t^{(H)}$-adapted processes $u : [0,T] \times \Omega \to U$ such that $X^{(u)}(t)$ exists and does not explode in $[0,T]$ and such that (9.17) holds. If $u \in \mathcal{A}$ and $X^{(u)}(t)$ is the corresponding state process, we call $(u, X^{(u)})$ an admissible pair.*

Consider the problem to find J^* and $u^* \in \mathcal{A}$ such that

$$J^* = \sup\{J(u); u \in \mathcal{A}\} = J(u^*). \tag{9.19}$$

If such $u^* \in \mathcal{A}$ exists, then u^* is called an *optimal control* and (u^*, X^*), where $X^* = X^{u^*}$, is called an *optimal pair*.

Let $C([0,T]; \mathbb{R}^{n \times m})$ be the set of continuous function from $[0,T]$ into $\mathbb{R}^{n \times m}$. Define the *Hamiltonian* $H : [0,T] \times \mathbb{R}^n \times U \times \mathbb{R}^n \times C([0,T]; \mathbb{R}^{n \times m}) \to \mathbb{R}$ by

$$H(t,x,u,p,q(\cdot)) = f(t,x,u) + b(t,x,u)^T p$$
$$+ \sum_{i=1}^{n} \sum_{k=1}^{m} \sigma_{ik}(t,x,u) \int_0^T q_{ik}(s) \phi_{H_k}(s,t)\, ds,$$

where $\phi_{H_k}(s,t)$ is defined in Chapter 3, (3.1). Consider the following *fractional backward stochastic differential equation* in the pair of unknown $\mathcal{F}_t^{(H)}$-adapted processes $p(t) \in \mathbb{R}^n$ and $q(t) \in \mathbb{R}^{n \times m}$ called the *adjoint processes*:

$$\begin{cases} dp(t) = -H_x(t, X(t), u(t), p(t), q(\cdot))\, dt + q(t) dB^{(H)}(t),\ t \in [0,T]. \\ p(T) = g_x(X(T)) + \lambda^T G_x(X(T)). \end{cases} \tag{9.20}$$

where $H_x = \nabla_x H = (\partial H/\partial x_1, \ldots, \partial H/\partial x_n)^T$ is the gradient of H with respect to x and similarly with g_x and G_x. $X(t) = X^{(u)}(t)$ is the process obtained by using the control $u \in \mathcal{A}$ and $\lambda \in \mathbb{R}_+^n$ is a constant. The equation (9.20) is called the *adjoint equation* and $p(t)$ is sometimes interpreted as the *shadow price* (of a resource).

Lemma 9.2.2. *Let $X(t)$ and $Y(t)$ be two processes of the form*

$$dX(t) = \mu(t,\omega)\, dt + \sigma(t,\omega)\, dB^{(H)}(t), \quad X(0) = x \in \mathbb{R}^n$$

and

$$dY(t) = \nu(t,\omega)\, dt + \theta(t,\omega)\, dB^{(H)}(t), \quad Y(0) = y \in \mathbb{R}^n,$$

where $\mu : [0,T] \times \Omega \to \mathbb{R}^n$, $\nu : [0,T] \times \Omega \to \mathbb{R}^n$, $\sigma : [0,T] \times \Omega \to \mathbb{R}^{n \times m}$, and $\theta : [0,T] \times \Omega \to \mathbb{R}^{n \times m}$ are given processes with rows $\sigma_i, \theta_i \in \mathcal{L}_\phi^{(m)}(0,T)$ for $1 \leq i \leq n$ and $B^H(\cdot)$ is an m-dimensional fBm.

1. *Then, for $T > 0$,*

$$E[X(T) \cdot Y(T)]$$
$$= x \cdot y + E\left[\int_0^T X(s)\, dY(s)\right] + E\left[\int_0^T Y(s)\, dX(s)\right]$$
$$+ E\left[\int_0^T \int_0^T \sum_{i=1}^n \sum_{k=1}^m \sigma_{ik}(s) \theta_{ik}(t) \phi_{H_k}(s,t)\, ds\, dt\right]$$
$$+ E\left[\sum_{i=1}^n \sum_{j,k=1}^m \left(\int_\mathbb{R} \int_\mathbb{R} D_{j,t}^\phi \sigma_{ik}(s) D_{k,s}^\phi \theta_{ij}(t)\, dt\, ds\right)\right]$$

provided that the first two integrals exist.

2. *In particular, if $\sigma(\cdot)$ or $\theta(\cdot)$ is deterministic, then*

$$E[X(T) \cdot Y(T)]$$
$$= x \cdot y + E\left[\int_0^T X(s)\, dY(s)\right] + E\left[\int_0^T Y(s)\, dX(s)\right]$$
$$+ E\left[\int_0^T \int_0^T \sum_{i=1}^n \sum_{k=1}^m \sigma_{ik}(s)\theta_{ik}(t)\phi_{H_k}(s,t)\, ds\, dt\right].$$

Theorem 9.2.3 (The fractional stochastic maximum principle). *[31]*
Suppose $\hat{u} \in \mathcal{A}$ and put $\hat{X} = X^{(\hat{u})}$. Let $\hat{p}(t), \hat{q}(t)$ be a solution of the corresponding adjoint equation (9.20) for some $\lambda \in \mathbb{R}_+^n$. Put Δ_4 equal to

$$E\left[\sum_{i=1}^n \sum_{j,k=1}^m \int_0^T \int_0^T D_{j,s}^{\phi_j}\left\{\sigma_{ik}(t, X(t), u(t)) - \sigma_{ik}(t, \hat{X}(t), \hat{u}(t))\right\} D_{k,t}^{\phi_k} \hat{q}_{ij}(s)\, dt\, ds\right].$$

Assume that (9.21) to (9.23) hold,

$$H(t, \cdot, \cdot, \hat{p}(t), \hat{q}(t)), \quad g(\cdot) \text{ and } G(\cdot) \text{ are concave, for all } t \in [0, T], \quad (9.21)$$

$$H(t, \hat{X}(t), \hat{u}(t), \hat{p}(t), \hat{q}(\cdot)) = \max_{v \in U} H(t, \hat{X}(t), v, \hat{p}(t), \hat{q}(\cdot)), \quad (9.22)$$

$$\Delta_4 \leq 0, \quad (9.23)$$

and that $[X(t) - \hat{X}(t)]\hat{q}(t)$ and $\hat{p}(t)^T \left[\sigma(t, \hat{X}(t), \hat{u}(t)) - \sigma(t, X(t), u(t))\right]$ are fWIS integrable for all $u \in \mathcal{A}$. Then if $\lambda \in \mathbb{R}_+^n$ is such that $(\hat{u}, \hat{X})$ is admissible [i.e., (9.17) holds], the pair $(\hat{u}, \hat{X})$ is an optimal pair for problem (9.19).

Proof. We first give a proof in the case when $G(x) = 0$, i.e., when there is no terminal condition.

If $dp(t) = \xi(t)\, dt + \eta(t)\, dB^{(H)}(t)$, then we put

$$\int_0^T \rho(t)\, dp(t) := \int_0^T \rho(t)\xi(t)\, dt + \int_0^T \rho(t)\eta(t)\, dB^{(H)}(t),$$

where $\int_0^T \rho(t)\eta(t)\, dB^{(H)}(t)$ denotes the fWIS stochastic integral defined in Chapter 3. With $(\hat{u}, \hat{X})$ as above, consider

$$\Delta =: E\left[\int_0^T f(t, \hat{X}(t), \hat{u}(t))\, dt - \int_0^T f(t, X(t), u(t))\, dt\right]$$
$$=: \Delta_1 + \Delta_2 + \Delta_3, \quad (9.24)$$

where

$$\Delta_1 := E\left[\int_0^T H(t, \hat{X}(t), \hat{u}(t), \hat{p}(t), \hat{q}(\cdot))\, dt - \int_0^T H(t, X(t), u(t), \hat{p}(t), \hat{q}(\cdot))\, dt\right]$$

$$\Delta_2 := -E\left[\int_0^T b(t, \hat{X}(t), \hat{u}(t))^T \hat{p}(t)\, dt - \int_0^T b(t, X(t), u(t))^T \hat{p}(t)\, dt\right]$$

and $-\Delta_3$ is equal to

$$E\left[\int_0^T \int_0^T \sum_{i=1}^n \sum_{k=1}^m \Psi_{ik}(s, \hat{X}(s), \hat{u}(s), X(s), u(s)) \hat{q}_{ik}(t) \phi_{H_k}(s,t)\, ds\, dt\right],$$

where $\Psi_{ik}(s, \hat{X}(s), \hat{u}(s), X(s), u(s)) := \sigma_{ik}(s, \hat{X}(s), \hat{u}(s)) - \sigma_{ik}(s, X(s), u(s))$. Since $(x, u) \to H(x, u) = H(t, x, u, p, q(\cdot))$ is concave, we have

$$H(x, u) - H(\hat{x}, \hat{u}) \leq H_x(\hat{x}, \hat{u}) \cdot (x - \hat{x}) + H_u(\hat{x}, \hat{u}) \cdot (u - \hat{u})$$

for all $(x, u), (\hat{x}, \hat{u})$. Since $v \to H(\hat{X}(t), v)$ is maximal at $v = \hat{u}(t)$, we have

$$H_u(\hat{x}, \hat{u}) \cdot (u(t) - \hat{u}(t)) \leq 0 \quad \forall t \in [0, T].$$

Therefore,

$$\Delta_1 \geq E\left[\int_0^T -H_x(t, \hat{X}(t), \hat{u}(t), \hat{p}(t), \hat{q}(\cdot)) \cdot (X(t) - \hat{X}(t))\, dt\right]$$

$$= E\left[\int_0^T (X(t) - \hat{X}(t))^T d\hat{p}(t) - \int_0^T (X(t) - \hat{X}(t))^T \hat{q}(t)\, dB^{(H)}(t)\right].$$

Since $E\left[\int_0^T (X(t) - \hat{X}(t))^T \hat{q}(t)\, dB^{(H)}(t)\right] = 0$, this gives

$$\Delta_1 \geq E\left[\int_0^T (X(t) - \hat{X}(t))^T d\hat{p}(t)\right].$$

By (9.17) we have

$$\Delta_2 = -E\left[\int_0^T \left\{b(t, \hat{X}(t), \hat{u}(t)) - b(t, X(t), u(t))\right\} \cdot \hat{p}(t)\, dt\right]$$

$$= -E\left[\int_0^T \hat{p}(t)\left(d\hat{X}(t) - dX(t)\right)\right]$$

$$\quad - E\left[\int_0^T \hat{p}(t)^T \left\{\sigma(t, \hat{X}(t), \hat{u}(t)) - \sigma(t, X(t), u(t))\right\} dB^{(H)}(t)\right]$$

$$= E\left[\int_0^T \hat{p}(t)\left(dX(t) - d\hat{X}(t)\right)\right].$$

Finally, since g is concave, we have

$$g(X(T)) - g(\hat{X}(T)) \leq g_x(\hat{X}(T)) \cdot (X(T) - \hat{X}(T)) \qquad (9.25)$$

Combining (9.24) through (9.25) with Lemma 9.2.2, we get, using (9.18), (9.20), and (9.23),

$$\begin{aligned}
J(\hat{u}) &- J(u) \\
&= \Delta + E\Big[g(\hat{X}(T)) - g(X(T))\Big] \\
&\geq \Delta + E\Big[g_x(\hat{X}(T)) \cdot (\hat{X}(T) - X(T))\Big] \\
&\geq \Delta - E\Big[\hat{p}(T) \cdot \big(X(T) - \hat{X}(T)\big)\Big] \\
&= \Delta - \Bigg\{ E\bigg[\int_0^T \big(X(t) - \hat{X}(t)\big) d\hat{p}(t)\bigg] + E\bigg[\int_0^T \hat{p}(t)\big(dX(t) - d\hat{X}(t)\big)\bigg] \\
&\quad + E\bigg[\int_0^T\int_0^T \sum_{i=1}^n \sum_{k=1}^m \Psi_{ik}(s, \hat{X}(s), \hat{u}(s), X(s), u(s))\hat{q}_{ik}(t)\phi_{H_k}(s,t)\, ds\, dt\bigg] \\
&\quad + E\bigg[\sum_{i=1}^n \sum_{j,k=1}^m \int_0^T \int_0^T D_{j,s}^{\phi_j}\big\{\Psi_{ik}(t, X(t), u(t), \hat{X}(t), \hat{u}(t))\big\} \\
&\qquad\qquad \cdot D_{k,t}^{\phi_k}\hat{q}_{ij}(s)\, dt\, ds\bigg]\Bigg\} \\
&\geq \Delta - (\Delta_1 + \Delta_2 + \Delta_3 + \Delta_4) \geq 0,
\end{aligned}$$

where $\Psi_{ik}(s, \hat{X}(s), \hat{u}(s), X(s), u(s)) = \sigma_{ik}(s, \hat{X}(s), \hat{u}(s)) - \sigma_{ik}(s, X(s), u(s))$. This shows that $J(\hat{u})$ is maximal among all admissible pairs $(u(\cdot), X(\cdot))$. This completes the proof in the case with no terminal conditions ($G = 0$).

Finally consider the general case with $G \neq 0$. Suppose that for some $\lambda_0 \in \mathbb{R}_+^n$ there exists $\hat{u}_{\lambda_0}$ satisfying (9.21) to (9.23). Then by the above argument we know that if we put

$$J_{\lambda_0}(u) = E\bigg[\int_0^T f(t, X(t), u(t))dt + g(X(T)) + \lambda_0^T G(X(T))\bigg],$$

then $J_{\lambda_0}(\hat{u}_{\lambda_0}) \geq J_{\lambda_0}(u)$ for all controls u (without terminal condition). If λ_0 is such that $\hat{u}_{\lambda_0}$ satisfies the terminal condition (i.e., $\hat{u}_{\lambda_0} \in \mathcal{A}$) and u is another control in $\mathcal{A}$, then

$$J(\hat{u}_{\lambda_0}) = J_{\lambda_0}(\hat{u}_{\lambda_0}) \geq J_{\lambda_0}(u) = J(u),$$

and hence $\hat{u}_{\lambda_0} \in \mathcal{A}$ maximizes $J(u)$ over all $u \in \mathcal{A}$. $\square$

Corollary 9.2.4. *Let $\hat{u} \in \mathcal{A}$, $\hat{X} = X^{(\hat{u})}$ and $(\hat{p}(t), \hat{q}(t))$ be as in Theorem 9.2.3. Assume that (9.21) and (9.22) hold and that condition (9.23) is replaced by the condition*

$$\hat{q}(\cdot) \quad or \quad \sigma(\cdot, \hat{X}(\cdot), \hat{u}(\cdot)) \quad is \ deterministic.$$

Suppose that $[X(t) - \hat{X}(t)]\hat{q}(t)$ and $\hat{p}(t)^T \left[\sigma(t, \hat{X}(t), \hat{u}(t)) - \sigma(t, X(t), u(t))\right]$ are fWIS integrable for all $u \in \mathcal{A}$. Then if $\lambda \in \mathbb{R}^n_+$ is such that $(\hat{u}, \hat{X})$ is admissible, the pair $(\hat{u}, \hat{X})$ is optimal for problem (9.19).

In the following sections we illustrate our main results with several examples.

9.3 Linear quadratic control

We start by extending the following classical control model to the *fBm* case. The controlled dynamics is given by the following fWIS-type SDE where the state is scalar-valued:

$$\begin{cases} dx_t = (a_t x_t + b_t u_t)\, dt + (c_t x_t + d_t u_t)\, dB_t^{(H)}, \\ x_0 \in \mathbb{R} \quad \text{is given and deterministic,} \end{cases} \quad (9.26)$$

where $a_t, b_t, c_t,$ and d_t, $0 \le t \le T$, are given essentially bounded deterministic functions of t. The process u_t is assumed to be a Markov linear feedback control, namely,

$$u_t = K_t x_t,$$

where K_t is an essentially bounded deterministic function of t. Such a control is also referred to as an *admissible (Markov linear feedback) control* in this section.

Under each admissible control $u_t = K_t x_t$, the system (9.26) reduces to the following linear SDE

$$\begin{cases} dx_t = (a_t + b_t K_t) x_t dt + (c_t + d_t K_t) x_t dB_t^{(H)}, \\ x_0 \in \mathbb{R} \quad \text{is given and deterministic.} \end{cases} \quad (9.27)$$

Hence K. itself, also known as the *feedback gain*, can be regarded as a control. For every initial state x_0 and admissible control $u_t = K_t x_t$, there is an associated cost

$$J(x_0, u.) \equiv J(x_0, K.) = E\left[\int_0^T (Q_t x_t^2 + R_t u_t^2)\, dt + G x_T^2\right], \quad (9.28)$$

9.3 Linear quadratic control

where $x_\cdot$ is the solution of (9.26) under the control $u_\cdot$ (or equivalently, $K_\cdot$), Q_t and R_t are given essentially bounded deterministic functions in t, and G is a given deterministic scalar. Our optimal stochastic control problem is to minimize the cost functional (9.28), for each given x_0, over the set of all admissible Markov linear feedback controls.

Theorem 9.3.1. *Assume that for almost every $t \in [0,T]$, $d_t = 0$, $Q_t \geq 0$, and $R_t > \delta$ for some given $\delta > 0$, and $G \geq 0$. Then the following Riccati equation*

$$\begin{cases} \dot{p}_t + 2p_t[a_t + c_t \int_0^t \phi(t,s) c_s \, ds] + Q_t - R_t^{-1} b_t^2 p_t^2 = 0 \\ p_T = G \end{cases} \tag{9.29}$$

admits a unique solution p_t over $[0,T]$ with $p_t \geq 0$ for all $t \in [0,T]$. Moreover, the optimal Markov linear feedback control for the problem in (9.27) and (9.28) is given by

$$\hat{u}_t = \hat{K}_t x_t \quad \text{with} \quad \hat{K}_t = -R_t^{-1} b_t p_t.$$

Finally, the optimal value is $p_0 x_0^2$.

Proof. The unique solvability of the (classical) Riccati equation (9.29) was proved in, e.g., [231, p. 297, Corollary 2.10]. Next, for any admissible control $u_t = K_t x_t$, applying the Itô formula to the equation (9.27) with $d_t = 0$, we get

$$d(p_t x_t^2) = x_t^2 \left[\dot{p}_t + 2p_t(a_t + b_t K_t) + 2p_t c_t \int_0^t \phi(t,s) c_s \, ds \right] dt + 2x_t^2 c_t p_t \, dB_t^{(H)}.$$

Taking integration from 0 to T, we get

$$p_T x_T^2 = p_0 x_0^2 + \int_0^T x_t^2 \left[\dot{p}_t + 2p_t(a_t + b_t K_t) + 2p_t c_t \int_0^t \phi(t,s) c_s \, ds \right] dt$$
$$+ 2 \int_0^T x_t^2 c_t p_t \, dB_t^{(H)}.$$

Denote $f_t = x_t^2 c_t p_t$. It is easy to see that

$$\int_0^T \int_0^T \phi(s,t) E\left[|f_s f_t|\right] ds \, dt < \infty.$$

On the other hand, $D_t^\phi x_t = x_t \int_0^t \phi(t,s) c_s \, ds$. It is straightforward to check by Corollary 9.2.2 that

$$\sup_{0 \leq s \leq t \leq T} E\left[|D_t^\phi x_t|^p\right] < \infty \quad \forall \, p \geq 1.$$

This implies that f_t is integrable and

$$E\left[\int_0^T x_t^2 c_t p_t \, dB_t^{(H)}\right] = 0.$$

Hence,

$$E\left[p_T x_T^2\right] = p_0 x_0^2 + E\left[\int_0^T x_t^2 \left[\dot{p}_t + 2p_t(a_t + b_t K_t) + 2p_t c_t \int_0^t \phi(t,s) c_s \, ds\right] dt\right].$$

Since $p_T = G$, we obtain

$$J(x_0, K.) = p_0 x_0^2 + E\Big[\int_0^T x_t^2(\dot{p}_t + 2p_t(a_t + b_t K_t)$$
$$+ (Q_t + R_t K_t^2) + 2p_t c_t \int_0^t \phi(t,s) c_s ds) dt\Big]$$
$$= p_0 x_0^2 + E\Big[\int_0^T x_t^2(\dot{p}_t + 2p_t a_t + 2p_t c_t \int_0^t \phi(t,s) c_s ds + Q_t$$
$$+ R_t(K_t + R_t^{-1} b_t p_t)^2 - R_t^{-1} b_t^2 p_t^2) dt\Big]$$
$$= p_0 x_0^2 + E\left[\int_0^T R_t(K_t + R_t^{-1} b_t p_t)^2 dt\right], \tag{9.30}$$

where the last equality is due to the Riccati equation (9.29). Equation (9.30) shows that the cost function achieves its minimum when $\hat{K}_t = -R_t^{-1} b_t p_t$, with the minimum value being $p_0 x_0^2$. This proves the theorem. □

Remark 9.3.2. It is interesting to note that the Riccati equation (9.29) corresponds to the following linear–quadratic control problem with (normal) Brownian motion: Minimize (9.28) subject to

$$\begin{cases} dx_t = (a_t x_t + b_t u_t) dt + \tilde{c}_t x_t dW_t, \\ x_0 \in \mathbb{R} \quad \text{is given and deterministic,} \end{cases}$$

where $\tilde{c}_t = \sqrt{2c_t \int_0^t \phi(t,s) c_s \, ds}$ if $c_t \int_0^t \phi(t,s) c_s \, ds \geq 0$ for all $t \geq 0$. This suggests that, in the current setting, the linear–quadratic control problem with *fBm* is equivalent (in the sense of sharing the same optimal feedback control and optimal value) to a linear–quadratic control problem with Brownian motion where the diffusion coefficient of the state is properly modified.

9.4 A minimal variance hedging problem

We move now to consider some problems from mathematical finance. The first one is a minimal variance hedging problem. For a discussion and an introduction to the use of *fBm* in finance we refer to Chapter 7.

9.4 A minimal variance hedging problem

Consider a financial market driven by two independent fBms $B_1(t) = B^{(H_1)}(t)$ and $B_2(t) = B^{(H_2)}(t)$, with $1/2 < H_i < 1$, $i = 1, 2$, according to the WIS integration model presented in Chapter 7, as follows:

Risk-free asset price: $\qquad dS_0(t) = 0,\qquad$ (9.31)
Generalized value of company 1: $\qquad dS_1(t) = dB_1(t),\qquad$ (9.32)
Generalized value of company 2: $\qquad dS_2(t) = dB_1(t) + dB_2(t),\qquad$ (9.33)

respectively, with initial values $S_0(0) = 1, S_1(0) = s_1, S_2(0) = s_2$. If $\theta(t) = (\theta_0(t), \theta_1(t), \theta_2(t)) \in \mathbb{R}^3$ is a WIS admissible portfolio (Definition 7.2.4) (giving the number of units of the risk-free asset, stock on company 1, and stock on company 2, respectively, held at time t), then the corresponding *generalized value process* is [see (7.14)]

$$V^\theta(t) = V^\theta(0) + \int_0^t \theta_1(s)\, dS_1(s)) + \int_0^t \theta_2(s)\, dS_2(s).$$

It follows from the fractional Clark–Haussmann–Ocone theorem (Theorem 3.10.8) that this market is *strong arbitrage free* and *complete*. The latter means that any bounded $\mathcal{F}_T^{(H)}$-measurable random variable F can be hedged (or replicated), in the sense that there exists a WIS admissible portfolio $\theta(t)$ and an initial value $z \in \mathbb{R}$ such that

$$F(\omega) = z + \int_0^T \theta_1(s)\, dS_1(s)) + \int_0^T \theta_2(s)\, dS_2(s) \qquad \text{for a.a. } \omega.$$

(See [233] for a general discussion about this.)

Let us now assume that we are not allowed to trade in stock 1, i.e., we must have $\theta_1(t) \equiv 0$. How close to, say, the value $F(\omega) = B_1(T, \omega)$ can we get at the terminal value if we must hedge under this constraint?

If we put $\theta_2(t) = u(t)$ and interpret "close" as having a small $L^2(\mathbb{P}^H)$ distance to F, then the problem can be stated as follows: Find $z \in \mathbb{R}$ and $u(t, \omega) \in \mathcal{L}_\phi^{(2)}(0, T)$, $\mathcal{F}_t^{(H)}$-adapted, such that

$$J(z, u) := E\left[\left(B_1(T) - \left\{z + \int_0^T u(t)[dB_1(t) + dB_2(t)]\right\}\right)^2\right]$$

$$= z^2 + E\left[\left\{\int_0^T [u(t) - 1]\, dB_1(t) + \int_0^T u(t)\, dB_2(t)\right\}^2\right]$$

is minimal. We see immediately that it is optimal to choose $z = 0$; so it remains to minimize over $u(t) = u(t, \omega)$ the functional

$$J(u) := E\left[\left\{\int_0^T [u(t) - 1]\, dB_1(t) + \int_0^T u(t)\, dB_2(t)\right\}^2\right].$$

If we apply the fractional Itô isometry (3.41), we get, after some simplifications,

$$J(u) = E\left[\int_0^T \int_0^T \{[u(s) - 1][u(t) - 1]\phi_1(s,t) + u(s)u(t)\phi_2(s,t)\}\,ds\,dt\right]$$

$$+ E\left[\int_0^T \int_0^T \{D_{1,t}^\phi u(s) - D_{2,t}^\phi u(s)\}\{D_{1,s}^\phi u(t) - D_{2,s}^\phi u(t)\}\,dt\,ds\right].$$

However, it is difficult to see from this what the minimizing $u(t)$ is.

To approach this problem by using the fractional maximum principle, we define the state process $X(t)$ by

$$dX(t) = (u(t) - 1)\,dB_1(t) + u(t)\,dB_2(t). \tag{9.34}$$

Then the problem is equivalent to maximizing

$$J_1(u) := E\left[-\frac{1}{2}X^2(T)\right].$$

The Hamiltonian for this problem is

$$\begin{aligned}
H(t,x,u,p,q(\cdot)) &= (u-1)\int_0^T q_1(s)\phi_1(s,t)\,ds + u\int_0^T q_2(s)\phi_2(s,t)\,ds \\
&= (u-1)\int_0^T q_1(s)\phi_1(s,t)\,ds + u\int_0^T q_2(s)\phi_2(s,t)\,ds \\
&= u\left[\int_0^T q_1(s)\phi_1(s,t)\,ds + \int_0^T q_2(s)\phi_2(s,t)\,ds\right] \\
&\quad - \int_0^T q_1(s)\phi_1(s,t)\,ds.
\end{aligned} \tag{9.35}$$

The adjoint equation is

$$dp(t) = q_1(t)dB_1(t) + q_2(t)dB_2(t), \quad t < T,$$
$$p(T) = -X(T).$$

Comparing with (9.34), we see that this equation has the solution

$$q_1(t) = 1 - u(t), \quad q_2 = -u_2(t), \quad p(t) = -X(t), \quad t \leq T.$$

Let $\hat{u}(t)$ be an optimal control candidate. Then by (9.35)

$$\begin{aligned}
&H(t, \hat{X}(t), v, \hat{p}(t), \hat{q}(\cdot)) \\
&= v\left[\int_0^T \hat{q}_1(s)\phi_1(s,t)\,ds + \int_0^T \hat{q}_2(s)\phi_2(s,t)\,ds\right] - \int_0^T \hat{q}_1(s)\phi_1(s,t)\,ds \\
&= v\left[\int_0^T (1 - \hat{u}(t))\phi_1(s,t)\,ds - \int_0^T \hat{u}(s)\phi_2(s,t)\,ds\right] \\
&\quad - \int_0^T \hat{q}_1(s)\phi_1(s,t)\,ds.
\end{aligned}$$

9.5 Optimal consumption and portfolio in a fractional BS market

The maximum principle requires that the maximum of this expression is attained at $v = \hat{u}(t)$. However, this is an affine function of v; so it is natural to guess that the coefficient of v must be 0, i.e.,

$$\int_0^T (1 - \hat{u}(s))\phi_1(s,t)\,ds - \int_0^T \hat{u}(s)\phi_2(s,t)\,ds = 0,$$

which gives

$$\int_0^T \hat{u}(s)[\phi_1(s,t) + \phi_2(s,t)]\,ds = \int_0^T \phi_1(s,t)\,ds. \quad (9.36)$$

This is a symmetric Fredholm integral equation of the first kind, and it is known that it has a unique solution $\hat{u}(t) \in L^2[0,T]$.

This choice of $\hat{u}(t)$ satisfies all the requirements of Theorem 9.2.3 (in fact, even those of Corollary 9.2.4) and we can conclude that this $\hat{u}(t)$ is optimal. Thus we have proved the following:

Theorem 9.4.1 (Solution of the minimal variance hedging problem). *The minimal value of*

$$J(z,u) = E\left[\left(B_1(T) - \left[z + \int_0^T u(t)[dB_1(t) + dB_2(t)]\right]\right)^2\right]$$

is attained when $z = 0$ and $u = \hat{u}(t)$ satisfies (9.36). The corresponding minimal value is

$$\inf_{z,u} J(z,u) = \int_0^T \int_0^T \{[\hat{u}(s) - 1][\hat{u}(t) - 1]\phi_1(s,t) + \hat{u}(s)\hat{u}(t)\phi_2(s,t)\}\,ds\,dt.$$

Remark 9.4.2. Note that if $\phi_1 = \phi_2$, then $\hat{u}(t) \equiv 1/2$, which is the same as the optimal value in the classical Brownian motion case ($H_1 = H_2 = 1/2$).

9.5 Optimal consumption and portfolio in a fractional Black and Scholes market

We address now the problem of finding the optimal consumption and portfolio in two market models affected in different ways by long-range dependence: in the first model the asset price is driven by a *fBm*; in the second one only the volatility is a function of $B_t^{(H)}$, $H > 1/2$.

We start by considering a WIS model of a fractional Black and Scholes (BS) financial market driven by a one-dimensional *fBm* as introduced in Chapter 7. Suppose now that we have the following two investment possibilities:

1. A *risk-free asset*, where the price $A(t)$ at time $t \geq 0$ is given by

$$dA(t) = rA(t)\,dt;\ A(0) = 1\ [\text{i.e.,}\ A(t) = e^{rt}],$$

where $r > 0$ is a constant, $0 \leq t \leq T$ (constant).

2. A stock on a company, whose generalized price process $S(t)$ at time $t \geq 0$ is given by

$$dS(t) = aS(t)\,dt + \sigma S(t)\,dB^{(H)}(t);\ S(0) = s > 0, \qquad (9.37)$$

where $a > r > 0$ and $\sigma \neq 0$ are constants, $0 \leq t \leq T$.

The solution of (9.37) is

$$S(t) = S(0)\exp\left(\sigma B^{(H)}(t) + at - \frac{1}{2}\sigma^2 t^{2H}\right), \qquad t \geq 0,$$

by Example 3.4.4. Suppose an investor chooses a WIS admissible portfolio (Definition 7.2.4) $\theta(t) = (\alpha(t), \beta(t))$ giving the number of units $\alpha(t), \beta(t)$ held at time t of risk-free assets and stocks, respectively. Suppose the investor is also free to choose a (t,ω)-measurable, adapted *consumption process* $c(t,\omega) \geq 0$. The *generalized wealth process* $Z(t) = Z^{c,\theta}(t)$ associated to a given assumption rate c and portfolio $\theta = (\alpha, \beta)$ is defined by [see (7.18)]

$$Z(t) = \alpha(t)A(t) + \beta(t)\diamond S(t). \qquad (9.38)$$

We say that θ is *Wick Itô Skorohod self-financing* with respect to c if

$$dZ(t) = \alpha(t)\,dA(t) + \beta(t)\,dS(t) - c(t)\,dt. \qquad (9.39)$$

From (9.38) we get

$$\alpha(t) = A^{-1}(t)[Z(t) - \beta(t)\diamond S(t)], \qquad (9.40)$$

which substituted into (9.39) gives, using (9.38),

$$dZ(t) = rZ(t)\,dt + (a-r)\beta(t)\diamond S(t)\,dt + \sigma\beta(t)\diamond S(t)\,dB^{(H)}(t) - c(t)\,dt$$

or

$$d(e^{-rt}Z(t)) + e^{-rt}c(t)\,dt = \sigma e^{-rt}\beta(t)\diamond S(t)\left[\frac{a-r}{\sigma}dt + dB^{(H)}(t)\right]. \qquad (9.41)$$

Define the measure $\hat{\mathbb{P}}^H$ on $\mathcal{F}_T^{(H)}$ by

$$\begin{aligned}\frac{d\hat{\mathbb{P}}^H}{d\mathbb{P}^H} &= \exp\left(-\int_0^T K(s)\,dB^{(H)}(s) - \frac{1}{2}\|K\|_H^2\right) \\ &:= \exp^\diamond\left(-\int_0^T K(s)\,dB^{(H)}(s)\right) =: \eta(T),\end{aligned} \qquad (9.42)$$

9.5 Optimal consumption and portfolio in a fractional BS market

where we have

$$K(s) = \frac{(a-r)(Ts-s^2)^{1/2-H}I_{[0,T]}(s)}{2\sigma H \cdot \Gamma(2H) \cdot \Gamma(2-2H)\cos[\pi(H-1/2)]} \quad (9.43)$$

and

$$\|K\|_H^2 = \int_0^T \int_0^T K(s)K(t)\phi(s,t)\,ds\,dt,$$

where Γ is the gamma function and ϕ is defined in (3.1).

Then by the fractional Girsanov formula (Theorem 3.2.4), the process

$$\hat{B}^{(H)}(t) := \frac{a-r}{\sigma}t + B^{(H)}(t) \quad (9.44)$$

is a fBm (with Hurst parameter H) with respect to $\hat{\mathbb{P}}^H$. In terms of $\hat{B}^{(H)}(t)$, we can write (9.41) as follows:

$$e^{-rt}Z(t) + \int_0^t e^{-ru}c(u)\,du = Z(0) + \int_0^t \sigma e^{-ru}\beta(u)\diamond S(u)\,d\hat{B}^{(H)}(u). \quad (9.45)$$

If $Z(0) = z > 0$, we write $Z_z^{c,\theta}(t)$ for the corresponding wealth process $Z(t)$ given by (9.45). We say that (c,θ) is *admissible* with respect to z and write $(c,\theta) \in \mathcal{A}(z)$ if $\theta = \theta(t) = (\alpha(t),\beta(t))$ with $\alpha(t)$ satisfying (9.40) and $\beta(t) = \beta(t,\omega)$ satisfying the condition

$$\beta(\cdot)\diamond S(\cdot) \in \mathcal{L}_\phi(0,T) \quad (9.46)$$

and in addition θ is (WIS) self-financing with respect to c and $Z_z^{c,\theta}(T) \geq 0$ almost surely.

Note that it follows from (9.46) and Theorem 3.2.4 that if we put

$$M(t) := \int_0^t \sigma e^{-ru}\beta(u)\diamond S(u)\,d\hat{B}^{(H)}(u), \quad 0 \leq t \leq T,$$

then $E_{\hat{\mathbb{P}}^H}[M(T)] = 0$, where $E_{\hat{\mathbb{P}}^H}$ is the expectation under $\hat{\mathbb{P}}^H$. Therefore, from (9.45) we get the *budget constraint*

$$E_{\hat{\mathbb{P}}^H}\left[e^{-rT}Z_z^{c,\theta}(T) + \int_0^T e^{-ru}c(u)\,du\right] = z,$$

valid for all $(c,\theta) \in \mathcal{A}(z)$.

Conversely, suppose $c(u) \geq 0$ is a given consumption rate and $F(\omega)$ is a given $\mathcal{F}_T^{(H)}$-measurable random variable such that $E_{\hat{\mathbb{P}}^H}[G^2] < \infty$, where

$$G(\omega) = e^{-rT}F(\omega) + \int_0^T e^{-ru}c(u,\omega)\,du. \quad (9.47)$$

Then by the fractional Clark–Haussmann–Ocone theorem (Theorem 3.10.8) applied to $(\hat{B}^{(H)}(\cdot), \hat{\mathbb{P}}^H)$, we get

$$G(\omega) = E_{\hat{\mathbb{P}}^H}[G] + \int_0^T \psi(t,\omega) \, d\hat{B}^{(H)}(t), \qquad (9.48)$$

where

$$\psi(t,\omega) := \tilde{E}_{\hat{\mathbb{P}}^H}\left[\hat{D}_t^{(H)} G \mid \mathcal{F}_t^{(H)}\right],$$

$\hat{D}_t^{(H)}$ denotes the Malliavin derivative with respect to $\hat{B}^{(H)}$, and

$$\psi(\cdot) \in \mathcal{L}_\phi(0,T).$$

Therefore, if $E_{\hat{\mathbb{P}}^H}[G] = z$ and we define

$$\beta(t) := \sigma^{-1} e^{rt} S^{-1}(t) \diamond \psi(t), \qquad (9.49)$$

then $\beta(t)$ satisfies (9.46) and with $\theta = (\alpha, \beta)$ with α as in (9.40) we have by comparing (9.45) and (9.48)

$$Z_z^{c,\theta}(T) = F \text{ a.s.}$$

We have proved the following:

Lemma 9.5.1. *Let $c(t) \geq 0$ be a given consumption rate, and let F be a given $\mathcal{F}_T^{(H)}$-measurable random variable such that the random variable*

$$G(\omega) := e^{-rT} F(\omega) + \int_0^T e^{-ru} c(u,\omega) \, du$$

satisfies

$$E_{\hat{\mathbb{P}}^H}[G^2] < \infty.$$

Then (9.50) and (9.51) are equivalent:

There exists a portfolio θ such that $(c,\theta) \in \mathcal{A}(x)$ and $Z_z^{c,\theta}(T) = F$ a.s. (9.50)

$$E_{\hat{\mathbb{P}}^H}[G] = z. \qquad (9.51)$$

Now let $D_1 > 0, D_2 > 0, T > 0$ and $\gamma \in (-\infty, 1) \setminus \{0\}$ be given constants. Consider the following quantity

$$J^{(c,\theta)}(z) = E\left[\int_0^T \frac{D_1}{\gamma} c^\gamma(t) \, dt + \frac{D_2}{\gamma} (Z_z^{c,\theta}(T))^\gamma\right], \qquad (9.52)$$

where $(c,\theta) \in \mathcal{A}(z)$ and we interpret Z^γ as $-\infty$ if $Z < 0$. We may regard $J^{(c,\theta)}(z)$ as the total expected utility obtained from the consumption rate $c(t) \geq 0$ and the terminal wealth $Z_z^{c,\theta}(T)$. We now seek $V(z)$ and $(c^*, \theta^*) \in \mathcal{A}(z)$ such that

9.5 Optimal consumption and portfolio in a fractional BS market

$$V(z) = \sup_{(c,\theta)\in\mathcal{A}(z)} J^{(c,\theta)}(z) = J^{c^*,\theta^*}(z), \quad z > 0.$$

By Lemma 9.5.1 we see that this problem is equivalent to the *constrained* optimization problem

$$V(z) = \sup_{c,F\geq 0} E\left[\int_0^T \frac{D_1}{\gamma} c^\gamma(t)dt + \frac{D_2}{\gamma} F^\gamma\right], \qquad (9.53)$$

given that

$$E_{\hat{\mathbb{P}}^H}\left[\int_0^T e^{-ru} c(u)du + e^{-rT} F\right] = z, \qquad (9.54)$$

where the supremum is taken over all consumption rates $c(t,\omega) \geq 0$ and $\mathcal{F}_T^{(H)}$-measurable $F(\omega) \geq 0$ such that

$$\int_0^T e^{-ru} c(u)\,du + e^{-rT} F \in L^2(\hat{\mathbb{P}}^H).$$

Consider for each $\lambda > 0$ the following related *unconstrained* optimization problem:

$$V_\lambda(z) = \sup_{c,F\geq 0} \left\{ E\left[\int_0^T \frac{D_1}{\gamma} c^\gamma(t)\,dt + \frac{D_2}{\gamma} F^\gamma\right] \right.$$
$$\left. - \lambda E_{\hat{\mathbb{P}}^H}\left[\int_0^T e^{-rt} c(t)\,dt + e^{-rT} F\right] \right\}. \qquad (9.55)$$

Suppose that for each $\lambda > 0$ we can find $V_\lambda(z)$ and corresponding $c_\lambda(t,\omega) \geq 0$, $F_\lambda \geq 0$. Moreover, suppose that there exists $\lambda^* > 0$ such that $c_{\lambda^*}, F_{\lambda^*}$ satisfies the constraint in (9.54):

$$E_{\hat{\mathbb{P}}^H}\left[\int_0^T e^{-ru} c_{\lambda^*}(u)du + e^{-rT} F_{\lambda^*}\right] = z. \qquad (9.56)$$

Then, $c_{\lambda^*}, F_{\lambda^*}$ actually solves the *constrained* problem (9.53) because if $c \geq 0$, $F \geq 0$ is another pair satisfying the constraint, then

$$E\left[\int_0^T \frac{D_1}{\gamma} c^\gamma(t)\,dt + \frac{D_2}{\gamma} F^\gamma\right]$$
$$= E\left[\int_0^T \frac{D_1}{\gamma} c^\gamma(t)\,dt + \frac{D_2}{\gamma} F^\gamma\right]$$
$$- \lambda^* E_{\hat{\mathbb{P}}^H}\left[\int_0^T e^{-ru} c(u)\,du + e^{-rT} F\right] + \lambda^* z$$

$$\leq E\left[\int_0^T \frac{D_1}{\gamma} c_{\lambda^*}^\gamma(t)\, dt + \frac{D_2}{\gamma} F_{\lambda^*}^\gamma\right]$$

$$-\lambda^* E_{\hat{\mathbb{P}}^H}\left[\int_0^T e^{-ru} c_{\lambda^*}(u)\, du + e^{-rT} F_{\lambda^*}\right] + \lambda^* z$$

$$= E\left[\int_0^T \frac{D_1}{\gamma} c_{\lambda^*}^\gamma(t)\, dt + \frac{D_2}{\gamma} F_{\lambda^*}^\gamma\right].$$

Finally, to solve the original problem (9.52), we use Lemma 9.5.1 to find θ^* such that $(c_{\lambda^*}, \theta^*) \in \mathcal{A}(z)$ and

$$Z_z^{c_{\lambda^*}, \theta^*}(T) = F_{\lambda^*} \text{ a.s..}$$

Then c_{λ^*}, θ^* are optimal for (9.52) and

$$V(z) = V_{\lambda^*}(z) = E\left[\int_0^T \frac{D_1}{\gamma} c_{\lambda^*}^\gamma(t)\, dt + \frac{D_2}{\gamma} (Z_z^{c_{\lambda^*}, \theta^*}(T))^\gamma\right].$$

In view of the above we now proceed to solve the unconstrained optimization problem (9.55). Note that with

$$\eta(t) = \exp^\diamond\left(-\int_0^t K(s)\, dB^{(H)}(s)\right) \tag{9.57}$$

as in (9.42), we can write

$$\begin{aligned} V_\lambda(z) &= \sup_{c, F \geq 0} E\left[\int_0^T \left(\frac{D_1}{\gamma} c^\gamma(t) - \lambda \eta(T) e^{-rt} c(t)\right) dt \right.\\ &\quad \left. + \frac{D_2}{\gamma} F^\gamma - \lambda \eta(T) e^{-rT} F\right] \\ &= \sup_{c, F \geq 0} E\left[\int_0^T \left(\frac{D_1}{\gamma} c^\gamma(t) - \lambda \rho(t) e^{-rt} c(t)\right) dt \right.\\ &\quad \left. + \frac{D_2}{\gamma} F^\gamma - \lambda \eta(T) e^{-rT} F\right], \end{aligned} \tag{9.58}$$

where

$$\rho(t) =: E\left[\eta(T) \mid \mathcal{F}_t^{(H)}\right].$$

In the above formula we have used that

$$E\left[\eta(T) c(t)\right] = E\left[E\left[\eta(T) c(t) \mid \mathcal{F}_t^{(H)}\right]\right] = E\left[c(t) E\left[\eta(T) \mid \mathcal{F}_t^{(H)}\right]\right]$$
$$= E\left[c(t) \rho(t)\right].$$

9.5 Optimal consumption and portfolio in a fractional BS market

The problem (9.58) can be solved by simply maximizing pointwise (for each t, ω) the two functions

$$g(c) = \frac{D_1}{\gamma} c^\gamma - \lambda \rho(t, \omega) e^{-rt} c, \quad c \geq 0$$

$$h(F) = \frac{D_2}{\gamma} F^\gamma - \lambda \eta(T, \omega) e^{-rT} F, \quad F \geq 0$$

for each $t \in [0, T]$ and $\omega \in \Omega$.
We have $g'(c) = 0$ for

$$c = c_\lambda(t, \omega) = \frac{1}{D_1} [\lambda e^{-rt} \rho(t, \omega)]^{1/(\gamma-1)} \tag{9.59}$$

and by concavity this is the maximum point of g. Similarly,

$$F = F_\lambda(\omega) = \frac{1}{D_2} \left[\lambda e^{-rT} \eta(T, \omega)\right]^{1/(\gamma-1)} \tag{9.60}$$

is the maximum point of h. We now seek λ^* such that (9.56) holds, i.e.,

$$E\left[\int_0^T e^{-rt} \rho(t) \frac{1}{D_1} [\lambda e^{-rt} \rho(t)]^{1/(\gamma-1)} \, dt \right.$$

$$\left. + e^{-rt} \eta(T) \frac{1}{D_2} [\lambda e^{-rT} \eta(T)]^{1/(\gamma-1)} \right] = z$$

or

$$\lambda^{1/(\gamma-1)} N = z,$$

where

$$N = E\left[\int_0^T \frac{1}{D_1} e^{r\gamma/(1-\gamma)t} \rho(t)^{\gamma/(\gamma-1)} dt + \frac{1}{D_2} e^{r\gamma/(1-\gamma)T} \eta(T)^{\gamma/(\gamma-1)}\right] > 0. \tag{9.61}$$

Hence,

$$\lambda^* = \left(\frac{z}{N}\right)^{\gamma-1}.$$

Substituted into (9.59) and (9.60) this gives

$$c_{\lambda^*}(t, \omega) = \frac{z}{D_1 N} e^{r/(1-\gamma)t} \rho(t, \omega)^{1/(\gamma-1)} \tag{9.62}$$

and

$$F_{\lambda^*}(\omega) = \frac{z}{D_2 N} e^{r/(1-\gamma)T} \eta(T, \omega)^{1/(\gamma-1)}. \tag{9.63}$$

This is the optimal c, F for the constrained problem (9.53), and we conclude that the solution of the original problem is

$$V(z) = V_{\lambda^*}(z) = E\left[\int_0^T \frac{D_1}{\gamma} c_{\lambda^*}^\gamma(t)\, dt + \frac{D_2}{\gamma} F_{\lambda^*}^\gamma\right]. \tag{9.64}$$

To find $V(z)$ we need to compute $E\left[\rho(t)^{\gamma/(\gamma-1)}\right]$. For $t = T$, this was done in (2.19) to (2.27) in [124].

Define $K^{(t)} = K \cdot I_{[0,t]}$, where $K(\cdot)$ is defined in (9.43). From (3.6) and (3.8) of [113], we obtain

$$\rho(t) = E\left[\eta(T) \mid \mathcal{F}_t^{(H)}\right] = \exp\left(\int_0^t \zeta(s)\, dB^{(H)}(s) - \frac{1}{2}\|\zeta\|_H^2\right),$$

where ζ is determined by the following equation:

$$(-\Delta)^{-(H-1/2)}\zeta(s) = -(-\Delta)^{-(H-1/2)}K^{(T)}(s), \quad 0 \le s \le t;$$
$$\zeta(s) = 0 \quad s < 0, \quad \text{or} \quad s > t.$$

We have

$$\zeta(s) = -\kappa_H s^{1/2-H} \frac{d}{ds} \int_s^t w^{2H-1}(w-s)^{1/2-H}\, dw$$
$$\cdot \frac{d}{dw} \int_0^w z^{1/2-H}(w-z)^{1/2-H} g(z)\, dz,$$

where $g(z) = -(-\Delta)^{-(H-1/2)} K^{(T)}(z)$ and

$$\kappa_H = \frac{2^{2H-2}\sqrt{\pi}\,\Gamma(-1/2)}{\Gamma(1-H)\Gamma^2(3/2-H)\cos[\pi(H-1/2)]}.$$

Hence,

$$E\left[\rho(t)^{\gamma/(\gamma-1)}\right] = E\left[\exp\left(-\frac{\gamma}{1-\gamma}\left[\int_0^t \zeta(s)\, dB^{(H)}(s) - \frac{1}{2}\|\zeta\|_H^2\right]\right)\right]$$
$$= E\left[\exp\left(-\frac{\gamma}{1-\gamma}\int_0^t \zeta(s)\, dB^{(H)}(s) - \frac{\gamma^2}{2(1-\gamma)^2}\|\zeta\|_H^2\right.\right.$$
$$\left.\left.+ \left(\frac{\gamma^2}{2(1-\gamma)^2} + \frac{\gamma}{2(1-\gamma)}\right)\|\zeta\|_H^2\right)\right]$$
$$= \exp\left(\frac{\gamma}{2(1-\gamma)^2}\|\zeta\|_H^2\right). \tag{9.65}$$

In the special case $t = T$ we see that $\zeta = K^{(T)} = K \cdot I_{[0,T]}$, where

$$\int_0^T K(s)\phi(s,t)\, ds = \frac{a-r}{\sigma} \quad \text{for } 0 \le t \le T.$$

Thus,

$$\|K^{(T)}\|_H^2 = \|K\|_H^2 = \frac{a-r}{\sigma} \int_0^T K(t)\, dt$$

$$= \frac{(a-r)^2}{2\sigma^2 H \cdot \Gamma(2H) \cdot \Gamma(2-2H)\cos(\pi(H-1/2))} \int_0^T (Tt - t^2)^{1/2-H} dt$$

$$= \frac{(a-r)^2}{\sigma^2} \Lambda_H \cdot T^{2-2H}, \tag{9.66}$$

where

$$\Lambda_H = \frac{\Gamma^2(3/2 - H)}{2H \cdot (2 - 2H) \cdot \Gamma(2H) \cdot \Gamma(2-2H)\cos[\pi(H-1/2)]}.$$

Substituting (9.65) and (9.66) into (9.61), we get

$$N = \frac{1}{D_1} \int_0^T \exp\left(\frac{r\gamma}{1-\gamma}t + \frac{\gamma}{2(1-\gamma)^2}\|\zeta\|_H^2\right) dt$$

$$+ \frac{1}{D_2} \exp\left(\frac{r\gamma}{1-\gamma}T + \frac{\gamma(a-r)^2 \Lambda_H}{2(1-\gamma)^2 \sigma^2} T^{2-2H}\right) \tag{9.67}$$

and (9.64) gives

$$V(z) = \frac{z^\gamma}{\gamma}\left[D_1^{-1\gamma} N^{-\gamma} \int_0^T \exp\left(\frac{r\gamma}{1-\gamma}t + \frac{\gamma}{2(1-\gamma)^2}\|\zeta\|_H^2\right) dt \right.$$

$$\left. + D_2^{-1\gamma} N^{-\gamma} \exp\left(\frac{r\gamma}{1-\gamma}T + \frac{\gamma(a-r)^2 \Lambda_H}{2(1-\gamma)^2 \sigma^2} T^{2-2H}\right)\right]. \tag{9.68}$$

We have proved the following:

Theorem 9.5.2. *The value function $V(z)$ of the optimal consumption and portfolio problem (9.52) is given by (9.67) and (9.68). The corresponding optimal consumption c_{λ^*} is given by (9.62), and the corresponding optimal terminal wealth $Z_z^{c_{\lambda^*}^*, \pi^*} = F_{\lambda^*}$ is given by (9.63).*

Remark 9.5.3. It is an interesting question how the value function $V(z) = V^{(H)}(z)$ of problem (9.52) depends on the Hurst parameter $H \in (1/2, 1)$. We will not pursue this question here, but simply note that since $\Lambda_{1/2} = 1$, we have

$$\lim_{H \to 1/2+} V^{(H)}(z) = V^{(\frac{1}{2})}(z)$$

where $V^{(\frac{1}{2})}(z)$ is the (well-known) value function in the standard Brownian motion case.

It remains to find the optimal portfolio $\theta^* = (\alpha^*, \beta^*)$ for problem (9.52). For this we use the fractional Clark–Haussmann–Ocone theorem (Theorem 3.10.8) with

$$G(\omega) = e^{-rt} F_{\lambda^*}(\omega) + \int_0^T e^{-ru} c_{\lambda^*}(u, \omega) \, du$$

as in (9.47). Then by (9.49)

$$\beta^*(t) = \sigma^{-1} e^{rt} S^{-1}(t) \tilde{E}_{\hat{\mathbb{P}}^H} \left[\hat{D}_t^{(H)} G \mid \mathcal{F}_t^{(H)} \right]. \tag{9.69}$$

To compute this we first note that by Theorem 3.2.4, (9.57) and (9.44), we have

$$\eta(t)^{\frac{1}{\gamma-1}} = \exp\left(\frac{1}{1-\gamma} \int_0^t K(s) \, dB^{(H)}(s) + \frac{1}{2(1-\gamma)} \|K^{(t)}\|_H^2 \right)$$

$$= \exp\left(\frac{1}{1-\gamma} \int_0^t K(s) \, d\hat{B}^{(H)}(s) - \frac{a-r}{\sigma(1-\gamma)} \int_0^t K(s) \, ds \right.$$
$$\left. + \frac{1}{2(1-\gamma)} \|K^{(t)}\|_H^2 \right)$$

$$= \exp\left(\frac{1}{1-\gamma} \int_0^t K(s) \, d\hat{B}^{(H)}(s) - \frac{1}{2(1-\gamma)^2} \|K^{(t)}\|_H^2 \right.$$
$$\left. + \frac{1}{2(1-\gamma)^2} \|K^{(t)}\|_H^2 - \frac{a-r}{\sigma(1-\gamma)} \int_0^t K(s) \, ds \right.$$
$$\left. + \frac{1}{2(1-\gamma)} \|K^{(t)}\|_H^2 \right)$$

$$= \exp^\diamond \left(\frac{1}{1-\gamma} \int_0^t K(s) \, d\hat{B}^{(H)}(s) \right) \cdot R(t),$$

where

$$R(t) = \exp\left(\frac{2-\gamma}{2(1-\gamma)^2} \|K^{(t)}\|_H^2 - \frac{a-r}{\sigma(1-\gamma)} \int_0^t K(s) \, ds \right).$$

Hence, by (9.63)

$$\tilde{E}_{\hat{\mathbb{P}}^H} \left[\hat{D}_t^{(H)} \left(e^{-rT} F_{\lambda^*} \right) \mid \mathcal{F}_t^{(H)} \right]$$

$$= \frac{z}{D_2 N} e^{-rT} e^{rT/(1-\gamma)} \tilde{E}_{\hat{\mathbb{P}}^H} \left[\hat{D}_t^{(H)} \left(\eta(T)^{1/(\gamma-1)} \right) \mid \mathcal{F}_t^{(H)} \right]$$

$$= \frac{z}{D_2 N} \exp\left(\frac{r\gamma T}{1-\gamma} \right) \tilde{E}_{\hat{\mathbb{P}}^H} \left[\frac{K(t)}{1-\gamma} \eta(T)^{1/(\gamma-1)} \mid \mathcal{F}_t^{(H)} \right]$$

$$= \frac{z}{D_2 N} \exp\left(\frac{r\gamma T}{1-\gamma} \right) \frac{K(t)}{1-\gamma} R(T)$$

$$\cdot \tilde{E}_{\hat{\mathbb{P}}^H} \left[\exp^\diamond \left(\frac{1}{1-\gamma} \int_0^T K(s) \, d\hat{B}^{(H)}(s) \right) \Big| \mathcal{F}_t^{(H)} \right]$$

$$= \frac{z}{D_2 N} \exp\left(\frac{r\gamma T}{1-\gamma} \right) \frac{K(t)}{1-\gamma} R(T) \exp^\diamond \left(\frac{1}{1-\gamma} \int_0^t K(s) \, d\hat{B}^{(H)}(s) \right)$$

9.5 Optimal consumption and portfolio in a fractional BS market

$$= \frac{z}{D_2 N} \exp\left(\frac{r\gamma T}{1-\gamma}\right) R(T) \cdot \frac{K(t)}{1-\gamma} \cdot \exp\left(\frac{1}{1-\gamma}\int_0^t K(s)\, d\widehat{B}^{(H)}(s)\right.$$
$$\left. - \frac{1}{1-\gamma} \|K^{(t)}\|_H^2 \right)$$

$$= \frac{zK(t)}{D_2 N(1-\gamma)} \exp\left(\frac{r\gamma T}{1-\gamma} + \frac{1}{1-\gamma}\int_0^t K(s)\, dB^{(H)}(s) - \frac{a-r}{\sigma(1-\gamma)}\right.$$
$$\left. \cdot \int_t^T K(s)\, ds + \frac{2-\gamma}{2(1-\gamma)^2}\|K^{(T)}\|_H^2 - \frac{1}{1-\gamma}\|K^{(t)}\|_H^2\right). \quad (9.70)$$

Similarly, by (9.62) and (9.70),

$$\tilde{E}_{\hat{\mathbb{P}}^H}\left[\hat{D}_t^{(H)}\left(\int_0^T e^{-ru}c_{\lambda^*}(u)\, du\right) \mid \mathcal{F}_t^{(H)}\right]$$

$$= \frac{z}{D_1 N}\tilde{E}_{\hat{\mathbb{P}}^H}\left[\int_0^T \hat{D}_t^{(H)}\left(e^{-ru}e^{ru/(1-\gamma)}\rho(u)^{1/(\gamma-1)}\right) du \mid \mathcal{F}_t^{(H)}\right]$$

$$= \frac{z}{D_1 N}\int_0^T \exp\left(\frac{r\gamma u}{1-\gamma}\right) \tilde{E}_{\hat{\mathbb{P}}^H}\left[\hat{D}_t^{(H)}\left(\rho(u)^{1/(\gamma-1)}\right) \mid \mathcal{F}_t^{(H)}\right] du$$

$$= \frac{zh(t)}{D_1 N(1-\gamma)} \int_0^T \exp\left(\frac{r\gamma u}{1-\gamma}\right) \exp\left(\frac{1}{1-\gamma}\int_0^t \zeta(s)\, dB^{(H)}(s)\right.$$
$$\left. - \frac{a-r}{\sigma(1-\gamma)}\int_t^u \zeta(s)\, ds + \frac{2-\gamma}{2(1-\gamma)^2}\|\zeta\|_H^2 - \frac{1}{1-\gamma}\|\zeta\|_H^2\right) du$$

$$= \frac{z\zeta(t)}{D_1 N(1-\gamma)} \exp\left(\frac{1}{1-\gamma}\int_0^t K(s)\, dB^{(H)}(s) - \frac{1}{1-\gamma}\|K^{(t)}\|_H^2\right) \quad (9.71)$$

$$\cdot \int_0^T \exp\left(\frac{r\gamma u}{1-\gamma} + \frac{2-\gamma}{2(1-\gamma)^2}\|K^{(u)}\|_H^2 - \frac{a-r}{\sigma(1-\gamma)}\int_t^{u\wedge t} K(s)\, ds\right) du.$$

Adding (9.70) and (9.71) and using (9.69), we get the following:

Theorem 9.5.4. *The optimal portfolio* $\theta^*(t) = (\alpha^*(t), \beta^*(t))$ *for problem* (9.52) *is given by*

$$\beta^*(t) = \sigma^{-1} e^{rt} S^{-1}(t)(Y_1 + Y_2),$$

where

$$Y_1 = \tilde{E}_{\hat{\mathbb{P}}^H}\left[\hat{D}_t^{(H)}\left(e^{-rT} F_{\lambda^*}\right) \mid \mathcal{F}_t^{(H)}\right]$$

is given by (9.70) *and*

$$Y_2 = \tilde{E}_{\hat{\mathbb{P}}^H}\left[\hat{D}_t^{(H)}\left(\int_0^T e^{-ru}c_{\lambda^*}(u)\, du\right) \mid \mathcal{F}_t^{(H)}\right]$$

is given by (9.71), *and*

$$\alpha^*(t) = e^{-rt}\left[Z^*(t) - \beta^*(t)S(t)\right]$$

with

$$e^{-rt}Z^*(t) + \int_0^t e^{-ru}c_{\lambda^*}(u)du = z + \int_0^t \sigma e^{-ru}\beta^*(u)S(u)\,d\widehat{B}^{(H)}(u)$$

and $c_{\lambda^*}(u)$ given by (9.62).

9.6 Optimal consumption and portfolio in presence of stochastic volatility driven by fBm

As a final example we compute the optimal consumption and portfolio in a market affected by *stochastic volatility driven by* fBm. Fix a terminal time $T > 0$ and consider a market with two securities.

1. A *risk-free asset* whose price per share A_t at time $t \geq 0$ is given by

$$dA_t = r_t A_t dt\,;\; A_0 = 1 \quad \text{(i.e., } A_t = e^{\int_0^t r_s ds}\text{)},$$

where $r_t > 0$ is a given adapted stochastic process satisfying

$$E\left[\int_0^T |r_t|\,dt\right] < \infty.$$

2. A stock whose price per share S_t at time $t \geq 0$ is given by the solution of a generalized "geometric Brownian motion" on $(\Omega, \mathcal{F}, \mathbb{P})$ with stochastic volatility

$$dS_t = \mu_t S_t dt + \sigma_t S_t dB_t\,,\; S_0 > 0,$$

where $\mu_t (> r_t)$ is a given adapted stochastic processes and $\sigma_t = f(t, Y_t)$ for a certain continuous function f and

$$dY_t = a(t, Y_t)\,dt + b(t, Y_t)\,dB_t + k(t, Y_t)\,dB_t^{(H)}. \tag{9.72}$$

Let $\mathcal{F}_t = \sigma(B_s,\, 0 \leq s \leq t)$ be the σ-algebra generated by $B_s, 0 \leq s \leq t$ and $\mathcal{G}_t = \sigma(B_s,\, 0 \leq s \leq t) \vee \sigma(B_s^{(H)}, 0 \leq s \leq T)$ be the σ-algebra generated by $B_s, 0 \leq s \leq t$ and $B_s^{(H)}, 0 \leq s \leq T$. A *portfolio* is a pair of $\mathcal{G}_t$-adapted process $\theta_t = (\alpha_t, \beta_t),\, 0 \leq t \leq T$, where α_t and β_t denote the numbers of shares in risk-free asset and stock, respectively, held by the investor at time t. Here we assume that the volatility is observable to simplify the setting. We first search the optimal portfolio among all portfolios which are adapted to the filtration $\mathcal{G}_t$. Then we will show that the optimal portfolio in this class is, in fact, $\mathcal{H}_t = \sigma(B_s,\, B_s^{(H)}, 0 \leq s \leq t)$ adapted. With this portfolio the investor's wealth at time instant t is

$$Z_t = Z_t^\theta = \alpha_t A_t + \beta_t S_t. \tag{9.73}$$

Let $c = (c_t, 0 \le t \le T)$ be a given $\mathcal{G}_t$-adapted process, denoting the investor's consumption rate. We assume that $c_t \ge 0$ and

$$\int_0^T c_t\, dt < \infty \quad \text{almost surely.}$$

The set of all such $(c_t, 0 \le t \le T)$ is denoted by $\mathcal{C}$.

A portfolio $\theta_t = (\alpha_t, \beta_t), 0 \le t \le T$ is called *self-financing* with respect to the consumption rate c if

$$dZ_t^\theta = \alpha_t dA_t + \beta_t dS_t - c_t dt, \quad 0 \le t \le T, \tag{9.74}$$

where Z_t^θ is given by (9.73). Denote $\xi_t = \exp\left(-\int_0^t r_s\, ds\right)$.

A self-financing $\mathcal{G}_t$-adapted portfolio $\theta = (\alpha_t, \beta_t)$ is called *admissible* with respect to c if

$$E\left[\int_0^T |\beta_t \sigma_t S_t \xi_t|^2 dt\right] < \infty. \tag{9.75}$$

The set of all admissible portfolios is denoted by $\mathcal{A}$. We shall denote by $Z_t^{\theta,c}$ an investor's wealth at time t that is associated with the portfolio θ and that is self-financing with respect to c. Namely, $Z_t^{\theta,c}$ satisfies (9.73), (9.74) and (9.75).

Let g and ψ be two given continuous concave functions. Define

$$J(\theta, c) = E\left[\int_0^T \psi(c_t)\, dt + g(Z_T^{\theta,c})\right].$$

The problem that will be studied in this section is the following:
Problem I: Find an admissible portfolio $\theta^* \in \mathcal{A}$ and a consumption rate $c^* \in \mathcal{C}$ such that

$$J(\theta^*, c^*) \ge J(\theta, c) \quad \forall \theta \in \mathcal{A}, \ c \in \mathcal{C}.$$

When r_t, μ_t and σ_t are constants, a similar problem was proposed and solved by R. Merton [162]. If Y_t is driven only by standard Brownian motion, namely, in the equation (9.72), $a_2(t, Y_t) = 0$, the problem has been discussed in the literature (see for example, [98]). Here we shall assume that the volatility process is observable and driven both by a standard Brownian motion and a fBm. Unlike in the only one standard Brownian motion case, we are no longer in a Markovian setup. We shall use the method which appeared in [63], [64] and [124].

Lemma 9.6.1. *Define*
$$\rho_s = \frac{a_s - r_s}{\sigma_s}$$
and let
$$E\left[\exp\left(\frac{1}{2}\int_0^T |\rho_s|^2 \, ds\right)\right] < \infty.$$
Let $\mathbb{Q}$ be a probability measure on $(\Omega, \mathcal{F})$ defined by
$$\frac{d\mathbb{Q}}{d\mathbb{P}} = \eta(T), \tag{9.76}$$
where
$$\eta(t) := \exp\left(\int_0^t \rho_s \, dB_s - \frac{1}{2}\int_0^t \rho_s^2 \, ds\right),$$
Denote $\xi_t = \exp\left(-\int_0^t r_s ds\right)$, $0 \le t \le T$. Assume that F is a given random variable which is $\mathcal{F}_T$ measurable. Then the following statements about F are equivalent:

1. *There is an admissible portfolio θ and a consumption rate c such that $Z_0^{\theta,c} = z$ and $Z_T^{\theta,c} = F$.*
2. *$G = \xi_T F + \int_0^T \xi_s c_s \, ds$ is square integrable and*
$$E_{\mathbb{Q}}[G] = z.$$

Proof. From (9.73) we see that a self-financing portfolio is uniquely determined by β_t, $0 \le t \le T$. In fact, we have
$$\alpha_t = \frac{Z_t - \beta_t S_t}{A_t} = \xi_t(z_t - \beta_t S_t), \tag{9.77}$$
which substituted into (9.74) yields
$$dZ_t = r_t Z_t \, dt - c_t \, dt + (\mu_t - r_t)\beta_t S_t \, dt + \sigma_t \beta_t S_t \, dB_t.$$
This we can also write as
$$dZ_t - r_t Z_t \, dt + c_t \, dt = \sigma_t \beta_t S_t \left(dB_t + \frac{\mu_t - r_t}{\sigma_t} dt\right)$$
$$= \sigma_t \beta_t S_t \left(dB_t + \rho_t \, dt\right).$$
Using the definition of ξ_t we may write the above equation as
$$d(\xi_t Z_t) + \xi_t c_t \, dt = \sigma_t \xi_t \beta_t S_t \, d\tilde{B}_t,$$
where
$$\tilde{B}_t = B_t + \frac{\mu_t - r_t}{\sigma_t}.$$

Hence
$$\xi_T Z_T + \int_0^T \xi_s c_s \, ds = z + \int_0^T \sigma_s \xi_s \beta_s S_s \, d\tilde{B}_s.$$

If $\mathbb{Q}$ is a probability measure on $(\Omega, \mathcal{F})$ defined by (9.76), then from the Girsanov theorem, $\tilde{B}_t$ is a Brownian motion on the probability space $(\Omega, \mathcal{F}, \mathbb{Q})$. Therefore, by (9.75) we have

$$E_{\mathbb{Q}}\left[\xi_T Z_T + \int_0^T \xi_s c_s \, ds\right] = z.$$

On the other hand, let

$$G(\omega) = \xi_T F + \int_0^T \xi_s c_s \, ds.$$

If $E_{\mathbb{Q}}[G] = z$ and G is square integrable, then there is a unique $\mathcal{G}_t$-adapted stochastic process f_t such

$$G = E_{\mathbb{Q}}[G] + \int_0^T f_t \, d\tilde{B}_t.$$

Define

$$\beta_t = \frac{f_t}{\sigma_t \xi_t S_t}$$

and α_t by (9.77). Then $\theta_t = (\alpha_t, \beta_t)$ is the portfolio that we seek. □

From this lemma we see that Problem I is equivalent to the following problem.

Problem II: *Find a $\mathcal{G}_t$-adapted nondecreasing process c_t and a $\mathcal{G}_T$-measurable nonnegative random variable F subject to*

$$E_{\mathbb{Q}}\left[\xi_T F + \int_0^T \xi_s c_s \, ds\right] = z, \tag{9.78}$$

which maximizes

$$E_{\mathbb{Q}}\left[g(F) + \int_0^T \psi(c_s) \, ds\right].$$

We shall use the Lagrange multiplier method to solve this constrained Problem II. Consider for each $\lambda > 0$ the following *unconstrained* optimization problem

$$V_\lambda(z) = \sup_{c, F \geq 0} \left\{ E\left[g(F) + \int_0^T \psi(c_s) \, ds\right] - \lambda E_{\mathbb{Q}}\left[\int_0^T \xi_t c_t \, dt + \xi_T F\right] \right\}.$$

9 Stochastic optimal control and applications

Lemma 9.6.2. *Suppose that for each $\lambda > 0$ one can find $V_\lambda(z)$ and corresponding optimal $c_\lambda(t, \omega) \geq 0, F_\lambda \geq 0$. If there exists $\lambda^* > 0$ such that $c^* = c_{\lambda^*}, F^* = F_{\lambda^*}$ satisfy the constraint (9.78), i.e.,*

$$E_{\mathbb{Q}}\left[\int_0^T \xi_s c_{\lambda^*}(s)\, ds + \xi_T F_{\lambda^*}\right] = z,$$

then $c_{\lambda^*}, F_{\lambda^*}$ solve the constrained Problem II.

Proof. If $c \geq 0, F \geq 0$ is any pair satisfying the constraint, then

$$E\left[\int_0^T \psi(c_t)\, dt + g(F)\right] = E\left[\int_0^T \psi(c_t)\, dt + g(F)\right]$$

$$- \lambda^* E_{\mathbb{Q}}\left[\int_0^T \xi_s c_s\, ds + \xi_T F\right] + \lambda^* z$$

$$\leq E\left[\int_0^T \psi(c_t^*)\, dt + g(F^*)\right]$$

$$- \lambda^* E_{\mathbb{Q}}\left[\int_0^T \xi_s c_s^*\, ds + \xi_T F^*\right] + \lambda^* z$$

$$= E\left[\int_0^T \xi_s c_s^*\, ds + g(F^*)\right].$$

This proves the lemma. □

From this lemma it follows that Problem I is equivalent to the following problem.
Problem III: Find λ^*, F_{λ^*} and c_{λ^*} such that $F = F_{\lambda^*}$ and $c = c_{\lambda^*}$ maximize

$$J_{\lambda^*}(F, c) = E\left[\int_0^T \psi(c_t)\, dt + g(F)\right] - \lambda^* E_{\mathbb{Q}}\left[\int_0^T \xi_s c_s\, ds + \xi_T F\right] \quad (9.79)$$

for the fixed λ^* and the following holds for λ^*:

$$E_{\mathbb{Q}}\left[\int_0^T \xi_s c_{\lambda^*}(s)\, ds + \xi_T F_{\lambda^*}\right] = z. \quad (9.80)$$

Now we outline the general method to solve the unconstrained optimization problem (9.79) and (9.80). Using the definition of η_t, we can write

$$V_\lambda(z) = \sup_{c, F \geq 0} E\left[\int_0^T (\psi(c_t) - \lambda \eta_T \xi_t c_t)\, dt + g(F) - \lambda \eta_T \xi_T F\right]$$
$$= \sup_{c, F \geq 0} E\left[\int_0^T (\psi(c_t) - \lambda \eta_t \xi_t c_t)\, dt + g(F) - \lambda \eta_T \xi_T F\right].$$

The problem (9.79) can be solved by maximizing the following two functions:

$$f_t(c) = \psi(c) - \lambda \eta_t \xi_t c, \ c \geq 0,$$

$$h_t(F) = g(F) - \lambda \eta_T \xi_T F, \ F \geq 0,$$

for each $t \in [0, T]$ and $\omega \in \Omega$. Since g and ψ are continuous concave functions for any given λ, the maximum $c_\lambda(t, \omega)$ and $F_\lambda(\omega)$ exist as a function of λ. Once this is done, one can substitute $c_\lambda(t, \omega)$ and $F_\lambda(\omega)$ into (9.78) to obtain an equation for λ^*. Then $c_{\lambda^*}(t, \omega)$ and $F_{\lambda^*}(\omega)$ is the solution to (9.79) and (9.80). We shall give more explicit solutions for some specific utility functions. Now we assume that

$$g(x) = \frac{x^\gamma}{D_1 \gamma} \quad \text{and} \quad \psi(x) = \frac{x^\gamma}{D_2 \gamma}.$$

In this case

$$f_t(c) = \frac{D_1}{\gamma} c^\gamma - \lambda \eta(t, \omega) \xi_t c, \ c \geq 0,$$

$$h(F) = \frac{D_2}{\gamma} F^\gamma - \lambda \eta(t, \omega) \xi_T F, \ F \geq 0,$$

for all $t \in [0, T]$ and $\omega \in \Omega$. We have $f_t'(c) = 0$ for

$$c = c_\lambda(t, \omega) = \left(\frac{\lambda \xi_t \eta_T}{D_1} \right)^{1/(\gamma-1)} \tag{9.81}$$

and by concavity this is the maximum point of f_t. Similarly,

$$F = F_\lambda(\omega) = \left(\frac{\lambda \xi_T \eta_T}{D_2} \right)^{1/(\gamma-1)} \tag{9.82}$$

is the maximum point of h_t.

We now seek λ^* such that (9.80) holds, i.e.,

$$E\left[\int_0^T \xi_t \eta_t \left(\frac{\lambda \xi_t \eta_t}{D_1} \right)^{1/(\gamma-1)} dt + \xi_T \eta_T \left(\frac{\lambda \xi_T \eta_T}{D_2} \right)^{1/(\gamma-1)} \right] = z$$

or

$$\lambda^{1/(\gamma-1)} N = z,$$

where

$$N = E\left[\int_0^T \xi_t \eta_t \left(\frac{\xi_t \eta_t}{D_1} \right)^{1/(\gamma-1)} dt + \xi_T \eta_T \left(\frac{\xi_T \eta_T}{D_2} \right)^{1/(\gamma-1)} \right]$$

$$= \int_0^T \frac{\xi_t^{\gamma/(\gamma-1)}}{D_1^{1/(\gamma-1)}} E\left[\eta_t^{1/(\gamma-1)} \right] dt + \frac{\xi_T^{\gamma/(\gamma-1)}}{D_2^{1/(\gamma-1)}} E\left[\eta_T^{1/(\gamma-1)} \right].$$

Hence,
$$\lambda^* = \left(\frac{z}{N}\right)^{\gamma-1}.$$

Substituted into (9.81) and (9.82) yields the optimal consumption rate

$$c_{\lambda^*}(t,\omega) = \frac{z}{N}\left(\frac{\xi_t \eta_t}{D_1}\right)^{1/(\gamma-1)}$$

and

$$F_{\lambda^*}(\omega) = \frac{z}{N}\left(\frac{\xi_T \eta_T}{D_1}\right)^{1/(\gamma-1)}.$$

If $g(x) = (\log x)/D_1$ and $\psi(x) = (\log x)/D_2$, then in similar way we have

$$\lambda^* = \frac{D_1 + D_2}{z},$$

$$F^* = \frac{D_2}{\lambda^* \xi_T \eta_T},$$

and

$$c^*(t,\omega) = \frac{D_2}{\lambda^* \xi_t \eta_t}.$$

It is easy to see that $c^*(t,\omega)$ is $\mathcal{H}_t$-adapted.

10
Local time for fractional Brownian motion

In this chapter we present the main results concerning the local time of the *fBm* and provide its chaos expansion. In addition, we investigate the definition and the properties of the weighted and renormalized self-intersection local time for *fBm* and present a Meyer Tanaka formula valid for every $H \in (0,1)$.

The main references for this part are [28], [62], [87], [100], [110], [120], [122], [123] and [177].

10.1 Local time for *fBm*

Fix a *d-dimensional Hurst parameter*

$$H = (H_1, H_2, \ldots, H_d) \in (0,1)^d,$$

and let $B^{(H)}(t) = (B_1^{(H_1)}(t), \ldots, B_d^{(H_d)}(t))$, $t \in \mathbb{R}$, be a *d-dimensional fBm* on $(\Omega, \mathcal{F}, \mathbb{P})$, where $\mathbb{P}$ is the measure defined in (4.41). We are considering d independent *fBm* with Hurst parameters $H_1, H_2, \ldots, H_d$, respectively, such that

$$E\left[B_j^{(H_j)}(t) B_k^{(H_k)}(s)\right] = \frac{1}{2}\left\{|s|^{2H_j} + |t|^{2H_j} - |s-t|^{2H_j}\right\} \delta_{jk}, \qquad 1 \le j, k \le d,$$

where

$$\delta_{jk} = \begin{cases} 0 & \text{when } j \ne k, \\ 1 & \text{when } j = k. \end{cases}$$

Given $x \in \mathbb{R}^d$, the local time of *fBm* $B^{(H)}(t)$ can be heuristically described by

$$\ell_T(x) = \ell_T^{(H)}(x) = \int_0^T \delta(B^{(H)}(t) - x)\, dt,$$

where δ is the Dirac delta function, which is approximated by

10 Local time for fractional Brownian motion

$$P_\varepsilon(x) = \frac{1}{(2\pi\varepsilon)^{d/2}} e^{-|x|^2/(2\varepsilon)} = \frac{1}{(2\pi)^d} \int_{\mathbb{R}^d} e^{ix\xi - 1/2\varepsilon\xi^2} d\xi, \qquad x \in \mathbb{R}^d, \quad (10.1)$$

where $i^2 = -1$. This means that formally

$$\delta(x) = \lim_{\varepsilon \to 0} P_\varepsilon(x) = \frac{1}{(2\pi)^d} \int_{\mathbb{R}^d} e^{ix\xi} d\xi.$$

The justification of this is similar to those in [4] and [112].
More rigorously we follow [123] and introduce the local time $\ell_T(x)$ of fBm $B^{(H)}(t) = (B_1^{(H_1)}(t), \ldots, B_d^{(H_d)}(t))$ at points $x \in \mathbb{R}^d$ up to time $T > 0$ by using the *(Donsker) delta function*

$$\delta_{B^{(H)}(t)}(x) = \delta(B^{(H)}(t) - x)$$

of $B^{(H)}(t)$ at $x \in \mathbb{R}^d$, a concept we will make precise using the fractional white noise theory introduced in Chapter 4. We now proceed as in [2] and define the *Donsker delta function* as follows:

Definition 10.1.1. *Let $Y : \Omega \to \mathbb{R}^d$ be a random variable which also belongs to $(\mathcal{S})^*$. Then a continuous function*

$$\delta_Y(\cdot) : \mathbb{R}^d \to (\mathcal{S})^*$$

is called a Donsker delta function *of Y if it has the property that*

$$\int_{\mathbb{R}^d} g(y)\delta_Y(y)\, dy = g(Y) \qquad a.s. \qquad (10.2)$$

for all (measurable) $g : \mathbb{R}^d \to \mathbb{R}$ such that the integral converges in $(\mathcal{S})^*$.

As in [2, Prop. 4.2] we can now prove the following:

Proposition 10.1.2. *Suppose $Y : \Omega \to \mathbb{R}^d$ is normally distributed with mean $m = (m_1, m_2, \ldots, m_d) = E[Y]$ and covariance matrix $C = [c_{jk}]_{1 \leq j,k \leq d}$. Suppose C is invertible with inverse*

$$A = C^{-1} = [a_{jk}]_{1 \leq j,k \leq d}.$$

Then $\delta_Y(y)$ exists, is unique, and, is given by

$$\delta_Y(y) = (2\pi)^{-d/2}\sqrt{|A|}\, \exp^\diamond\Big(-\frac{1}{2}\sum_{j,k=1}^d a_{jk}(Y_j + m_j - y_j) \diamond (Y_k + m_k - y_k)\Big),$$
$$(10.3)$$

where $|A|$ is the determinant of A.

Proof. The proof of Proposition 4.2 in [2] applies. We omit the details. □

Proposition 10.1.3. *Let $Y, C = [c_{jk}]$, $A = C^{-1} = [a_{jk}]$ be as in Proposition 10.1.2. Then*

$$\delta_Y(y) = (2\pi)^{-d} \int_{\mathbb{R}^d} \exp^\diamond \left(i \sum_{j=1}^d \xi_j (Y_j + m_j - y_j) - \frac{1}{2} \sum_{j,k=1}^d c_{jk} \xi_j \xi_k \right) d\xi, \quad (10.4)$$

where $i^2 = -1$ is the imaginary unit.

Proof. Recall the well-known formula

$$\int_{\mathbb{R}^d} \exp \left(b^T \xi - \frac{1}{2} \xi^T C \xi \right) d\xi = (2\pi)^{d/2} |A|^{1/2} \exp \left[\frac{1}{2} b^T A b \right] \quad (10.5)$$

valid for all $b \in \mathbb{C}^d$. Using the *fractional Hermite transform* (see [109] for the standard case), we obtain the following Wick analogue of (10.5):

$$\int_{\mathbb{R}^d} \exp^\diamond \left(b^T \xi - \frac{1}{2} \xi^T C \xi \right) d\xi = (2\pi)^{d/2} |A|^{1/2} \exp^\diamond \left(\frac{1}{2} b^T \diamond A \diamond b \right),$$

valid for every d-dimensional square-integrable random variable b. In particular, if we apply (10.5) with $b = i(Y + m - y)$, we get from (10.3) that

$$\delta_Y(y) = (2\pi)^{-d/2} |A|^{1/2} \exp^\diamond \left(\frac{1}{2} b^T \diamond A \diamond b \right)$$

$$= (2\pi)^{-d} \int_{\mathbb{R}^d} \exp^\diamond \left(b^T \xi - \frac{1}{2} \xi^T C \xi \right) d\xi,$$

which is (10.4). □

From this we deduce the *Hu formula* for the Donsker delta function proved in [110].

Theorem 10.1.4.

$$\delta_Y(y) = (2\pi)^{-d} \int_{\mathbb{R}^d} \exp \left(i \sum_{j=1}^d \xi_j (Y_j + m_j - y_j) \right) d\xi, \quad (10.6)$$

where the integral exists in $(\mathcal{S})^*$.

Proof. Recall the following connection between the Wick exponential and the ordinary exponential of a Gaussian random variable Z with values in $\mathbb{R}^d$ and with $E[Z] = 0$:

$$\exp^\diamond(b^T Z) = \exp \left(b^T Z - \frac{1}{2} b^T E[ZZ^T] b \right)$$

for all (deterministic) $b \in \mathbb{C}^d$. If we apply this to (10.4) with $Z = Y - m$ and $b = i\xi$ we get (10.6). □

242   10 Local time for fractional Brownian motion

We now focus on the case when $Y = B^{(H)}(t)$:

Corollary 10.1.5. *The Donsker delta function $\delta_{B^{(H)}(t)}(x)$ of fractional Brownian $B^{(H)}(t) \in \mathbb{R}^d$ is given by*

$$\delta_{B^{(H)}(t)}(x) = (2\pi)^{-d/2} \left(\prod_{j=1}^{d} t^{-H_j}\right) \exp^{\diamond}\left(-\frac{1}{2}\sum_{j=1}^{d} t^{-2H_j}(B_j^{(H_j)}(t) - x_j)^{\diamond 2}\right).$$

Proof. The random variable $Y := B^{(H)}(t)$ is normally distributed with mean 0 and with covariance matrix $C = [c_{jk}]$ given by

$$c_{jk} = E\left[B_j^{(H_j)}(t) B_k^{(H_k)}(t)\right] = t^{2H_j} \delta_{jk}, \qquad 1 \leq j,\ k \leq d. \tag{10.7}$$

Thus C is diagonal with inverse $A = [a_{jk}]$ given by

$$a_{jk} = t^{-2H_j} \delta_{jk}.$$

Then

$$|A| = \prod_{j=1}^{d} t^{-2H_j}.$$

Hence the result follows from Proposition 10.1.2. $\square$

Corollary 10.1.6. *The Donsker delta function is also given by*

$$\delta_{B^{(H)}(t)}(x) = (2\pi)^{-d} \int_{\mathbb{R}^d} \exp^{\diamond}\left(i\sum_{j=1}^{d} \xi_j(B_j^{(H_j)}(t) - x_j)\right.$$
$$\left. - \frac{1}{2}\sum_{j=1}^{d} t^{2H_j}\xi_j^2(B_j^{(H_j)}(t) - x_j)^{\diamond 2}\right) d\xi.$$

Proof. This follows from Proposition 10.1.3 and (10.7). $\square$

Remark 10.1.7. Note that this integral converges in the fractional distribution space $(\mathcal{S})^*$. This follows by considering the fractional Hermite transforms. See [109].

Corollary 10.1.8.

$$\delta_{B^{(H)}(t)}(x) = (2\pi)^{-d} \int_{\mathbb{R}^d} \exp\left(i\sum_{j=1}^{d} \xi_j[B_j^{(H_j)}(t) - x_j]\right) d\xi,$$

where the integral converges in $(\mathcal{S})^*$.

Proof. This is a direct consequence of Theorem 10.1.4. $\square$

10.1 Local time for fBm

We now proceed to define *fractional local time*:

Definition 10.1.9. *Fix $T > 0$ and a point $x \in \mathbb{R}^d$. The* local time *of $B^{(H)}(t)$ up to time T at the point x is defined by*

$$\ell_T^{(H)}(x) = \ell_T^{(H)}(\omega, x) = \lim_{\varepsilon \to 0} \frac{1}{|D(x,\varepsilon)|} \lambda_1(\{s \in [0,T] : B^{(H)}(s,\omega) \in D(x,\varepsilon)\}),$$

where the limit is taken in $(\mathcal{S})^*$,

$$D(x,\varepsilon) = \{y \in \mathbb{R}^d : |y - x| < \varepsilon\},$$

$|D(x,\varepsilon)| = \lambda_d(D(x,\varepsilon))$ is the volume of $D(x,\varepsilon)$, and λ_d denotes the Lebesgue measure in $\mathbb{R}^d$.

Remark 10.1.10. This definition is natural from the point of view that local time at x is the *amount of time spent at the point*.

Proposition 10.1.11. *1. The local time of $B^{(H)}(t) \in \mathbb{R}^d$ at the point $x \in \mathbb{R}^d$ exists in $(\mathcal{S})^*$ and is given by*

$$\ell_T^{(H)}(x) = \int_0^T \delta_{B^{(H)}(s)}(x)\, ds.$$

2. Hence, for d-dimensional $B^{(H)}(t)$ we have

$$\ell_T^{(H)}(x) = (2\pi)^{-d/2} \int_0^T \left(\prod_{j=1}^d t^{-H_j} \right) \cdot \exp^\diamond \left(-\frac{1}{2} \sum_{j=1}^d t^{-2H_j} [B_j^{(H_j)}(t) - x_j]^{\diamond 2} \right) dt \quad (10.8)$$

and

$$\ell_T^{(H)}(x) = (2\pi)^{-d} \int_0^T \left[\int_{\mathbb{R}^d} \exp\left(i \sum_{j=1}^d \xi_j [B_j^{(H_j)}(t) - x_j] \right) d\xi \right] dt. \quad (10.9)$$

3. In particular, for $d = 1$ this gives

$$\ell_T^{(H)}(x) = (2\pi)^{-1/2} \int_0^T t^{-H} \exp^\diamond \left(-\frac{1}{2} t^{-2H} (B^{(H)}(t) - x)^{\diamond 2} \right) dt \quad (10.10)$$

and

$$\ell_T^{(H)}(x) = (2\pi)^{-1} \int_0^T \left[\int_{\mathbb{R}} \exp(i\xi[B^{(H)}(t) - x]) d\xi \right] dt.$$

Proof. If we apply (10.2) to the function

$$g_\varepsilon(y) = I_{D(x,\varepsilon)}(y) = \begin{cases} 1 & \text{if } y \in D(x,\varepsilon), \\ 0 & \text{otherwise}, \end{cases}$$

we get

$$\lambda_1(\{s \in [0,T] : B^{(H)}(s) \in D(x,\varepsilon)\}) = \int_0^T I_{D(x,\varepsilon)}[B^{(H)}(s)]\,ds$$

$$= \int_0^T \left[\int_{\mathbb{R}^d} I_{D(x,\varepsilon)}(y)\delta_{B^{(H)}(s)}(y)\,dy\right] ds$$

$$= \int_{D(x,\varepsilon)} \left[\int_0^T \delta_{B^{(H)}(s)}(y)\,ds\right] dy.$$

Hence

$$\ell_T^{(H)}(x) = \lim_{\varepsilon \to 0} \frac{1}{|D(x,\varepsilon)|} \lambda_1(\{s \in [0,T] : B^{(H)}(s) \in D(x,\varepsilon)\})$$

$$= \lim_{\varepsilon \to 0} \frac{1}{|D(x,\varepsilon)|} \int_{D(x,\varepsilon)} \left[\int_0^T \delta_{B^{(H)}(s)}(y)\,ds\right] dy = \int_0^T \delta_{B^{(H)}(s)}(x)\,ds.$$

Formulas (10.8) and (10.9) now follow from Corollary 10.1.5 and Corollary 10.1.6. □

The *generalized expectation* operator E can be defined on $(\mathcal{S})^*$ in exactly the same way as in the standard case (for example, see [109]): If $X \in (\mathcal{S})^*$, then $E[X]$ is the zero order element in the chaos expansion of X. In particular, E has the properties

$$E[Y \diamond Z] = E[Y]E[Z] \quad \text{for } Y, Z \in (\mathcal{S})^*$$

and

$$E[\exp^\diamond Y] = \exp E[Y] \quad \text{if } \exp^\diamond(Y) \in (\mathcal{S})^*.$$

Therefore we obtain the following directly from (10.8) and (10.10):

Corollary 10.1.12. *1. The generalized expectation of fractional local time of $B^{(H)}(t)$, $0 \leq t \leq T$ at $x \in \mathbb{R}^d$ is*

$$E\left[\ell_T^{(H)}(x)\right] = (2\pi)^{-d/2} \int_0^T \left(\prod_{j=1}^d t^{-H_j}\right)$$

$$\exp\left(-\frac{1}{2}\sum_{j=1}^d t^{-2H_j} x_j^2\right) dt. \tag{10.11}$$

2. In particular, for $d = 1$ we have

$$E\left[\ell_T^{(H)}(x)\right] = (2\pi)^{-1/2} \int_0^T t^{-H} \exp\left(-\frac{1}{2}t^{-2H}x^2\right) dt. \quad (10.12)$$

In the next section we will find the whole chaos expansion of $\ell_T^{(H)}(x)$. Then we will find conditions which ensure that $\ell_T^{(H)}(x, \cdot) \in L^2(\mathbb{P})$. If this is the case, then (10.11) and (10.12) give the usual expectation of $\ell_T^{(H)}(x, \cdot)$.

It is well-known that $L^2(\mathbb{P})$ is a dense subset of $(\mathcal{S})^*$. Thus $\ell_T^{(H)}(x)$ can be approximated by elements in $L^2(\mathbb{P})$. It is easy to verify that

Proposition 10.1.13. *Let P_ε be defined as in (10.1), and let*

$$\ell_{\varepsilon,T}^{(H)}(x) = \int_0^T P_\varepsilon(B^{(H)}(t) - x)\, dt = \frac{1}{(2\pi)^d} \int_0^T \int_{\mathbb{R}^d} e^{i(B^{(H)}(t)-x)\xi - 1/2\varepsilon\xi^2}\, d\xi\, dt$$

Then

$$\lim_{\varepsilon \to 0} \ell_{\varepsilon,T}^{(H)}(x) = \ell_T^{(H)}(x) \qquad \text{in } (\mathcal{S})^*.$$

10.2 The chaos expansion of local time for fBm

In this section we will use the following formula for the Donsker delta function,

$$\delta_{B^{(H)}(t)}(x) = (2\pi)^{-d} \int_{\mathbb{R}^d} \exp\left(i\sum_{j=1}^d \xi_j[B_j^{(H_j)}(t) - x_j]\right) d\xi \in (\mathcal{S})^*, \quad (10.13)$$

obtained in Corollary 10.1.8, to find the chaos expansion of the fractional local time

$$\ell_T^{(H)}(x) = \int_0^T \delta_{B^{(H)}(t)}(x)\, dt. \quad (10.14)$$

For simplicity we will first assume that $x = 0$ (the case $x \neq 0$ is similar), and we will put

$$\delta(B^{(H)}(t)) = \delta_{B^{(H)}(t)}(0).$$

Let $f(s) = (f_1(s), \ldots, f_d(s))$ be a (complex) deterministic function belonging to $L_H^{2,(d)}(\mathbb{R}) = L_{H_1}^2(\mathbb{R}) \times \cdots \times L_{H_d}^2(\mathbb{R})$, $f = 0$ outside $[0, T]$. Define

$$\mathcal{E}[f](t) := \exp\left(\int_0^t f(s)\, dB^{(H_j)}(s) - \frac{1}{2}\|fI_{[0,t]}\|_H^2\right),$$

where $\|fI_{[0,t]}\|_H^2$ is defined in (3.59). Then it is easy to see that

$$\mathcal{E}[f](t) = 1 + \sum_{j=1}^d \int_0^t \mathcal{E}[f](s) f_j(s)\, dB_j^{(H)}(s).$$

By iteration of this identity it follows that

$$\exp\left(\int_0^T f(s)dB^{(H)}(s) - \frac{1}{2}\|f\|_H^2\right)$$
$$= 1 + \sum_{n=1}^\infty \sum_{1\leq j_1,\cdots,j_n \leq d} \int_0^T \int_0^{s_{n-1}}$$
$$\cdots \int_0^{s_2} f_{j_1}(s_1) \cdots f_{j_n}(s_n) \, dB_{j_1}^{(H_{j_1})}(s_1) \cdots dB_{j_n}^{(H_{j_n})}(s_n).$$

Note that

$$iB^{(H)}(t)\xi = i\sum_{j=1}^d \xi_j \int_0^T I_{[0,t]}(s) \, dB_j^{(H_j)}(s) \qquad \text{for all } \xi \in \mathbb{R}^d$$

and

$$I_{[0,t]}(s_1) \cdots I_{[0,t]}(s_n) = I_{[\max\{s_1,\ldots,s_n\},\infty)}(t).$$

Thus for any $\xi \in \mathbb{R}^d$,

$$\exp\left(i\xi B^{(H)}(t)\right)$$
$$= \exp\left(i\xi B^{(H)}(t) + \frac{1}{2}\sum_{j=1}^d t^{2H_j}\xi_j^2\right) \exp\left(-\frac{1}{2}\sum_{j=1}^d t^{2H_j}\xi_j^2\right)$$
$$= \sum_{n=0}^\infty i^n \exp\left(-\frac{1}{2}\sum_{j=1}^d t^{2H_j}\xi_j^2\right)$$
$$\cdot \int_0^T \int_0^{s_{n-1}} \cdots \int_0^{s_2} \sum_{1\leq j_1,\ldots,j_n\leq d} \xi_{j_1}\cdots\xi_{j_n} I_{[0,t]}(s_1)\cdots I_{[0,t]}(s_n)$$
$$\cdot dB_{j_1}^{(H_{j_1})}(s_1) \cdots dB_{j_n}^{(H_{j_n})}(s_n)$$
$$= \sum_{n=0}^\infty i^n \exp\left(-\frac{1}{2}\sum_{j=1}^d t^{2H_j}\xi_j^2\right)$$
$$\cdot \int_0^T \int_0^{s_{n-1}} \cdots \int_0^{s_2} \sum_{1\leq j_1,\ldots,j_n\leq d} \xi_{j_1}\cdots\xi_{j_n} I_{[\max\{s_1,\ldots,s_n\},\infty)}(t)$$
$$\cdot dB_{j_1}^{(H_{j_1})}(s_1) \cdots dB_{j_n}^{(H_{j_n})}(s_n).$$

Therefore by Proposition 10.1.11 item 2,

$$\ell_T^{(H)}(0) = \sum_{n=0}^\infty i^n \sum_{1\leq j_1,\ldots,j_n\leq d} \int_0^T \int_0^{s_{n-1}} \cdots \int_0^{s_2} f_{j_1,\ldots,j_n}(s_1,\ldots,s_n)$$
$$\cdot dB_{j_1}^{(H_{j_1})}(s_1) \cdots dB_{j_n}^{(H_{j_n})}(s_n),$$

where, for $n \geq 1$,

$$f_{j_1,\ldots,j_n}(s_1,\ldots,s_n) = \frac{1}{(2\pi)^d} \int_0^T \int_{\mathbb{R}^d} \exp\left(-\frac{1}{2}\sum_{k=1}^d t^{2H_k}\xi_k^2\right) \xi_{j_1}\cdots\xi_{j_n}$$
$$\cdot I_{[\max\{s_1,\ldots,s_n\},\infty]}(t)\, dt\, d\xi_1\cdots d\xi_d.$$

To compute $f_{j_1,\ldots,j_n}(s_1,\ldots,s_n)$, let us introduce

$$j = (j_1, j_2, \ldots, j_n)$$

and

$$\nu(j,k) := \#\{l \in \{1,2,\ldots,n\} : j_l = k\}$$

for $k = 1, 2, \ldots, d$. Thus $\nu(j,1) + \cdots + \nu(j,d) = n$ for $j = 1,\ldots,n$. With this notation, we obtain

$$\int_{\mathbb{R}^d} \exp\left(-\frac{1}{2}\sum_{k=1}^d t^{2H_k}\xi_k^2\right)\xi_{j_1}\cdots\xi_{j_n}\, d\xi_1\cdots d\xi_d$$
$$= \int_{\mathbb{R}^d} \exp\left(-\frac{1}{2}\sum_{k=1}^d t^{2H_k}\xi_k^2\right)\xi_1^{\nu(j,1)}\cdots\xi_d^{\nu(j,d)}\, d\xi_1\cdots d\xi_d$$
$$= t^{-K(j)}\int_{\mathbb{R}^d} e^{-1/2|\eta|^2}\eta_1^{\nu(j,1)}\cdots\eta_d^{\nu(j,d)}\, d\eta_1\cdots d\eta_d$$
$$= t^{-K(j)}(2\pi)^d C(j),$$

where

$$C(j) = \begin{cases} 0 & \text{when one of } \nu(j,k) \text{ is odd,} \\ (2\pi)^{-d/2}\prod_{k=1}^d \dfrac{\nu(j,k)!}{2^{\nu(j,k)/2}(\nu(j,k)/2)!} & \text{when all } \nu(j,k) \text{ are even.} \end{cases}$$

and

$$K(j) = \sum_{k=1}^d H_k(1+\nu(j,k)).$$

Thus

$$f_{j_1,\ldots,j_n}(s_1,\ldots,s_n) = C(j)\int_{\max\{s_1,\ldots,s_n\}}^T t^{-K(j)}\, dt$$
$$= \frac{C(j)}{1-K(j)}\left(T^{1-K(j)} - \max\{s_1,\ldots,s_n\}^{1-K(j)}\right).$$

Consequently,

$$\ell_T^{(H)}(0) = \sum_{n \text{ even}} \sum_{1 \le j_1,\ldots,j_n \le d} \int_0^T \int_0^{s_{n-1}} \cdots \int_0^{s_2} g_n(T, \max\{s_1,\ldots,s_n\})$$
$$\cdot dB_{j_1}^{(H)}(s_1) \cdots dB_{j_n}^{(H)}(s_n),$$

where

$$g_n(T, u) = (-1)^n \frac{C(j)}{1 - K(j)} \left(T^{1-K(j)} - u^{1-K(j)} \right) \quad \text{for } n \ge 1. \quad (10.15)$$

(When $n = 0$, we assume that $\max\{s_1,\ldots,s_n\} = 0$.) Thus we obtain

Theorem 10.2.1. *The chaos expansion of the fractional local time $\ell_T^{(H)}(0) = \int_0^T \delta(B^{(H)}(t)) \, dt$ at $x = 0$ is given by*

$$\ell_T^{(H)}(0) = \int_0^T \delta(B^{(H)}(t)) \, dt$$
$$= \sum_{n \text{ even}} \sum_{1 \le j_1,\ldots,j_n \le d} \int_0^T \int_0^{s_{n-1}} \cdots \int_0^{s_2} g_n(T, \max\{s_1,\ldots,s_n\})$$
$$\cdot dB_{j_1}^{(H)}(s_1) \cdots dB_{j_n}^{(H)}(s_n), \quad (10.16)$$

where $g(T, u)$ is given by (10.15).

Another proof of the chaos expansion for the *fBm* can be found in Proposition 4 of [87].

Remark 10.2.2. If the series in (10.16) converges in $L^2(\mathbb{P})$, then the expectation is the zero-order term in the chaos expansion. Thus, by choosing $n = 0$ in (10.15) we have, for $d = 1$ (see Corollary 10.1.12),

$$E\left[\int_0^T \delta(B^{(H)}(t)) \, dt\right] = \frac{T^{1-H}}{\sqrt{2\pi}(1-H)}.$$

Now we compute the L^2 norm of the local time of the *fBm*. We still use expressions (10.13) and (10.14) for the Donsker delta function. It suffices to show that $\ell_{\varepsilon,T}^{(H)}(0)$ is a bounded sequence in $L^2(\mathbb{P})$ as $\varepsilon \to 0+$. For the sake of simplicity, we let $\varepsilon = 0$. Thus we need to estimate

$$E\left[(\ell_T^{(H)}(x))^2\right] = E\left[(\int_0^T \delta(B^{(H)}(t) - x) \, dt)^2\right]$$
$$= \frac{1}{(2\pi)^{2d}} \int_{[0,T]^2} \int_{\mathbb{R}^{2d}} e^{-i\xi x + i\eta x} E\left[e^{-i\xi B^{(H)}(t) + i\eta B^{(H)}(s)}\right] d\xi \, d\eta \, ds \, dt$$
$$\le \frac{1}{(2\pi)^{2d}} \int_{[0,T]^2} \int_{\mathbb{R}^{2d}} e^{-1/2 \operatorname{Var}(\xi B^{(H)}(t) - \eta B^{(H)}(s))} d\xi \, d\eta \, ds \, dt$$

$$\leq \frac{2}{(2\pi)^{2d}} \int_0^T \int_0^{s_{n-1}} \cdots \int_0^{s_2} \int_{\mathbb{R}^{2d}} e^{-1/2\,\mathrm{Var}\left(\xi B^{(H)}(t) - \eta B^{(H)}(s)\right)} \, d\xi \, d\eta \, ds \, dt.$$

Here we have used that for any Gaussian random variable X,

$$E\left[e^{iX}\right] = e^{-1/2\,\mathrm{Var}(X)}.$$

Using the nondeterminism property of fBm we have that when $0 \leq s < t \leq T$, there is a positive constant $k > 0$ such that

$$\mathrm{Var}\left(\xi B^{(H)}(t) - \eta B^{(H)}(s)\right) = \mathrm{Var}\left(\xi[B^{(H)}(t) - B^{(H)}(s)] + (\xi - \eta)B^{(H)}(s)\right)$$
$$\geq k\left[\xi^2 |t-s|^{2H_0} + (\eta - \xi)^2 s^{2H_0}\right],$$

where

$$H_0 = \max(H_1, \ldots, H_d).$$

See [25], [26], [27], [28], [29], [119], [197] for the use of this property. Therefore, we have when $H_0 d < 1$,

$$E\left[(\ell_T^{(H)}(x))^2\right] \leq \frac{2}{(2\pi)^{2d}} \int_{0 \leq s < t \leq T} \int_{\mathbb{R}^{2d}} e^{-k/2[\xi^2|t-s|^{2H_0} + (\eta - \xi)^2 s^{2H_0}]} \, d\xi \, d\eta \, ds \, dt$$

$$= \frac{2}{(2\pi)^d k^{d/2}} \int_{0 \leq s < t \leq T} \frac{1}{s^{H_0 d}(t-s)^{H_0 d}} \, ds \, dt$$

$$= \frac{2}{(2\pi)^d k^{d/2}} \int_0^T \left(\int_0^t \frac{1}{s^{H_0 d}(t-s)^{H_0 d}} \, ds\right) dt$$

$$= \frac{2\Gamma(1-H_0 d)^2}{(2\pi)^d k^{d/2} \Gamma(2-2H_0 d)} \int_0^T t^{1-2H_0 d} \, dt$$

$$= \frac{2\Gamma(1-H_0 d)^2}{(2\pi)^d k^{d/2} \Gamma(2-2H_0 d)(2-2H_0 d)} T^{2-2H_0 d}$$

$$= \frac{2\Gamma(1-H_0 d)^2}{(2\pi)^d k^{d/2} \Gamma(3-2H_0 d)} T^{2-2H_0 d}.$$

Summarizing the above, we obtain

Theorem 10.2.3. *Assume that $H_0 d < 1$. Then the local time $\ell_T^{(H)}(x) = \int_0^T \delta(B(t) - x) \, dt$ is square integrable, and for any $x \in \mathbb{R}$ we have*

$$\lim_{T \to \infty} E\left[\left(\int_0^T \delta(B^{(H)}(t) - x) \, dt\right)^2\right] \leq \frac{2\Gamma(1-H_0 d)^2 T^{2-2H_0 d}}{(2\pi)\sqrt{k}\,\Gamma(3-2H_0 d)},$$

where k is a constant depending on H_0.

Moreover, by [87] we obtain the following further regularity result concerning $\ell_T^{(H)}$.

Theorem 10.2.4. *The local time $\ell_T^{(H)}$ of the one-dimensional fBm $B^{(H)}$ belongs to the space $\mathbb{D}_H^{\alpha,2}$ for every $\alpha < (1-H)/(2H)$.*

This result can be obtained by using the Wiener chaos expansion for $\ell_T^{(H)}$ (see, for example [177]). For further details on the proof of this Theorem, we refer to Theorem 5 of [87].

By Theorem 1.6.1 of Chapter 1 we know that the *fBm* has β-Hölder continuous trajectories for all $\beta < H$. Thus, when H becomes smaller, the paths of $B^{(H)}$ become less regular, but the regularity of its local time increases. If $H = 1/2$, we obtain the regularity result for the standard Brownian motion that holds for $\alpha < 1/2$.

10.3 Weighted local time for *fBm*

Let $d = 1$ and consider a one-dimensional $B^{(H)}$ for $H \in (0,1)$. Now we introduce the *weighted local time* as

$$\mathcal{L}_t^{(B^{(H)})}(x) = \int_0^t \delta(B_s^{(H)} - x) s^{2H-1} \, ds. \tag{10.17}$$

For $H = 1/2$, the usual local time is the same as the weighted local time. In [62] the weighted local time $\mathcal{L}_t^{(B^{(H)})}$ is introduced as the density of the occupation measure

$$m_t^w(\Gamma) = 2H \int_0^t I_\Gamma(B_s^{(H)}) s^{2H-1} \, ds,$$

where $\Gamma \in \mathcal{B}(\mathbb{R})$. By [28] and [100] it follows that the occupation measure

$$m_t(\Gamma) = \int_0^t I_\Gamma(B_s^{(H)}) \, ds$$

has a density $\lambda_t^{(x)}$ that has a continuous version in t and x. In particular, by [100] we have that $\lambda_t^{(x)}$ is Hölder continuous respectively of order $\gamma < 1 - H$ in t and of order $\alpha < (1-H)/(2H)$ in x. Since we have

$$\mathcal{L}_t^{(B^{(H)})}(x) = 2H \int_0^t s^{2H-1} \lambda_t^{(x)} \, ds,$$

the weighted local time $\mathcal{L}_t^{(B^{(H)})}(x)$ inherits the continuity properties of $\lambda_t^{(x)}$. In particular, for any continuous function $g : \mathbb{R} \to \mathbb{R}$ the following holds:

$$\int_0^t g(B_s^{(H)}) s^{2H-1} \, ds = \int_0^t g(y) \mathcal{L}_t^{(B^{(H)})}(y) \, dy.$$

Following the proof of [29], it can also be easily seen that the local time $\ell_t^{(B^{(H)})}(x)$ is a jointly continuous function of t and x for almost all $\omega \in \Omega$. We can now prove the following:

Proposition 10.3.1.
$$E\left[\mathcal{L}_t^{(B^{(H)})}(x)\right] = (2\pi)^{-1/2} \int_0^t r^{H-1} \exp\left(-\frac{1}{2}r^{-2H}x^2\right) dr. \tag{10.18}$$

Proof. To prove (10.18), use integration by parts to write
$$\mathcal{L}_t^{(B^{(H)})}(x) = \int_0^t \delta(B_s^{(H)} - x)s^{2H-1} ds$$
$$= \ell_t^{(B^{(H)})}(x)t^{2H-1} - (2H-1)\int_0^t s^{2H-2}\ell_s^{(B^{(H)})}(x) ds. \tag{10.19}$$

By using equation (10.11), we have
$$E\left[\int_0^t s^{2H-2}\ell_s^{(B^{(H)})}(x) ds\right]$$
$$= \int_0^t s^{2H-2}\left[\frac{1}{\sqrt{2\pi}}\int_0^s r^{-H}\exp\left(-\frac{1}{2}r^{-2H}x^2\right) dr\right] ds$$
$$= \frac{1}{\sqrt{2\pi}}\int_0^t \left(\int_r^t s^{2H-2} ds\right) r^{-H}\exp\left(-\frac{1}{2}r^{-2H}x^2\right) dr$$
$$= \frac{1}{\sqrt{2\pi}}\int_0^t \frac{1}{2H-1}\left(t^{2H-1} - r^{2H-1}\right) r^{-H}\exp\left(-\frac{1}{2}r^{-2H}x^2\right) dr$$
$$= \frac{t^{2H-1}}{2H-1}E\left[\ell_t^{(B^{(H)})}(x)\right]$$
$$- \frac{1}{(2H-1)\sqrt{2\pi}}\int_0^t r^{H-1}\exp\left(-\frac{1}{2}r^{-2H}x^2\right) dr. \tag{10.20}$$

Combining (10.19) and (10.20), we obtain (10.18). □

The exact second moment is harder to compute. However, we are able to obtain an upper bound for the second moment.

Proposition 10.3.2. *The weighted local time* $\mathcal{L}_T^{(B^{(H)})}(x) = \int_0^T \delta(B^{(H)}(t) - x)t^{2H-1} dt$ *is square integrable, and for any* $x \in \mathbb{R}$, *we have*
$$E\left[(\mathcal{L}_T^{(B^{(H)})}(x))^2\right] \leq \frac{\Gamma(H)\Gamma(1-H)T^{2H}}{2H\pi\sqrt{k}},$$
where k is a constant depending on H and is defined by (10.21) below.

Proof. Using the representation (10.17), we follow the proof of Theorem 10.2.3 and get
$$E\left[\left(\mathcal{L}_T^{(B^{(H)})}(x)\right)^2\right] = E\left[\left(\int_0^T \delta(B^{(H)}(t) - x)t^{2H-1} dt\right)^2\right]$$

$$= \frac{1}{(2\pi)^2} \int_{[0,T]^2} \int_{\mathbb{R}^2} e^{-(i\xi x + i\eta x)}$$
$$\cdot E\left[e^{i\xi B^{(H)}(t) + i\eta B^{(H)}(s)}\right] (st)^{2H-1} d\xi\, d\eta\, ds\, dt$$
$$\leq \frac{1}{(2\pi)^2} \int_{[0,T]^2} \int_{\mathbb{R}^2} e^{-1/2 \operatorname{Var}(\xi B^{(H)}(t) - \eta B^{(H)}(s))} (st)^{2H-1} d\xi\, d\eta\, ds\, dt$$
$$\leq \frac{2}{(2\pi)^2} \int_{0 \leq s < t \leq T} \int_{\mathbb{R}^2} e^{-1/2 \operatorname{Var}(\xi B^{(H)}(t) - \eta B^{(H)}(s))} (st)^{2H-1} d\xi\, d\eta\, ds\, dt.$$

As in the proof of Theorem 10.2.1 by using the nondeterminism property of fBm (see [29], [119], [122], and the references therein), we find that, when $0 \leq s < t \leq T$, there is a positive constant $k > 0$ such that

$$\operatorname{Var}\left(\xi B^{(H)}(t) - \eta B^{(H)}(s)\right) = \operatorname{Var}\left(\xi [B^{(H)}(t) - B^{(H)}(s)] + (\xi - \eta) B^{(H)}(s)\right)$$
$$\geq k \left[\xi^2 |t-s|^{2H} + (\eta - \xi)^2 s^{2H}\right], \qquad (10.21)$$

Therefore, we have

$$E\left[\left(\mathcal{L}_T^{(B^{(H)})}(x)\right)^2\right]$$
$$\leq \frac{2}{(2\pi)^2} \int_{0 \leq s < t \leq T} \int_{\mathbb{R}^2} e^{-k/2[\xi^2 |t-s|^{2H} + (\eta-\xi)^2 s^{2H}]} (st)^{2H-1} d\xi\, d\eta\, ds\, dt$$
$$= \frac{2}{2\pi\sqrt{k}} \int_0^T \int_0^{s_{n-1}} \cdots \int_0^{s_2} \frac{(st)^{2H-1}}{s^H (t-s)^H} ds\, dt$$
$$= \frac{2}{2\pi\sqrt{k}} \int_0^T \left(\int_0^t \frac{s^{H-1} t^{2H-1}}{(t-s)^H} ds\right) dt$$
$$= \frac{2\Gamma(H)\Gamma(1-H)}{2\pi\sqrt{k}} \int_0^T t^{2H-1} dt = \frac{\Gamma(H)\Gamma(1-H)}{2H\pi\sqrt{k}} T^{2H}.$$

This proves the proposition. □

By [62] we obtain the Wiener chaos expansion for the weighted local time.

Proposition 10.3.3. *Let $H \in (0,1)$ and $B^{(H)}$ a one-dimensional fBm. The weighted local time $\mathcal{L}_t^{(B^{(H)})}$ has the following Wiener chaos expansion:*

$$\mathcal{L}_t^{(B^{(H)})} = 2H \sum_{n=0}^{\infty} \int_0^t s^{(2-n)H-1} p_{s^{2H}}(x) h_n\left(\frac{x}{s^{2H}}\right) I_n(K_H(s,\cdot)^{\otimes n}) ds.$$

where $p_{s^{2H}}(x) = 1/(s^H \sqrt{2\pi}) \exp\left(-x^2/(2s^{2H})\right)$; h_n is the nth Hermite polynomial, and $K_H(s,t)$ is the reproducing kernel introduced in Chapter 2.

10.4 A Meyer Tanaka formula for *fBm*

Here we provide a generalized Itô formula for convex functions valid for every $H \in (0,1)$ following [123]. At this purpose we consider the stochastic integral defined in Chapter 4. Let $f : \mathbb{R} \to \mathbb{R}$ be a convex function. Then it is well-known that its left derivative $D^- f(x)$ exists and is finite for every $x \in \mathbb{R}$, where we have

$$D^- f(x) := \lim_{h \to 0, h > 0} \frac{f(x-h) - f(x)}{h}. \qquad (10.22)$$

Define the *second derivative measure* of f by $\nu_f(dx)$ by

$$\nu_f([a,b)) := D^- f(b) - D^- f(a), \qquad -\infty < a < b < \infty. \qquad (10.23)$$

Theorem 10.4.1. *Let $H \in (0,1)$, $B^{(H)}$ a one-dimensional fBm and f be a convex function of polynomial growth. Then*

$$f(B_t^{(H)}) = f(0) + \int_0^t D^- f(B_s^{(H)}) \, dB_s^{(H)} + H \int_{\mathbb{R}} \mathcal{L}_t^{(B^{(H)})}(x) \, \nu_f(dx),$$

where here the stochastic integral in terms of $B^{(H)}$ is to be interpreted as the one defined in Chapter 4.

Proof. Define a function ρ by

$$\rho(x) := c \exp\left(\frac{1}{(x-1)^2 - 1}\right),$$

for $x \in (0,2)$, and $\rho(x) = 0$ elsewhere, where c is a normalizing constant such that $\int_{\mathbb{R}} \rho(x) \, dx = 1$. Let

$$\rho_n(x) := n\rho(nx).$$

If

$$f_n(x) = \int_{\mathbb{R}} \rho_n(x - y) f(y) \, dy, \qquad n \geq 1,$$

then it is well-known that

$$\lim_{n \to \infty} f_n(x) = f(x) \quad \text{and} \quad \lim_{n \to \infty} f'_n(x) = D^- f(x)$$

for every $x \in \mathbb{R}$. Moreover, if g is of class C^1 and has compact support, then

$$\lim_{n \to \infty} \int_{\mathbb{R}} g(x) f''_n(x) \, dx = \int_{\mathbb{R}} g(x) \, \nu_f(dx).$$

Since f_n is C^2, we have by Theorem 4.2.6 in Chapter 4

$$f_n(B_t^{(H)}) = f_n(0) + \int_0^t f'_n(B_s^{(H)}) \, dB_s^{(H)} + H \int_0^t s^{2H-1} f''_n(B_s^{(H)}) \, ds. \qquad (10.24)$$

If $n \to \infty$, then it is easy to see that $f_n(B_t^{(H)})$ converges to $f(B_t^{(H)})$ almost surely, and $f_n(0)$ converges to $f(0)$.

Next, we consider the limit of the last term. Since

$$H \int_0^t s^{2H-1} f_n''(B_s^{(H)}) \, ds = H \int_0^t s^{2H-1} \int_{\mathbb{R}} f_n''(x) \delta(B_s^{(H)} - x) \, dx \, ds$$

$$= H \int_{\mathbb{R}} f_n''(x) \int_0^t s^{2H-1} \delta(B_s^{(H)} - x) \, ds \, dx$$

$$= H \int_{\mathbb{R}} f_n''(x) \mathcal{L}_t^{(B^{(H)})}(x) \, dx$$

we have

$$H \int_0^t s^{2H-1} f_n''(B_s^{(H)}) \, ds \longrightarrow H \int_{\mathbb{R}} \mathcal{L}_t^{(B^{(H)})}(x) \nu_f(dx)$$

as n goes to infinity. Finally we see that

$$\int_0^t f_n'(B_s^{(H)}) \diamond W_s^{(H)} \, ds \xrightarrow{n \to \infty} \int_0^t D^- f(B_s^{(H)}) \diamond W_s^{(H)} \, ds \quad \text{in } (\mathcal{S})^*$$

and also almost surely, by the convergence of the other terms in (10.24). So, the result follows. $\square$

The function f defined by $f(x) = |x - z|$ for $x \in \mathbb{R}$ is convex, $D^- f(x) = \text{sign}\,(x - z)$, and $\nu_f(dx) = 2\delta_z(dx)$. Thus we obtain the following representation for the reflection of fBm.

Corollary 10.4.2 (A Meyer Tanaka formula for fBm). Let $H \in (0,1)$ and $B^{(H)}$ a one-dimensional fBm. For any $z \in \mathbb{R}$,

$$|B_t^{(H)} - z| = |z| + \int_0^t \text{sign}\,(B_s^{(H)} - z) \, dB_s^{(H)} + 2H \mathcal{L}_t^{(B^{(H)})}(z). \tag{10.25}$$

In [62] a Meyer Tanaka formula for fBm is obtained for $H > 1/3$. Moreover in [54] a Meyer Tanaka formula for fBm is provided for all $H < 1/2$ by using the extended divergence operator δ introduced in Chapter 2, Definition 2.2.7. We recall it for the sake of completeness.

Theorem 10.4.3. Let $H < 1/2$ and $x \in \mathbb{R}$. Then $I_{(x,\infty)}(B^{(H)}(s))I_{(0,t)}(s) \in \text{dom}^* \delta$ and

$$\delta(I_{(x,\infty)}(B^{(H)}(s))I_{(0,t)}(s)) = [B^{(H)}(t) - x]^+ - (-x^+) - \frac{1}{2}\mathcal{L}_t^{(B^{(H)})}(x).$$

For the proof of this result, we refer to Theorem 4.4 of [54].

10.5 A Meyer Tanaka formula for geometric *fBm*

Let X_t be a *geometric fBm*, defined by

$$dX_t = \mu X_t\, dt + \sigma X_t\, dB_t^{(H)}, \qquad t \geq 0,\ X_0 = x > 0,$$

with x a constant, where $B^{(H)}$ is a one-dimensional *fBm* and $H \in (0,1)$. Using the Wick calculus (see Example 4.2.4), the solution of this equation is found to be

$$X_t = x \exp\left(\sigma B_t^{(H)} + \mu t - \frac{1}{2}\sigma^2 t^{2H}\right). \tag{10.26}$$

For this process, we define *the local time* $\ell_t^{(X)}(z)$ of X at the point $z > 0$ by

$$\ell_t^{(X)}(z) = \int_0^t \delta(X_s - z)\, ds = \lim_{\varepsilon \to 0} \frac{1}{2\varepsilon}\lambda_1\left(\{s \in [0,t] : |X_s - z| < \varepsilon\}\right),$$

where λ_1 is the Lebesgue measure on $\mathbb{R}$, and the *weighted local time* $\mathcal{L}_t^{(X)}(z)$ of X at z by

$$\mathcal{L}_t^{(X)}(z) = \int_0^t \delta(X_s - z) s^{2H-1}\, ds.$$

Since X_t is a functional of $B_t^{(H)}$, we expect that the local time of X will also be related to the local time of $B^{(H)}$. The approach that we are going to develop is applicable more generally to any process of the form $Y_t = f(t, B^{(H)}(t))$. Let $f(t, x)$ be a continuous function of t and x such that as a function of x it is invertible with continuously differentiable inverse function $f^{-1}(t, x)$. Then for any smooth function ψ of compact support we have, fixing t and writing $f(y) = f(t, y)$,

$$\int_{\mathbb{R}} \delta(f(y) - z)\psi(z)\, dz = \psi(f(y)).$$

On the other hand, if we make the substitution $u = f^{-1}(z)$, we get

$$\int_{\mathbb{R}} \delta(y - f^{-1}(z)) \frac{d}{dz}\left[f^{-1}(z)\right]\psi(z)\, dz = \int_{\mathbb{R}} \delta(y - u)\psi(f(u))\, du$$
$$= \psi(f(y)).$$

Thus we have

$$\delta(f(y) - z) = \delta(y - f^{-1}(z)) \frac{d}{dz}\left[f^{-1}(z)\right]$$

(in the distribution sense). Now we consider the local time of $Y_t = f(t, B^{(H)}(t))$ with

$$f(t, y) = x \exp\left(\sigma y + \mu t - \frac{1}{2}\sigma^2 t^{2H}\right).$$

Denote the inverse of $y \to f(t, y)$ by $h(t, y)$. Then

10 Local time for fractional Brownian motion

$$h(t, z) = \frac{1}{\sigma}\left(\log \frac{z}{x} - \mu t + \frac{\sigma^2}{2}t^{2H-1}\right). \qquad (10.27)$$

The derivative of $h(t, z)$ with respect to z is

$$\frac{d}{dz}h(t, z) = \frac{d}{dz}\left[f^{-1}(t, z)\right] = \frac{1}{\sigma z}.$$

Thus

$$\mathcal{L}_t^{(X)}(z) = \int_0^t \delta(f(s, B^{(H)}(s)) - z)s^{2H-1}\, ds$$

$$= \int_0^t \delta(B^{(H)}(s) - h(s, z))\frac{1}{\sigma z}s^{2H-1}\, ds$$

$$= \frac{1}{2\sigma z\pi}\int_0^t \int_{\mathbb{R}} e^{i\xi(B^{(H)}(s) - h(s,z))}\, d\xi\, s^{2H-1}\, ds.$$

Now it is easy to see from our previous calculations of the first and second moments of the local time of fBm (see Propositions 10.3.1 and 10.3.2) that we have the following result.

Proposition 10.5.1. Let $H \in (0, 1)$ and X_t given in (10.26). Then

$$E\left[\mathcal{L}_t^{(X)}(z)\right] = \frac{1}{\sqrt{2\pi}\sigma z}\int_0^t s^{H-1}\exp\left(-\frac{h(s, z)^2}{2s^{2H}}\right)\, ds,$$

where $h(s, z)$ is defined by (10.27), and

$$E\left[\left(\mathcal{L}_t^{(X)}(z)\right)^2\right] \le \frac{\Gamma(H)\Gamma(1-H)t^{2H}}{2H\pi\sqrt{k}\sigma^2 z^2},$$

where Γ is the gamma function.

Remark 10.5.2. Similarly, we can obtain

$$E\left[\ell_t^{(X)}(z)\right] = \frac{1}{\sqrt{2\pi}\sigma z}\int_0^t s^{-H}\exp\left(-\frac{h(s, z)^2}{2s^{2H}}\right)\, ds.$$

It is interesting to note that when f is independent of t, i.e., $Y_t = f(B^{(H)}(t))$, then the local time of Y_t is given by

$$\mathcal{L}_t^{(Y)}(f(y)) = \int_0^t \delta(f(B^{(H)}(s)) - f(y))s^{2H-1}\, ds$$

$$= \int_0^t \delta(B^{(H)}(s) - y)s^{2H-1}\frac{1}{f'(y)}\, ds$$

$$= \frac{1}{f'(y)}\int_0^t \delta(B^{(H)}(s) - y)s^{2H-1}\, ds$$

$$= \frac{1}{f'(y)} \mathcal{L}_t^{(B^{(H)})}(y).$$

This means that we have

$$\mathcal{L}_t^{(f(B^{(H)}))}(f(y)) = \frac{1}{f'(y)} \mathcal{L}_t^{(B^{(H)})}(y).$$

It is interesting to compare this formula to those in the semimartingale case (see [194, p. 234]).

The above calculations also prove the following theorem.

Theorem 10.5.3. *The weighted local time $\mathcal{L}_t^{(X)}(z)$ exists as an element of L^2, and it is a positive, jointly continuous function of $t > 0$ and $z > 0$.*

Furthermore, we have the following theorem and its corollaries.

Theorem 10.5.4. *Let $f : \mathbb{R}^+ \to \mathbb{R}$ be a convex function having polynomial growth. Then*

$$f(X_t) = f(X_0) + \int_0^t D^- f(X_s)\, dX_s + \sigma^2 H \int_{\mathbb{R}} x^2 \mathcal{L}_t^{(X)}(x)\, d\nu_f(x), \quad (10.28)$$

where, as before, $D^- f$ denotes the left derivative of f and ν_f denotes the second derivative measure of f, as introduced in (10.22) and (10.23), respectively.

Proof. We proceed as in the proof of Theorem 10.4.1. Let $\{f_n\}$ be a sequence of smooth functions converging to f (as in the proof of Theorem 10.4.1). Define $g_n(t, y) = f_n(x \exp(\sigma y - 1/2\sigma^2 t^{2H} + \mu t))$. Then by the fractional Itô formula (see Theorem 4.2.6), we have

$$f_n(X_t) = g_n(t, B_t^{(H)})$$

$$= g_n(0, 0) + \int_0^t \frac{\partial g_n}{\partial s}(s, B_s^{(H)})\, ds$$

$$+ \int_0^t \frac{\partial g_n}{\partial y}(s, B_s^{(H)})\, dB_s^{(H)} + H \int_0^t s^{2H-1} \frac{\partial^2 g_n}{\partial y^2}(s, B_s^{(H)})\, ds$$

$$= f_n(x) + \int_0^t f_n'(X_s) X_s (-H\sigma^2 s^{2H-1}\, ds + \mu\, ds + \sigma\, dB_s^{(H)})$$

$$+ H \int_0^t s^{2H-1} [f_n''(X_s) X_s^2 \sigma^2 + f_n'(X_s) X_s \sigma^2]\, ds$$

$$= f_n(x) + \int_0^t f_n'(X_s)\, dX_s + H \int_0^t s^{2H-1} f_n''(X_s) X_s^2 \sigma^2\, ds. \quad (10.29)$$

Proceeding as in the proof of Theorem 10.4.1, we have

$$\int_0^t f_n''(X_s) \sigma^2 X_s^2 H s^{2H-1}\, ds = \int_0^t \left[\int_{\mathbb{R}} f_n''(X_s) \sigma^2 x^2 \delta(X_s - x)\, dx \right] H s^{2H-1}\, ds$$

$$= \sigma^2 H \int_{\mathbb{R}} f_n''(x) x^2 \mathcal{L}_t^{(X)}(x) \, dx, \qquad (10.30)$$

which converges to $\sigma^2 H \int_{\mathbb{R}} x^2 \mathcal{L}_t^{(X)}(x) \, d\nu_f(x)$ as $n \to \infty$. Letting $n \to \infty$ in (10.29) and using (10.30), we obtain (10.28). □

Corollary 10.5.5 (A Meyer Tanaka formula for geometric *fBm*). Let $H \in (0,1)$ and X_t given in (10.26). For any $z > 0$ we have

$$|X_t - z| = |X_0 - z| + \int_0^t \operatorname{sign}(X_s - z) \, dX_s + 2\sigma^2 H z^2 \mathcal{L}_t^{(X)}(z).$$

Proof. In this case $f(x) = |x - z|$ in Theorem 10.5.4; so $D^- f(x) = \operatorname{sign}(x - z)$ and $\nu_f(x) = 2\delta_z(x)$. □

Corollary 10.5.6. Let $H \in (0,1)$ and X_t given in (10.26). For any $z > 0$ we have

$$(X_t - z)^+ = (X_0 - z)^+ + \int_0^t I_{[z,\infty)}(X_s) \, dX_s + \sigma^2 H z^2 \mathcal{L}_t^{(X)}(z).$$

Proof. In this case $f(x) = (x - z)^+$ in Theorem 10.5.4; so $D^- f(x) = I_{[z,\infty)}$ and $\nu_f(x) = \delta_z(x)$. □

A convergence of functionals of weighted sums of independent random variables to local times of *fBm* can be found in [132], where discrete approximations of $\mathcal{L}_t^{(B^{(H)})}(x)$ are established in the L^2 sense.

10.6 Renormalized self-intersection local time for *fBm*

Let $B^{(H)} = \{B^{(H)}(t), t \geq 0\}$ be a d-dimensional *fBm* of Hurst parameter $H \in (0,1)$. The *self-intersection local time* of $B^{(H)}$ is formally defined as

$$I = \int_0^T \int_0^t \delta_0(B^{(H)}(t) - B^{(H)}(s)) \, ds \, dt,$$

where $\delta_0(x)$ is the Dirac delta function. It measures the amount of time that the process spends intersecting itself on the time interval $[0,T]$ and has been an important topic of the theory of stochastic process. A rigorous definition of this random variable may be obtained by approximating the Dirac function by the heat kernel

$$p_\varepsilon(x) = (2\pi\varepsilon)^{-d/2} \exp\left(-\frac{|x|^2}{2\varepsilon}\right),$$

as $\varepsilon > 0$ tends to zero. We denote the approximated self-intersection local time by

$$I_\varepsilon = \int_0^T \int_0^t p_\varepsilon(B^{(H)}(t) - B^{(H)}(s))\,ds\,dt, \qquad (10.31)$$

and a natural question is to study the behavior of I_ε as ε tends to zero. At this purpose, we present some of the results of [120].

The following result extends a result from [197] to the case of arbitrary dimensions and with Hurst parameter $H < 3/4$.

Theorem 10.6.1. *Let I_ε be the random variable defined in (10.31). We have*

1. *If $H < 1/d$, then I_ε converges in L^2 as ε tends to zero.*
2. *If $1/d < H < 3/(2d)$, then*

$$I_\varepsilon - TC_{H,d}\varepsilon^{-d/2 + 1/(2H)}, \qquad (10.32)$$

converges in L^2 as ε tends to zero, where

$$C_{H,d} = (2\pi)^{-d/2} \int_0^\infty \left(z^{2H} + 1\right)^{-d/2} dz.$$

3. *If $1/d = H < 3/(2d)$, then*

$$I_\varepsilon - \frac{T}{2H(2\pi)^{d/2}} \log\left(\frac{1}{\varepsilon}\right), \qquad (10.33)$$

converges in L^2 as ε tends to zero.
4. *If $H \geq 3/(2d)$, then the difference $I_\varepsilon - E[I_\varepsilon]$ does not converge in L^2.*

That means, if $H < 3/(2d)$, the difference $I_\varepsilon - E[I_\varepsilon]$ converges in L^2 as ε tends to zero to the *renormalized self-intersection local time*. In the case $H \geq 3/(2d)$ we know from the above (10.33) that $I_\varepsilon - E[I_\varepsilon]$ does not converge in L^2. However, we have the following in this case.

Theorem 10.6.2. *Suppose $3/(2d) \leq H < 3/4$. Then the random variables*

$$\begin{cases} (\log(\frac{1}{\varepsilon}))^{-1/2} (I_\varepsilon - E[I_\varepsilon]) & \text{if } H = 3/(2d) \\ \varepsilon^{d/2 - 3/(4H)} (I_\varepsilon - E[I_\varepsilon]) & \text{if } H > 3/(2d) \end{cases}$$

converge as ε tends to zero in distribution to a normal law $N(0, T\sigma^2)$, where σ^2 is a constant depending on d and H.

We shall prove Theorem 10.6.1. The proof of Theorem 10.6.2 is more involved, and the reader is referred to the paper [120].

Proof (Theorem 10.6.1). Let $B^{(H)} = \{B^{(H)}(t), t \geq 0\}$ be a d-dimensional *fBm* of Hurst parameter $H \in (0,1)$.

Consider the approximation I_ε of the self-intersection local time introduced in (10.31). From the equality

$$p_\varepsilon(x) = (2\pi)^{-d} \int_{\mathbb{R}^d} \exp\left(i\langle \xi, x\rangle_{L^2(\mathbb{R}^d)}\right) \exp\left(-\frac{\varepsilon \|\xi\|^2_{L^2(\mathbb{R}^d)}}{2}\right) d\xi$$

and the definition of I_ε, we obtain

$$I_\varepsilon = (2\pi)^{-d} \int_0^T \int_0^t \int_{\mathbb{R}^d} \exp\left(i\langle \xi, B^{(H)}(t) - B^{(H)}(s)\rangle_{L^2(\mathbb{R}^d)}\right)$$
$$\cdot \exp\left(-\frac{\varepsilon \|\xi\|^2_{L^2(\mathbb{R}^d)}}{2}\right) d\xi \, ds \, dt.$$

Therefore,

$$E\left[I_\varepsilon^2\right] = \int_{\mathcal{J}} \int_{\mathbb{R}^{2d}} E\left[\exp\left(i\langle \xi, B_t^{(H)} - B_s^{(H)}\rangle_{L^2(\mathbb{R}^d)}\right.\right.$$
$$\left.\left. + i\langle \eta, B_{t'}^{(H)} - B_{s'}^{(H)}\rangle_{L^2(\mathbb{R}^d)}\right)\right] \qquad (10.34)$$
$$\cdot \frac{1}{(2\pi)^{2d}} \exp\left(-\left(\varepsilon\|\xi\|^2_{L^2(\mathbb{R}^d)} + \varepsilon\|\eta\|^2_{L^2(\mathbb{R}^d)}\right)/2\right)$$
$$\cdot d\xi \, d\eta \, ds \, dt \, ds' \, dt',$$

where

$$\mathcal{J} = \{(s,t,s',t') : 0 < s < t < T, 0 < s' < t' < T\}. \qquad (10.35)$$

Throughout this part we will make use of the following notation: for any $\tau = (s,t,s',t')$,

$$\lambda(\tau) = |t-s|^{2H}, \quad \rho(\tau) = |t'-s'|^{2H}, \qquad (10.36)$$

and

$$\mu(\tau) = \frac{1}{2}\left(|s-t'|^{2H} + |s'-t|^{2H} - |t-t'|^{2H} - |s-s'|^{2H}\right). \qquad (10.37)$$

Notice that λ is the variance of $B_t^{H,1} - B_s^{H,1}$, ρ is the variance of $B_{t'}^{H,1} - B_{s'}^{H,1}$, and μ is the covariance between $B_t^{H,1} - B_s^{H,1}$ and $B_{t'}^{H,1} - B_{s'}^{H,1}$, where $B^{H,1}$ denotes a one-dimensional *fBm* with Hurst parameter H.

With this notation, for any $\xi, \eta \in \mathbb{R}^d$, we can write

$$E\left[\left(\langle \xi, B^{(H)}(t) - B^{(H)}(s)\rangle_{L^2(\mathbb{R}^d)} + \langle \eta, B^{(H)}(t') - B^{(H)}(s')\rangle_{L^2(\mathbb{R}^d)}\right)^2\right]$$
$$= \lambda\|\xi\|^2_{L^2(\mathbb{R}^d)} + \rho\|\eta\|^2_{L^2(\mathbb{R}^d)} + 2\mu\langle \xi,\eta\rangle_{L^2(\mathbb{R}^d)}. \qquad (10.38)$$

As a consequence, from (10.34) and (10.38) we deduce for all $\varepsilon > 0$,

$$E\left(I_\varepsilon^2\right) = (2\pi)^{-2d}$$
$$\cdot \int_{\mathcal{J}} \int_{\mathbb{R}^{2d}} e^{-\left[(\lambda+\varepsilon)\|\xi\|^2_{L^2(\mathbb{R}^d)} + 2\mu\langle \xi,\eta\rangle_{L^2(\mathbb{R}^d)} + (\rho+\varepsilon)\|\eta\|^2_{L^2(\mathbb{R}^d)}\right]/2} d\xi \, d\eta \, d\tau$$

$$= (2\pi)^{-d} \int_{\mathcal{J}} \left[(\lambda + \varepsilon)(\rho + \varepsilon) - \mu^2 \right]^{-d/2} d\tau. \tag{10.39}$$

On the other hand, the expectation of the random variable I_ε is given by

$$E[I_\varepsilon] = \int_0^T \int_0^t p_{\varepsilon + |t-s|^{2H}}(0) \, ds \, dt$$

$$= (2\pi)^{-d/2} \int_0^T \int_0^t \left(\varepsilon + |t-s|^{2H} \right)^{-d/2} ds \, dt$$

$$= (2\pi)^{-d/2} \int_0^T (T-s) \left(\varepsilon + s^{2H} \right)^{-d/2} ds. \tag{10.40}$$

Assertion 1 follows easily from (10.39) and (10.40). From (10.40), making the change of variables $s = z\varepsilon^{1/(2H)}$, we obtain, if $1/d < H < 3/(2d)$

$$E[I_\varepsilon] = \frac{\varepsilon^{1/(2H)-d/2}}{(2\pi)^{d/2}} \int_0^{T\varepsilon^{1/(2H)}} (T - z\varepsilon^{1/(2H)}) \left(z^{2H} + 1 \right)^{-d/2} dz$$

$$= \varepsilon^{1/(2H)-d/2} T C_{H,d} + o(\varepsilon).$$

For $H = 1/d$ we get

$$E[I_\varepsilon] = \frac{T \log(1/\varepsilon)}{2H(2\pi)^{d/2}} + o(\varepsilon).$$

Hence, the convergence in L^2 of the random variables (10.32) and (10.33) is equivalent to the convergence of $I_\varepsilon - E[I_\varepsilon]$.

From (10.39) and (10.40) we obtain

$$E[I_\varepsilon I_\eta] - E[I_\varepsilon] E[I_\eta] = (2\pi)^{-d} \int_{\mathcal{J}} \left[((\lambda + \varepsilon)(\rho + \eta) - \mu^2)^{-d/2} \right.$$

$$\left. - ((\lambda + \varepsilon)(\rho + \eta))^{-d/2} \right] d\tau.$$

Therefore, a necessary and sufficient condition for the convergence in L^2 of $I_\varepsilon - E[I_\varepsilon]$ is that

$$\Xi_T =: \int_{\mathcal{J}} \left[(\lambda\rho - \mu^2)^{-d/2} - (\lambda\rho)^{-d/2} \right] d\tau < \infty. \tag{10.41}$$

This is a purely analysis problem and is solved by the following lemma. □

Lemma 10.6.3. *Let Ξ_T be defined by (10.41). Then $\Xi_T < \infty$ if and only if $dH < 3/2$.*

Proof. We divide the proof into three steps.
Step 1. We will denote by k a generic constant which may be different from one formula to another one. We will decompose the region $\mathcal{J}$ defined in (10.35) as follows:

$$\mathcal{T} \cap \{s < s'\} = \mathcal{T}_1 \cup \mathcal{T}_2 \cup \mathcal{T}_3,$$

where

$$\mathcal{T}_1 = \{(t, s, t', s') : 0 < s < s' < t < t' < T\},$$
$$\mathcal{T}_2 = \{(t, s, t', s') : 0 < s < s' < t' < t < T\},$$
$$\mathcal{T}_3 = \{(t, s, t', s') : 0 < s < t < s' < t' < T\}.$$

We will make use of the notation:

i. If $(t, s, t', s') \in \mathcal{T}_1$, we put $a = s' - s$, $b = t - s'$ and $c = t' - t$. On this region, the functions λ, ρ and μ defined in (10.36) and (10.37) take the following values

$$\lambda = \lambda_1 := \lambda_1(a, b, c) := (a + b)^{2H}, \rho = \rho_1 := (b + c)^{2H},$$
$$\mu = \mu_1 := \mu_1(a, b, c) := \frac{1}{2}\left[(a + b + c)^{2H} + b^{2H} - c^{2H} - a^{2H}\right].$$

ii. If $(t, s, t', s') \in \mathcal{T}_2$, we put $a = s' - s$, $b = t' - s'$ and $c = t - t'$. On this region we will have

$$\lambda = \lambda_2 := b^{2H}, \rho = \rho_2 := (a + b + c)^{2H},$$
$$\mu = \mu_2 := \frac{1}{2}\left[(b + c)^{2H} + (a + b)^{2H} - c^{2H} - a^{2H}\right].$$

iii. If $(t, s, t', s') \in \mathcal{T}_3$, we put $a = t - s$, $b = s' - t$ and $c = t' - s'$. On this region we will have

$$\lambda = \lambda_3 := a^{2H}, \rho = \rho_3 := c^{2H},$$
$$\mu = \mu_3 := \frac{1}{2}\left[(a + b + c)^{2H} + b^{2H} - (b + c)^{2H} - (a + b)^{2H}\right].$$

For $i = 1, 2, 3$ we set $\delta_i = \lambda_i \rho_i - \mu_i^2$, $\Theta_i = \delta_i^{-d/2} - (\lambda_i \rho_i)^{-d/2}$. Note that λ_i, ρ_i, μ_i, and so on, $i = 1, 2, 3$, are functions of a, b, and c.

The following lower bounds for the determinant of the covariance matrix of $B_t^{H,1} - B_s^{H,1}$ and $B_{t'}^{H,1} - B_{s'}^{H,1}$, were obtained in [119] using the local nondeterminism property of the fBm (see [29]).

i.
$$\delta_1 \geq k\left[(a + b)^{2H} c^{2H} + (b + c)^{2H} a^{2H}\right]. \tag{10.42}$$

ii. For $i = 2, 3$
$$\delta_i \geq k \lambda_i \rho_i. \tag{10.43}$$

Step 2. The following estimates are important.

$$k(a + b + c)^{2H-2} ac \leq \mu_3 \leq k b^{2H-2} ac. \tag{10.44}$$

For $i = 2, 3$, we have

10.6 Renormalized self-intersection local time for fBm

$$\Theta_i \leq k\mu_i^2 \, (\lambda_i \rho_i)^{-d/2-1} \tag{10.45}$$

and

$$\Theta_i \leq k \, (\lambda_i \rho_i)^{-d/2}. \tag{10.46}$$

In fact, the inequalities in formula (10.44) follow from

$$\mu_3 = \frac{1}{2}\left[(a+b+c)^{2H} + b^{2H} - (a+b)^{2H} - (b+c)^{2H}\right]$$

$$= H(2H-1)ac \int_0^1 \int_0^1 (b+vc+ua)^{2H-2} \, du \, dv.$$

We have, for $i = 2, 3$,

$$\Theta_i = \left[\left(1 - \frac{\mu_i^2}{\lambda_i \rho_i}\right)^{-d/2} - 1\right] (\lambda_i \rho_i)^{-d/2}.$$

The estimate (10.43), assuming $k < 1$, implies $\mu_i^2/(\lambda_i \rho_i) \leq 1 - k$ and (10.45) holds. Moreover, (10.43) also implies (10.46).

Step 3. Suppose $dH < 3/2$. We claim that

$$\int_{[0,T]^3} \Theta_i \, da \, db \, dc < \infty \tag{10.47}$$

for $i = 1, 2, 3$. From (10.42) we deduce

$$\delta_1 \geq k(a+b)^H (b+c)^H a^H c^H$$
$$\geq k(abc)^{4H/3}. \tag{10.48}$$

Then, (10.48) together with the estimate

$$\lambda_1 \rho_1 = (a+b)^{2H}(b+c)^{2H} \geq (abc)^{4H/3}.$$

implies (10.47) for $i = 1$.

To handle the case $i = 2$, we decompose the integral over the regions $\{b \geq \eta a\}$, $\{b \geq \eta c\}$, and $\{b < \eta a, b < \eta c\}$ for some fixed but arbitrary $\eta > 0$. We have, using (10.46)

$$\Xi = \int_{b \geq \eta a} \Theta_2 \, da \, db \, dc \leq k \int_{b \geq \eta a} \frac{da \, db \, dc}{(a+b+c)^{dH} b^{dH}}.$$

If $dH < 1$, then this integral is finite. If $1 < dH$, then

$$\Xi \leq k \int_0^T \int_0^T \frac{da \, dc}{(a+c)^{dH}} \int_{\eta a}^T b^{-Hd} \, db$$

$$\leq k \int_0^T \int_0^T a^{-4dH/3+1} c^{-2dH/3} \, da \, dc < \infty.$$

It is also easy to show that $\Xi < \infty$ in the case $1 = dH$. The case $b \geq \eta c$ can be treated in a similar way.

To deal with the case both $b < \eta a$ and $b < \eta c$, we make use of the estimate (10.45) and the following upper bound for μ_2.

$$\mu_2 = \frac{1}{2}\left[(a+b)^{2H} - a^{2H} + (b+c)^{2H} - c^{2H}\right]$$
$$\leq k(a^{2H-1} + c^{2H-1})b$$

for η small enough. In this way we obtain

$$\Theta_2 \leq k\left(a^{4H-2} + c^{4H-2}\right)(a+b+c)^{-2H-dH} b^{2-2H-dH}$$
$$\leq k\left(a^{(2-d/3)H}b^{dH/3} + c^{(2-d/3)H}b^{dH/3}\right)(a+b+c)^{-2H-dH} b^{-dH}.$$

Hence,

$$\int_{b<\eta a, b\leq \eta c} \Theta_2 \, da \, db \, dc$$
$$\leq k \int_{b<\eta a, b\leq \eta c} b^{-dH}(a+b+c)^{-2H-dH}$$
$$\quad \cdot \left[a^{(2-d/3)H}b^{dH/3} + c^{(2-d/3)H}b^{dH/3}\right] da \, db \, dc$$
$$\leq k \int_{[0,T]^3} b^{-dH}(a+b+c)^{-2H-dH} a^{(2-d/3)H}b^{dH/3} \, da \, db \, dc$$
$$\leq k \int_{[0,T]^3} b^{-2dH/3} a^{-2dH/3} c^{-2dH/3} \, da \, db \, dc$$

which is finite if $dH < 3/2$.

To handle the case $i = 3$, we decompose the integral over the regions $\{a \geq \eta_1 b, c \geq \eta_2 b\}$, $\{a < \eta_1 b, c < \eta_2 b\}$, $\{a \geq \eta_1 b, c < \eta_2 b\}$, and $\{a < \eta_1 b, c \geq \eta_2 b\}$. By symmetry it suffices to consider the first three regions. We have, using (10.46)

$$\int_{a \geq \eta_1 b, c \geq \eta_2 b} \Theta_3 \, da \, db \, dc \leq k \int_0^T db \int_{\eta_1 b}^T \frac{da}{a^{dH}} \int_{\eta_2 b}^T \frac{dc}{c^{dH}}$$
$$\leq k \int_0^T \frac{db}{b^{2dH-2}} < \infty.$$

Let us now suppose both $a < \eta_1 b$ and $c < \eta_2 b$. Using (10.44) and (10.45) and that $H < 3/4$ yields

$$\Theta_3 \leq kb^{4H-4}a^{2-2H-dH}c^{2-2H-dH} \leq ka^{-2/3dH}c^{-2/3dH}b^{-2/3dH},$$

which implies that the integral over this region is finite. Finally, let us consider the case $c < \eta_1 b$ and $a \geq \eta_2 b$. If $Hd > 1$, then (10.46) yields

10.6 Renormalized self-intersection local time for fBm

$$\Theta_3 \leq k(ac)^{-Hd}$$

which is integrable. So we can assume $H \leq 1/d \leq 1/2$. Then

$$\mu_3 = \frac{1}{2}\left[(a+b+c)^{2H} - (a+b)^{2H} - (c+b)^{2H} + b^{2H}\right]$$
$$\leq kb^{2H-1}c \qquad (10.49)$$

if η_2 is small enough. Hence, using (10.45) and (10.49) we get

$$\Theta_3 \leq kb^{4H-2}a^{-2H-dH}c^{2-2H-dH}.$$

Consequently, if $-dH + 2H + 1 < 0$

$$\int_{c<\eta_1 b, a\geq\eta_2 b} \Theta_3 \, da\, db\, dc \leq k \int_{c<\eta_1 b, a\geq\eta_2 b} b^{4H-2}a^{-2H-dH}c^{2-2H-dH} \, dc\, db\, da$$

$$\leq k \int_{a\geq\eta_2 b} a^{-dH-2H} b^{-dH+2H+1} \, db\, da$$

$$\leq k \int_0^T a^{-2dH+2} \, da,$$

which is finite if $dH < 3/2$. The case $-dH + 2H + 1 \geq 0$ is easier. Assume that $H = 3/(2d)$. We note that

$$E\left[(I_{2m}(f_{2m,\varepsilon}))^2\right] = (2m)! \, \|f_{2m,\varepsilon}\|^2_{H^{\otimes(2m)}} \qquad (10.50)$$

$$= (2m)! \sum_{m_1+\cdots+m_d=m} \frac{(2m)!}{(2m_1)!\cdots(2m_d)!} \frac{(2\pi)^{-d}}{((2m)!)^2} \alpha(\mathbf{i}_{2m})^2$$

$$\cdot \int_{\mathcal{J}} (\varepsilon+\lambda)^{-d/2-m}(\varepsilon+\rho)^{-d/2-m} \mu^{2m} \, d\tau$$

$$= \frac{\alpha_m}{(2\pi)^d 2^{2m}} \int_{\mathcal{J}} (\varepsilon+\lambda)^{-d/2-m}(\varepsilon+\rho)^{-d/2-m} \mu^{2m} d\tau,$$

where

$$\alpha_m = \sum_{m_1+\cdots+m_d=m} \frac{(2m_1)!\cdots(2m_d)!}{(m_1!)^2\cdots(m_d!)^2}.$$

Let us show that $\Xi_T = \infty$. It suffices to prove that

$$\frac{d}{2(2\pi)^d} \int_{\mathcal{J}} \mu^2(\lambda\rho)^{-d/2-1} \, ds\, dt\, ds'\, dt' = \infty, \qquad (10.51)$$

because from the identity (10.50) this is the second moment of the second chaos of the renormalized self-intersection local time. In order to check (10.51) we will show that

$$A := \int_{0<a+b+c<T} (T-a-b-c)\mu_3^2(\lambda_3\rho_3)^{-d/2-1} \, da\, db\, dc = \infty.$$

With the above notation, we have, using (10.44), for $\varepsilon > 0$ small enough,

$$A \geq k \int_{[0,\varepsilon]^3} (b+c+a)^{4H-4}(ac)^{2-Hd-2H} \, da \, db \, dc := B.$$

If $d = 2$, we get

$$B = k \int_{[0,\varepsilon]^3} \frac{1}{(a+b+c)ac} \, da \, db \, dc = \infty.$$

For $d > 2$, we have $2 - Hd - 2H = 1/2 - 3/d > -1$. Hence,

$$B = \frac{k}{3-4H} \int_{[0,\varepsilon]^2} \left[(c+a)^{4H-3} - (\varepsilon + c + a)^{4H-3} \right] (ac)^{2-Hd-2H} \, da \, dc,$$

and

$$\int_{0 < a < c < \varepsilon} (c+a)^{4H-3}(ac)^{2-Hd-2H} \, da \, dc$$
$$\geq 2^{4H-3} \int_{0 < a < c < \varepsilon} a^{2-Hd-2H} c^{2H-1-Hd} \, da \, dc$$
$$\geq k \int_0^\varepsilon a^{2-2Hd} \, da = \infty$$

because $2H - 1 - Hd < -1$. $\square$

10.7 Application to finance

To show an application of the previous results, we now use Corollary 10.5.6 to prove that the fractional analogue of the *stop-loss–start-gain* (SLSG) portfolio is not self-financing in a fractional Black Scholes market. Our result includes the classical Black Scholes market ($H = 1/2$), which was first proved in [49]. For further information on this portfolio from a p-variation approach, see [205] and the references therein.

We adopt here the WIS model of a financial market driven by fBm, as described in Chapter 7. The fractional Black Scholes market consists of a *bank account* or a *bond*, and a *stock*. The generalized price process A_t of the bond, at time t is given by

$$dA_t = \rho A_t dt, \qquad A_0 = 1, \qquad 0 \leq t \leq T,$$

where $\rho \geq 0$, $T > 0$ are constants. The generalized price process X_t of the stock is given by a geometric fBm

$$dX_t = \mu X_t \, dt + \sigma X_t \, dB_t^{(H)}, \qquad X_0 = x > 0, \qquad (10.52)$$

where μ and $\sigma \neq 0$ are constants. The solution of (10.52) is given in Example 4.2.4 of Chapter 4. This financial market is complete, and there exists an explicit fractional Black Scholes formula for the price of *European call option* (see Section 9.5). By normalizing the market (i.e., by considering all prices in units of A_t), we may assume without loss of generality that

$$\rho = 0, \quad \text{i.e. } A_t = 1 \, \forall t.$$

A European call option in this market gives the (generalized) payoff

$$F(\omega) = (X_T(\omega) - q)^+$$

at time T, for some given constant $q > 0$ (the exercise price of the option). The (generalized WIS) SLSG portfolio $\theta_t = (\alpha_t, \beta_t)$ for this payoff is defined by

$$\alpha_t = -q I_{[q,\infty)}(X_t),$$

i.e., the number of bonds or monetary units held at time t, and

$$\beta_t = I_{[q,\infty)}(X_t),$$

i.e., the number of stocks held at time t. We may describe this as follows: Assume that we start with initial fortune 0 and $X_0 < q$. As long as $X_t < q$, we do nothing. However, as soon as $X_t \geq q$, we immediately borrow the amount q from the bank to buy one stock at the price q. We keep this stock until its price drops below q. Then we sell it immediately and pay q back to the bank. The (generalized) value V_t^θ at time t of this portfolio θ is given by

$$\begin{aligned}
V_t^\theta &= \alpha_t A_t + \beta_t \diamond X_t \\
&= -q I_{[q,\infty)}(X_t) + I_{[q,\infty)}(X_t) \diamond X_t \\
&= (X_t - q)^+.
\end{aligned} \quad (10.53)$$

Thus this portfolio replicates the (generalized) payoff $(X_t - q)^+$ *at all times* $t \leq T$ with initial fortune 0 when $X_0 < q$. Recall that θ is self-financing (in the WIS sense of Definition 7.2.3) if and only if

$$dV_t^\theta = \alpha_t \, dA_t + \beta_t \diamond dX_t.$$

That is, with $\rho = 0$, we have

$$dV_t^\theta = I_{[q,\infty)}(X_t) \, dX_t.$$

By (10.53), this implies that

$$(X_t - q)^+ = (X_0 - q)^+ + \int_0^t I_{[q,\infty)}(X_s) \, dX_s. \quad (10.54)$$

However, Corollary 10.25 states that

$$(X_t - q)^+ = (X_0 - q)^+ + \int_0^t I_{[q,\infty)}(X_s) \, dX_s + \sigma^2 H q^2 \mathcal{L}_t^{(X)}(q). \tag{10.55}$$

Since $\mathcal{L}_t^{(X)}(q) > 0$ for all $t > 0$, we see that (10.55) contradicts (10.54). Hence θ is not self-financing (in the WIS sense).

Thus in this model, we are able to extend the SLSG portfolio from the geometric Brownian motion case of [49] to a geometric *fBm*. The fact that θ is not self-financing is in agreement with the result that our fractional Black Scholes model does not allow for strong arbitrage opportunities (Theorem 7.2.6). Further, we obtain new expressions for the value of European call and put options in our *fBm* model. Like in [49], our expressions decompose option prices into their *intrinsic* and *time* values. That is, we have

Proposition 10.7.1 (An alternative call valuation formula). *Consider a European call option with market value $C(0)$, maturity date T, and exercise price q. Assume the WIS fBm model for $H \in (0,1)$ (Chapter 7). Then*

$$C(0) = (X_0 - q)^+ + \sigma^2 H q^2 \widehat{E}[\mathcal{L}_T^{(X)}(q)],$$

where $\widehat{E}$ denotes the expectation under the risk-free measure $\widehat{\mathbb{P}}$ defined in (7.20).

Proof. By Corollary 5.6 in [121] and Section 6 in [89], we know that

$$C(0) = \widehat{E}[(X_T - q)^+].$$

Combining this with (10.55), we get

$$C(0) = \widehat{E}\left[(X_0 - q)^+ + \int_0^T I_{[q,\infty)}(X_s) \, dX_s + \sigma^2 H q^2 \mathcal{L}_T^{(X)}(q)\right]$$

$$= (X_0 - q)^+ + \sigma^2 H q^2 \widehat{E}[\mathcal{L}_T^{(X)}(q)]$$

since the expectation of the middle term, a quasi-martingale with respect to $\widehat{\mathbb{P}}$, is zero. □

Following [49], we call $(X_0 - q)^+$ the *intrinsic* value of the option and $\sigma^2 H q^2 \widehat{E}^{(H)}[\mathcal{L}_T^{(X)}(q)]$ the *time* value of the option.

Note that, letting $H \to 1/2$ in (10.55), we get the price process of a geometric Brownian motion. Taking the expectation of (10.55), under the risk-free measure for a geometric Brownian motion, the middle term again vanishes, and the expectation of the local time remains in agreement with Lemma A5 of [49]. This case is Proposition 2 of [49]. On the other hand, letting $H \to 1$ in (10.55), we get the case of a continuous price process of bounded variation. Then the local time term disappears since $\mathcal{L}_t^{(X)}(q)$ is zero for bounded variation processes. This complements the results of [49] and [205] using Lebesgue and Riemann Stieltjes integrals. In the latter paper, the

SLSG portfolio is shown to be self-financing for continuous price process of bounded variation, and so arbitrage opportunities exist for this process.

The put–call parity implies that puts and calls have the same time value. Hence, by the above proposition, we have a value for European put options. Recall that we assume the bond process $A_t = 1$ for all t.

Corollary 10.7.2 (Put Valuation). *Let $P(0)$ be the market price of the European put with maturity date T and exercise price q. Assume the WIS fBm model for $H \in (0,1)$ (Chapter 7). Then*

$$P(0) = (q - X_0)^+ + \sigma^2 H q^2 \widehat{E}\big[\mathcal{L}_T^{(X)}(q)\big].$$

Part IV

Appendixes

A
Classical Malliavin calculus

For the reader's convenience we recall the standard setup for the classical white noise probability space. For further details, we refer to [1], [108], [109], [144], [176] and [179].

A.1 Classical white noise theory

Definition A.1.1. *Let $\mathcal{S}(\mathbb{R})$ denote the Schwartz space of rapidly decreasing smooth functions on $\mathbb{R}$, and let $\Omega := \mathcal{S}'(\mathbb{R})$ be its dual, usually called the* space of tempered distributions. *Let $\mathbb{P}$ be the probability measure on the Borelian σ-algebra $\mathcal{F} := \mathcal{B}(\mathcal{S}'(\mathbb{R}))$ defined by the property that*

$$\int_{\mathcal{S}'(\mathbb{R})} exp(i<\omega,f>)\,d\mathbb{P}(\omega) = exp\left(-\frac{1}{2}\|f\|^2_{L^2(\mathbb{R})}\right), \quad f \in \mathcal{S}(\mathbb{R}), \tag{A.1}$$

where $i = \sqrt{-1}$ and $<\omega,f> = \omega(f)$ is the action of $\omega \in \Omega = \mathcal{S}'(\mathbb{R})$ on $f \in \mathcal{S}(\mathbb{R})$.

The measure $\mathbb{P}$ is called the white noise probability measure. Its existence follows from the Bochner Minlos theorem (see, for example, [109] or [144]).

Using (A.1) one can prove that

$$E[<\omega,f>] = 0 \quad \forall f \in \mathcal{S}(\mathbb{R}),$$

where in general

$$E[F(\omega)] = \int_\Omega F(\omega)\,d\mathbb{P}(\omega)$$

is the expectation of F with respect to $\mathbb{P}$. Moreover, we have the isometry

$$E[<\omega,f>^2] = \|f\|^2_{L^2(\mathbb{R})} \quad \forall f \in \mathcal{S}(\mathbb{R}). \tag{A.2}$$

Based on this we can now define $<\omega, f>$ for an arbitrary $f \in L^2(\mathbb{R})$ as follows:
$$<\omega, f> = \lim_{n \to \infty} <\omega, f_n> \quad [\text{limit in } L^2(\mathbb{P})],$$
where $f_n \in \mathcal{S}(\mathbb{R})$ is a sequence converging to f in $L^2(\mathbb{R})$. In particular, this makes
$$\tilde{B}(t) := \tilde{B}(t, \omega) := <\omega, I_{[0,t]}(\cdot)> \tag{A.3}$$
well-defined as an element of $L^2(\mathbb{P})$ for all $t \in \mathbb{R}$, where
$$I_{[0,t]}(s) = \begin{cases} 1 & \text{if } 0 \leq s \leq t, \\ -1 & \text{if } t \leq s \leq 0, \text{ except } t = s = 0, \\ 0 & \text{otherwise.} \end{cases}$$

By Kolmogorov's continuity theorem the process $\tilde{B}(t)$ has a continuous version, which we will denote by $B(t)$. It can now be verified that $B(t)$ is a Gaussian process and
$$E[B(t_1)B(t_2)] = \int_{\mathbb{R}} I_{[0,t_1]}(s)I_{[0,t_2]}(s)\,ds = \begin{cases} \min(|t_1|, |t_2|) & \text{if } t_1, t_2 > 0, \\ 0 & \text{otherwise.} \end{cases}$$

Therefore, $B(t)$ is a Brownian motion with respect to the probability law $\mathbb{P}$. It follows from (A.3) that
$$<\omega, f> = \int_{\mathbb{R}} f(t)\,dB(t) \text{ for all deterministic } f \in L^2(\mathbb{R}). \tag{A.4}$$

Let $\hat{L}^2(\mathbb{R}^n)$ be the set of all symmetric deterministic functions $f \in L^2(\mathbb{R}^n)$. If $f \in \hat{L}^2(\mathbb{R}^n)$, the *iterated Itô integral* of f is defined by
$$I_n(f) := \int_{\mathbb{R}^n} f(t)\,dB^{\otimes n}(t)$$
$$:= n! \int_{\mathbb{R}} \left\{ \int_{-\infty}^{t_n} \cdots \left[\int_{-\infty}^{t_2} f(t_1, \ldots, t_n)\,dB(t_1) \right] dB(t_2) \cdots dB(t_n) \right\}.$$

We now recall the following fundamental result:

Theorem A.1.2 (The Wiener Itô chaos expansion theorem I). *Let $F \in L^2(\mathbb{P})$. Then there exists a unique sequence $\{f_n\}_{n=0}^{\infty}$ of functions $f_n \in \hat{L}^2(\mathbb{R}^n)$ such that*
$$F(\omega) = \sum_{n=0}^{\infty} I_n(f_n) \quad [convergence \text{ in } L^2(\mathbb{P})],$$
where $I_0(f_0) := E[F]$.

Moreover, we have the isometry

$$E[F^2] = \sum_{n=0}^{\infty} n! \|f_n\|_{L^2(\mathbb{R}^n)}^2.$$

By convention we put $I_0(f_0) = f_0$ for constants f_0, and then $\|f_0\|^2 = |f_0|^2$.
Define the *Hermite polynomials* by

$$h_n(x) = (-1)^n e^{x^2/2} \frac{d^n}{dx^n}(e^{-x^2/2}), \quad n = 0, 1, 2, \ldots \quad (A.5)$$

For example, the first Hermite polynomials are

$$h_0(x) = 1, \quad h_1(x) = x, \quad h_2(x) = x^2 - 1,$$
$$h_3(x) = x^3 - 3x, \quad h_4(x) = x^4 - 6x^2 + 3, \ldots$$

The generating function is

$$\exp\left(tx - \frac{x^2}{2}\right) = \sum_{n=0}^{\infty} \frac{t^n}{n!} h_n(x), \quad \forall\, t,\, x \in \mathbb{R}.$$

In the following we let

$$\xi_n(x) = \pi^{-1/4}((n-1)!)^{-1/2} h_{n-1}(\sqrt{2}x) e^{-x^2/2}; \quad n = 1, 2, \ldots, \quad (A.6)$$

be the *Hermite functions* (see [109] and [223]). Then $\xi_n \in \mathcal{S}(\mathbb{R})$ and there exist constants C and γ such that

$$|\xi_n(x)| \leq \begin{cases} C n^{-1/12} & \text{if } |x| \leq 2\sqrt{n} \\ Ce^{-\gamma x^2} & \text{if } |x| > 2\sqrt{n} \end{cases} \quad (A.7)$$

for all n (see, for example, [223, Lemma 1.5.1]). It is also proved in [223] that $\{\xi_n\}_{n=1}^{\infty}$ constitutes an orthonormal basis for $L^2(\mathbb{R})$.

Let $\mathcal{J}$ be the set of all multi-indices $\alpha = (\alpha_1, \alpha_2, \ldots)$ of finite *length* $l(\alpha) = \max\{i; \alpha_i \neq 0\}$, with $\alpha_i \in \mathbb{N}_0 = \{0, 1, 2, \ldots\}$ for all i. For $\alpha = (\alpha_1, \ldots, \alpha_n) \in \mathcal{J}$ we put $\alpha! = \alpha_1! \alpha_2! \cdots \alpha_n!$ and $|\alpha| = \alpha_1 + \cdots + \alpha_n$, and we define

$$\mathcal{H}_\alpha(\omega) = h_{\alpha_1}(<\omega, \xi_1>) h_{\alpha_2}(<\omega, \xi_2>) \cdots h_{\alpha_n}(<\omega, \xi_n>). \quad (A.8)$$

Thus, for example,

$$\mathcal{H}_{(2,0,3,1)}(\omega) = h_2(<\omega,\xi_1>) h_0(<\omega,\xi_2>) h_3(<\omega,\xi_3>) h_1(<\omega,\xi_4>)$$
$$= (<\omega,\xi_1>^2 - 1)(<\omega,\xi_3>^3 - 3<\omega,\xi_3>)<\omega,\xi_4>$$

since

$$h_0(x) = 1, h_1(x) = x, h_2(x) = x^2 - 1, h_3(x) = x^3 - 3x.$$

Important special cases are the unit vectors

with 1 on the kth entry, 0 otherwise, $k = 1, 2, \ldots$ Note that

$$\mathcal{H}_{\varepsilon^{(k)}}(\omega) = h_1(<\omega, \xi_k>) = <\omega, \xi_k> = \int_{\mathbb{R}} \xi_k(t)\, dB(t).$$

More generally we have, by a result of Itô [I]:

$$\mathcal{H}_\alpha(\omega) = \int_{\mathbb{R}^{|\alpha|}} \xi^{\hat{\otimes}\alpha}(x)\, dB^{\otimes|\alpha|}(x)$$

where $\hat{\otimes}$ denotes symmetrized tensor product, i.e., $\xi^{\hat{\otimes}\alpha}(x)$ is the symmetrization with respect to the $l(\alpha)$ variables $x_1, \ldots, x_{l(\alpha)}$ of the tensor product

$$\xi^{\otimes\alpha}(x) := \xi_1^{\otimes\alpha_1}(x_1, \ldots, x_{\alpha_1}) \cdots \xi_m^{\otimes\alpha_m}(x_{l(\alpha)-\alpha_m+1}, \ldots, x_{l(\alpha)}),$$

where $x = (x_1, \ldots, x_{l(\alpha)})$ and $\alpha = (\alpha_1, \ldots, \alpha_{l(\alpha)}) \in \mathcal{J}, \alpha_m \neq 0$. This is the link between Theorem A.1.2 and the following result.

Theorem A.1.3 (The Wiener Itô chaos expansion theorem II). *Let $F \in L^2(\mathbb{P})$. Then there exists a unique family $\{c_\alpha\}_{\alpha \in \mathcal{J}}$ of constants $c_\alpha \in \mathbb{R}$ such that*

$$F(\omega) = \sum_{\alpha \in \mathcal{J}} c_\alpha \mathcal{H}_\alpha(\omega) \quad [convergence\ in\ L^2(\mathbb{P})].$$

Moreover, we have the isometry

$$E[F^2] = \sum_{\alpha \in \mathcal{J}} c_\alpha^2 \alpha!.$$

We now use Theorem A.1.2 and Theorem A.1.3 to define the following space $(\mathcal{S})$ of stochastic test functions and the dual space $(\mathcal{S})^*$ of stochastic distributions:

Definition A.1.4. *1. We define the* Hida space $(\mathcal{S})$ *of stochastic test functions to be all $\psi \in L^2(\mathbb{P})$ whose expansion*

$$\psi(\omega) = \sum_{\alpha \in \mathcal{J}} a_\alpha \mathcal{H}_\alpha(\omega)$$

satisfies

$$\|\psi\|_k^2 := \sum_{\alpha \in \mathcal{J}} a_\alpha^2 \alpha! (2\mathbb{N})^{k\alpha} < \infty \quad \forall k = 1, 2, \ldots,$$

where

$$(2\mathbb{N})^\gamma = (2 \cdot 1)^{\gamma_1}(2 \cdot 2)^{\gamma_2} \cdots (2 \cdot m)^{\gamma_m} \quad if\ \gamma = (\gamma_1, \ldots, \gamma_m) \in \mathcal{J}.$$

2. We define the Hida space $(S)^*$ of stochastic distributions *to be the set of formal expansions*

$$G(\omega) = \sum_{\alpha \in \mathcal{J}} b_\alpha \mathcal{H}_\alpha(\omega)$$

such that

$$\|G\|_q^2 := \sum_{\alpha \in \mathcal{J}} b_\alpha^2 \alpha! (2\mathbb{N})^{-q\alpha} < \infty \quad \text{for some } q < \infty.$$

We equip (S) with the projective topology and $(S)^*$ with the inductive topology. Convergence in (S) means convergence in $\|\cdot\|_k$ for every $k = 1, 2, \ldots$, while convergence in $(S)^*$ means convergence in $\|\cdot\|_q$ for some $q < \infty$. Then $(S)^*$ can be identified with the dual of (S), and the action of $G \in (S)^*$ on $\psi \in (S)$ is given by

$$\langle G, \psi \rangle_{(S)^*,(S)} := \sum_{\alpha \in \mathcal{J}} \alpha! a_\alpha b_\alpha.$$

In the sequel, we will denote the action $\langle \cdot, \cdot \rangle_{(S)^*,(S)}$ simply with symbol $\langle \cdot, \cdot \rangle$. In particular, if G belongs to $L^2(\mathbb{P}) \subset (S)^*$ and $\psi \in (S) \subset L^2(\mathbb{P})$, then

$$\langle G, \psi \rangle = \langle G, \psi \rangle_{L^2(\mathbb{P})} = E[G\psi].$$

We can in a natural way define $(S)^*$-valued integrals as follows:

Definition A.1.5. *Suppose that* $Z : \mathbb{R} \longrightarrow (S)^*$ *is a given function with property that*

$$\langle Z(t), \psi \rangle \in L^1(\mathbb{R}, dt) \quad \forall \psi \in (S). \tag{A.9}$$

Then $\int_\mathbb{R} Z(t) \, dt$ *is defined to be the unique element of* $(S)^*$ *such that*

$$\langle \int_\mathbb{R} Z(t) \, dt, \psi \rangle = \int_\mathbb{R} \langle Z(t), \psi \rangle \, dt \quad \forall \psi \in (S). \tag{A.10}$$

Just as in [109, Proposition 8.1], one can show that (A.10) defines $\int_\mathbb{R} Z(t) \, dt$ as an element of $(S)^*$. If (A.9) holds, we say that $Z(t)$ is *dt-integrable* in $(S)^*$.

Example A.1.6 (White noise).
For given $t \in \mathbb{R}$ the random variable $B(t) \in L^2(\mathbb{P})$ has the expansion

$$B(t) = <\omega, I_{[0,t]}(\cdot)> = <\omega, \sum_{k=1}^\infty \langle I_{[0,t]}, \xi_k \rangle_{L^2(\mathbb{R})} \xi_k(\cdot) >$$

$$= \sum_{k=1}^\infty \int_0^t \xi_k(s) \, ds <\omega, \xi_k> = \sum_{k=1}^\infty \int_0^t \xi_k(s) \, ds \mathcal{H}_{\varepsilon(k)}(\omega).$$

From this and (A.7) we see that regarded as a map $B(\cdot) : \mathbb{R} \to (S)^*$, $B(t)$ is differentiable with respect to t and

278   A Classical Malliavin calculus

$$\frac{d}{dt}B(t) = \sum_{k=1}^{\infty} \xi_k(t)\mathcal{H}_{\varepsilon^{(k)}}(\omega) \quad \text{in } (\mathcal{S})^*.\tag{A.11}$$

The expansion on the right-hand side of (A.11) is called *white noise* and denoted by $W(t)$.

The space $(\mathcal{S})^*$ is convenient for the *Wick product*.

Definition A.1.7. *If $F_i(\omega) = \sum_{\alpha \in \mathcal{J}} c_\alpha^{(i)} \mathcal{H}_\alpha(\omega); i = 1, 2$, are two elements of $(\mathcal{S})^*$ we define their* Wick product $(F_1 \diamond F_2)(\omega)$ *by*

$$(F_1 \diamond F_2)(\omega) = \sum_{\alpha,\beta \in \mathcal{J}} c_\alpha^{(1)} c_\beta^{(2)} \mathcal{H}_{\alpha+\beta}(\omega) = \sum_{\gamma \in \mathcal{J}} (\sum_{\alpha+\beta=\gamma} c_\alpha^{(1)} c_\beta^{(2)}) \mathcal{H}_\gamma(\omega).$$

The Wick product is a commutative, associative, and distributive (over addition) binary operation on each of the spaces $(\mathcal{S})$ and $(\mathcal{S})^*$.

Example A.1.8. 1. If F is deterministic, then $F \diamond G = F \cdot G$.
2. If $f \in L^2(\mathbb{R})$ is deterministic, then

$$\int_{\mathbb{R}} f(t)\,dB(t) = <\omega, f> = \sum_{k=1}^{\infty} \langle f, \xi_k \rangle_{L^2(\mathbb{R})} <\omega, \xi_k>$$

$$= \sum_{k=1}^{\infty} \langle f, \xi_k \rangle_{L^2(\mathbb{R})} \mathcal{H}_{\varepsilon^{(k)}}(\omega).$$

Moreover, if $\|f\|_2 = 1$, then $<\omega, f>^{\diamond n} = h_n(<\omega, f>)$.
3. If also $g(t) \in L^2(\mathbb{R})$ is deterministic, we have

$$\left[\int_{\mathbb{R}} f(t)\,dB(t)\right] \diamond \left[\int_{\mathbb{R}} g(t)\,dB(t)\right]$$

$$= \sum_{i,j=1}^{\infty} \langle f, \xi_i \rangle_{L^2(\mathbb{R})} \langle g, \xi_j \rangle_{L^2(\mathbb{R})} \mathcal{H}_{\varepsilon^{(i)}+\varepsilon^{(j)}}(\omega) \tag{A.12}$$

$$= \left[\int_{\mathbb{R}} f(t)\,dB(t)\right] \cdot \left[\int_{\mathbb{R}} g(t)\,dB(t)\right] - \langle f, g \rangle_{L^2(\mathbb{R})}$$

A fundamental property of the Wick product is the following relation to Itô Skorohod integration.

A.2 Stochastic integration

We recall the definition of Skorohod integral for the standard Brownian motion and its relation with the Wick product. For further details we refer to [109], [154], [179] and [176].

A.2 Stochastic integration

This integral may be regarded as an extension of the Itô integral to integrands that are not necessarily adapted. We first introduce some convenient notation.

Let $u(t,\omega)$, $\omega \in \Omega$, $t \in [0,T]$, be a stochastic process [always assumed to be (t,ω)-measurable] such that

$$u(t,\cdot) \quad \text{is } \mathcal{F}\text{-measurable for all } t \in [0,T] \quad (A.13)$$

and

$$E[u^2(t,\omega)] < \infty \quad \forall t \in [0,T]. \quad (A.14)$$

Then for each $t \in \mathbb{R}$ we can apply the Wiener Itô chaos expansion to the random variable $\omega \to u(t,\omega)$ and obtain functions $f_{n,t}(t_1,\ldots,t_n) \in \widehat{L}^2(\mathbb{R}^n)$ such that

$$u(t,\omega) = \sum_{n=0}^{\infty} I_n(f_{n,t}(\cdot)).$$

The functions $f_{n,t}(\cdot)$ depend on the parameter t, so we can write

$$f_{n,t}(t_1,\ldots,t_n) = f_n(t_1,\ldots,t_n,t).$$

Hence, we may regard f_n as a function of $n+1$ variables $t_1,\ldots,t_n,t$. Since this function is symmetric with respect to its first n variables, its *symmetrization* $\widetilde{f}_n$ as a function of $n+1$ variables $t_1,\ldots,t_n,t$ is given by, with $t_{n+1}=t$,

$$\widetilde{f}_n(t_1,\ldots,t_{n+1}) = \frac{1}{n+1}[f_n(t_1,\ldots,t_{n+1}) + \cdots$$
$$+ f_n(t_1,\ldots,t_{i-1},t_{i+1},\ldots,t_{n+1},t_i) + \cdots$$
$$+ f_n(t_2,\ldots,t_{n+1},t_1)],$$

where we only sum over those permutations σ of the indices $(1,\ldots,n+1)$ which interchange the *last* component with one of the others and leave the rest in place.

Definition A.2.1. *Suppose $u(t,\omega)$ is a stochastic process satisfying* (A.13) *and* (A.14) *and with Wiener Itô chaos expansion*

$$u(t,\omega) = \sum_{n=0}^{\infty} I_n(f_n(\cdot,t)).$$

Then we define the Skorohod integral *of u by*

$$\delta(u) := \int_{\mathbb{R}} u(t,\omega)\delta B(t) := \sum_{n=0}^{\infty} I_{n+1}(\widetilde{f}_n) \quad \text{(when convergent)} \quad (A.15)$$

where $\widetilde{f}_n$ is the symmetrization of $f_n(t_1,\ldots,t_n,t)$ as a function of $n+1$ variables $t_1,\ldots,t_n,t$.

We say u is Skorohod integrable *and write* $u \in \text{dom}(\delta)$ *if the series in* (A.15) *converges in* $L^2(\mathbb{P})$. *This occurs if and only if*

$$E[\delta(u)^2] = \sum_{n=0}^{\infty} (n+1)! \|\widetilde{f}_n\|_{L^2(\mathbb{R}^{n+1})}^2 < \infty.$$

Theorem A.2.2. *Suppose* $Y(t,\omega) : \mathbb{R} \times \Omega \to \mathbb{R}$ *is Skorohod integrable. Then* $Y(t,\cdot) \diamond W(t)$ *is dt-integrable in* $(\mathcal{S})^*$ *and*

$$\int_{\mathbb{R}} Y(t,\omega) \delta B(t) = \int_{\mathbb{R}} Y(t,\omega) \diamond W(t) \, dt, \qquad \text{(A.16)}$$

where the integral on the left is the Skorohod integral defined in (A.15).

Proof. See [109] for details. □

The Skorohod integral is an extension of the classical Itô integral in the sense that if $Y(t,\omega)$ is adapted to the filtration $\mathcal{F}_t$ generated by B and

$$E\Big[\int_{\mathbb{R}} Y^2(t,\omega) dt\Big] < \infty,$$

then

$$\int_{\mathbb{R}} Y(t) \delta B(t) = \int_{\mathbb{R}} Y(t) \, dB(t), \qquad \text{the classical Itô integral.}$$

The integral on the right-hand side of (A.16) may exist even if Y is not Skorohod integrable. Therefore, we may regard the right-hand side of (A.16) as an extension of the Skorohod integral, and we call it the *extended Skorohod integral*. We will use the same notation

$$\int_{\mathbb{R}} Y(t) \delta B(t)$$

for the extended Skorohod integral.

If $[a,b] \subset \mathbb{R}$ is an interval and u is Skorohod integrable, we define

$$\int_a^b u(t) \, \delta B(t) := \int_{\mathbb{R}} u(t) I_{[a,b]}(t) \, \delta B(t).$$

More generally, if $L \subset \mathbb{R}$ is a Borel set, we define

$$\int_L u(t) \, \delta B(t) := \int_{\mathbb{R}} u(t) I_L(t) \, \delta B(t).$$

Example A.2.3. Using Wick calculus in $(\mathcal{S})^*$, we get

$$\int_0^T B(T) \delta B(t) = \int_0^T B(T) \diamond W(t) \, dt = B(T) \diamond \int_0^T W(t) \, dt$$
$$= B(T) \diamond B(T) = B^2(T) - T,$$

by Example A.1.8 with $f = g = I_{[0,T]}$.

The following result gives a useful interpretation of the Skorohod integral as a limit of Riemann sums:

Theorem A.2.4. *Let $Y : [0, T] \to (\mathcal{S})^*$ be a càglàd function, i.e., $Y(t)$ is left-continuous with right-sided limits. Then Y is Skorohod integrable over $[0, T]$ and*

$$\int_{\mathbb{R}} Y(t)\, \delta B(t) = \lim_{\Delta t_j \to 0} \sum_{j=0}^{N-1} Y(t_j) \diamond (B(t_{j+1}) - B(t_j)),$$

where the limit is taken in $(\mathcal{S})^*$ and $0 = t_0 < t_1 < \cdots < t_n = T$ is a partition of $[0, T]$, $\Delta t_j = t_{j+1} - t_j$, $j = 0, \ldots, N-1$.

Proof. This is an easy consequence of Theorem A.2.2. □

We also note the following:

Theorem A.2.5. *Let $Y : \mathbb{R} \to (\mathcal{S})^*$. Suppose $Y(t)$ has the expansion*

$$Y(t) = \sum_{\alpha \in \mathcal{J}} c_\alpha(t) \mathcal{H}_\alpha(\omega), \qquad t \in \mathbb{R},$$

where

$$c_\alpha \in L^2(\mathbb{R}) \qquad \forall \alpha \in \mathcal{J}.$$

Then

$$\int_{\mathbb{R}} Y(t)\, \delta B(t) = \sum_{\alpha \in \mathcal{J}} \sum_{k \in \mathbb{N}} \langle c_\alpha, \xi_k \rangle_{L^2(\mathbb{R})} \mathcal{H}_{\alpha + \varepsilon^{(k)}}(\omega),$$

provided that the right-hand side converges in $(\mathcal{S})^*$. In particular, if $\int_{\mathbb{R}} Y(t)\, \delta B(t)$ belongs to $L^2(\mathbb{P})$, then

$$E\left[\int_{\mathbb{R}} Y(t) \delta B(t) \right] = 0.$$

A.3 Malliavin derivative

We recall here the main results concerning stochastic derivatives and the main relations between integral and differential stochastic calculus.

Definition A.3.1. *1. Let $F : \Omega \to \mathbb{R}$, $\gamma \in L^2(\mathbb{R})$. Then the directional derivative of F in the direction γ is defined by*

$$D_\gamma F(\omega) = \lim_{\varepsilon \to 0} \frac{F(\omega + \varepsilon \gamma) - F(\omega)}{\varepsilon}$$

provided that the limit exists in $(\mathcal{S})^*$.

2. Suppose there exists a function $\psi : \mathbb{R} \to (\mathcal{S})^*$ such that

$$D_\gamma F(\omega) = \int_\mathbb{R} \psi(t)\gamma(t)\,dt \qquad \forall \gamma \in L^2(\mathbb{R}).$$

Then we say that F is *differentiable* and we call $\psi(t)$ the *stochastic gradient* of F (or the *Hida Malliavin derivative of F*). We use the notation

$$D_t F = \psi(t)$$

for the stochastic gradient of F at $t \in \mathbb{R}$.

Note that – in spite of the notation – $D_t F$ is *not* a derivative with respect to t but is a (kind of) derivative with respect to $\omega \in \Omega$.

Example A.3.2. Suppose $F(\omega) = \langle \omega, f \rangle = \int_\mathbb{R} f(s)\,dB(s)$ for some $f \in L^2(\mathbb{R})$. Then by linearity

$$D_\gamma F(\omega) = \lim_{\varepsilon \to 0} \frac{1}{\varepsilon}(\langle \omega + \varepsilon\gamma, f \rangle - \langle \omega, f \rangle) = \langle \gamma, f \rangle_{L^2(\mathbb{R})} = \int_\mathbb{R} f(t)\gamma(t)\,dt$$

for all $\gamma \in L^2(\mathbb{R})$. We conclude that F is differentiable and

$$D_t\left[\int_\mathbb{R} f(s)\,dB(s)\right] = f(t) \qquad \text{for a.e. } t.$$

(Note that this is only valid for *deterministic* integrands f. See Theorem A.3.6 for the general case.)

We note two useful chain rules for stochastic differentiation.

Theorem A.3.3 (Chain rule I). *Let $\phi : \mathbb{R}^n \to \mathbb{R}$ be a Lipschitz continuous function, i.e., there exists $C < \infty$ such that*

$$|\phi(x) - \phi(y)| \leq C|x - y| \qquad \forall x, y \in \mathbb{R}^n.$$

Let $X = (X_1, \ldots, X_n)$, where each $X_i : \Omega \to \mathbb{R}$ is differentiable. Then $\phi(X)$ is differentiable and

$$D_t \phi(X) = \sum_{k=1}^n \frac{\partial \phi}{\partial x_k}(X) D_t X_k. \qquad (A.17)$$

We refer to [176] for a proof.

If $f(x) = \sum_{m=0}^\infty a_m x^m$ is a real analytic function and $X \in (\mathcal{S})^*$ we put

$$f^\diamond(X) = \sum_{m=0}^\infty a_m X^{\diamond m},$$

provided the sum converges in $(\mathcal{S})^*$.

We call $f^\diamond(X)$ the *Wick version* of $f(X)$. A similar definition applies to real analytic functions on $\mathbb{R}^n$.

A.3 Malliavin derivative

Theorem A.3.4 (The Wick chain rule). *Let $f : \mathbb{R}^n \to \mathbb{R}$ be real analytic and let $X = (X_1, \ldots, X_n) \in ((\mathcal{S})^*)^n$. Then if $f^\diamond(X) \in (\mathcal{S})^*$,*

$$D_t[f^\diamond(X)] = \sum_{k=1}^{n} \left(\frac{\partial f}{\partial x_k}\right)^\diamond (X) \diamond D_t X_k, \qquad t \in \mathbb{R}.$$

We refer to [179] for a proof. Note that by Example A.3.2 and the chain rule (A.17) we have

$$D_t \mathcal{H}_\alpha(\omega) = \sum_{i=1}^{m} \alpha_i \mathcal{H}_{\alpha - \varepsilon^{(i)}}(\omega) \xi_i(t) \in (\mathcal{S})^* \qquad \forall t.$$

where $\alpha = (\alpha_1, \ldots, \alpha_m) \in \mathcal{J}$. In fact, using the topology for $(\mathcal{S})^*$ one can prove, by using the approach of [1, Lemma 3.10], the following:

Theorem A.3.5. *Let $F \in (\mathcal{S})^*$. Then F is differentiable, and if F has the expansion*

$$F(\omega) = \sum_{\alpha \in \mathcal{J}} c_\alpha \mathcal{H}_\alpha(\omega),$$

then

$$D_t F(\omega) = \sum_{\alpha, i} c_\alpha \alpha_i \mathcal{H}_{\alpha - \varepsilon^{(i)}}(\omega) \xi_i(t) \qquad \forall t \in \mathbb{R}.$$

We now mention without proofs some of the most fundamental results from stochastic differential and integral calculus. For proofs we refer to [179].

Theorem A.3.6 (Fundamental theorem of stochastic calculus). *Suppose $Y(\cdot) : \mathbb{R} \to (\mathcal{S})^*$ and $D_t Y(\cdot) : \mathbb{R} \to (\mathcal{S})^*$ are Skorohod integrable. Then*

$$D_t \left[\int_\mathbb{R} Y(s) \, \delta B(s) \right] = \int_\mathbb{R} D_t Y(s) \, \delta B(s) + Y(t).$$

Theorem A.3.7 (Relation between the Wick product and the ordinary product). *Suppose $g \in L^2(\mathbb{R})$ is deterministic and that $F \in (\mathcal{S})^*$. Then*

$$F \diamond \int_\mathbb{R} g(t) \, dB(t) = F \cdot \int_\mathbb{R} g(t) \, dB(t) - \int_\mathbb{R} g(t) D_t F \, dt.$$

Corollary A.3.8. *Let $g \in L^2(\mathbb{R})$ be deterministic and $F \in L^2(\mathbb{P})$. Then*

$$E\left[F \cdot \int_\mathbb{R} g(t) \, dB(t)\right] = E\left[\int_\mathbb{R} g(t) D_t F \, dt\right]$$

provided that the integrals converge.

Theorem A.3.9 (Integration by parts). *Let $F \in L^2(\mathbb{P})$ and assume that $Y : \mathbb{R} \times \Omega \to \mathbb{R}$ is Skorohod integrable with $\int_\mathbb{R} Y(t) \delta B(t) \in L^2(\mathbb{P})$. Then*

$$F \int_\mathbb{R} Y(t) \, \delta B(t) = \int_\mathbb{R} FY(t) \, \delta B(t) + \int_\mathbb{R} Y(t) D_t F \, dt$$

provided that the integral on the extreme right converges in $L^2(\mathbb{P})$.

This immediately gives the following generalization of Corollary A.3.8:

Corollary A.3.10. *Let F and $Y(t)$ be as in Theorem A.3.9. Then*

$$E\Big[F \int_{\mathbb{R}} Y(t)\, \delta B(t)\Big] = E\Big[\int_{\mathbb{R}} Y(t) D_t F\, dt\Big].$$

Theorem A.3.11 (The Itô Skorohod isometry). *Suppose $Y : \mathbb{R} \times \Omega \to \mathbb{R}$ is Skorohod integrable with $\int_{\mathbb{R}} Y(t)\, \delta B(t) \in L^2(\mathbb{P})$. Then*

$$E\Big[\Big(\int_{\mathbb{R}} Y(t)\, \delta B(t)\Big)^2\Big] = E\Big[\int_{\mathbb{R}} Y^2(t)\, dt\Big] + E\Big[\int_{\mathbb{R}} \int_{\mathbb{R}} D_t Y(s) D_s Y(t)\, ds\, dt\Big].$$

B
Notions from fractional calculus

In this appendix we briefly recall the main features of the classical theory of fractional calculus by following [237]. For a complete treatment of this subject, we refer to [206].

B.1 Fractional calculus on an interval

Definition B.1.1. *Let f be a deterministic real-valued function that belongs to $L^1(a,b)$, where (a,b) is a finite interval of $\mathbb{R}$. The fractional Riemann Liouville integrals of order $\alpha > 0$ are determined at almost every $x \in (a,b)$ and defined as the*

1. *Left-sided version:*

$$I_{a+}^{\alpha}f(x) := \frac{1}{\Gamma(\alpha)}\int_a^x (x-y)^{\alpha-1} f(y)\,dy. \tag{B.1}$$

2. *Right-sided version:*

$$I_{b-}^{\alpha}f(x) := \frac{1}{\Gamma(\alpha)}\int_x^b (y-x)^{\alpha-1} f(y)\,dy. \tag{B.2}$$

For $\alpha = n \in \mathbb{N}$ one obtains the n-order integrals

$$I_{a+}^{n}f(x) = \int_a^x \int_a^{x_{n-1}} \cdots \int_a^{x_2} f(x_1)\,dx_1\,dx_2\ldots dx_n$$

and

$$I_{b-}^{n}f(x) = \int_x^b \int_{x_{n-1}}^b \cdots \int_{x_2}^b f(x_1)\,dx_1\,dx_2\ldots dx_n.$$

Definition B.1.2. *Consider $\alpha < 1$. We define the fractional Liouville derivatives as*

$$D_{a+}^\alpha f := \frac{d}{dx} I_{a+}^{1-\alpha} f$$

and

$$D_{b-}^\alpha f := \frac{d}{dx} I_{b-}^{1-\alpha} f,$$

if the right-hand sides are well-defined (or determined).

For any $f \in L^1(a,b)$ one obtains

$$D_{a+}^\alpha I_{a+}^\alpha f = f, \quad D_{b-}^\alpha I_{b-}^\alpha f = f. \tag{B.3}$$

In order to guarantee that also the opposite order of operation holds, we consider the following family of functions.

Definition B.1.3. *We denote by $I_{a+}^\alpha(L^p(a,b))$ [respectively, $I_{b-}^\alpha(L^p(a,b))$] the family of functions f that can be represented as an I_{a+}^α-integral (respectively, I_{b-}^α-integral) of some function $\phi \in L^p(a,b)$, $p \geq 1$. Such ϕ is unique (in L^p sense) and coincides with $D_{a+}^\alpha f$ (respectively, with $D_{b-}^\alpha f$).*

In particular we denote by I_{a+}^α (respectively, I_{b-}^α) the map from $L^p(a,b)$ into $I_{a+}^\alpha(L^p(a,b))$ [respectively, $I_{b-}^\alpha(L^p(a,b))$].

This means that if $f \in I_{a+}^\alpha(L^p(a,b))$, we have

$$I_{a+}^\alpha D_{a+}^\alpha f = f.$$

Moreover, given $f \in L^p(a,b)$ the following *Weyl representation* holds:

$$D_{a+}^\alpha f(x) = \frac{1}{\Gamma(1-\alpha)} \left[\frac{f(x)}{(x-a)^\alpha} + \alpha \int_a^x \frac{f(x)-f(y)}{(x-y)^{\alpha-1}} dy \right]$$

and

$$D_{b-}^\alpha f(x) = \frac{1}{\Gamma(1-\alpha)} \left[\frac{f(x)}{(b-x)^\alpha} + \alpha \int_x^b \frac{f(x)-f(y)}{(y-x)^{\alpha-1}} dy \right],$$

for almost every $x \in (a,b)$. The convergence of the integrals at the singularity $y = x$ holds pointwise for almost every $x \in (a,b)$ and moreover in L^p sense if $1 < p < \infty$.

It is easy to verify that the fractional integrals satisfy the composition formulas

$$I_{a+}^\alpha(I_{a+}^\beta f) = I_{a+}^{\alpha+\beta} f$$

and

$$I_{b-}^\alpha(I_{b-}^\beta f) = I_{b-}^{\alpha+\beta} f.$$

Moreover, the following integration by parts formula holds

$$\int_a^b I_{a+}^\alpha f(x) g(x)\, dx = \int_a^b f(x) I_{b-}^\alpha g(x)\, dx \tag{B.4}$$

if $f \in L^p, g \in L^q(a,b)$, $1/p + 1/q \leq 1 + \alpha$, where $p > 1$ and $q > 1$ if $1/p + 1/q = 1 + \alpha$. This implies the following relations for the fractional derivatives

$$D_{a+}^\alpha(D_{a+}^\beta f) = D_{a+}^{\alpha+\beta} f$$

and

$$D_{b-}^\alpha(D_{b-}^\beta f) = D_{b-}^{\alpha+\beta} f$$

if $f \in I_{a+}^{\alpha+\beta}(L^p(a,b))$ [respectively, $f \in I_{b-}^{\alpha+\beta}(L^p(a,b))$], $\alpha > 0$, $\beta > 0$, $\alpha + \beta < 1$. We introduce the *boundary orders of differentiation* 0 or 1 by taking the limits in $L^p(a,b)$

$$\lim_{\alpha \to 1^-} D_{a+}^\alpha f = f',$$

if f is differentiable in the L^p sense with derivative f', and

$$\lim_{\alpha \to 0^+} D_{a+}^\alpha f = f.$$

If $p = 1$, the latter limit is to be interpreted in pointwise convergence. The corresponding integration by parts formula is

$$\int_a^b D_{a+}^\alpha f(x) g(x)\, dx = \int_a^b f(x) D_{b-}^\alpha g(x)\, dx, \tag{B.5}$$

for $f \in I_{a+}^\alpha(L^p(a,b))$, $g \in I_{b-}^\alpha(L^q(a,b))$, $1/p + 1/q \leq 1 + \alpha$, $0 \leq \alpha \leq 1$. By using these relations, we introduce the concept of *fractal integral* following [238]. In the following we denote by C^α, $0 < \alpha < 1$, the space of α-Hölder continuous functions on (a,b).

Definition B.1.4. *Consider the "corrected" functions*

$$f_{a+}(x) := I_{(a,b)}(f(x) - f(a+)),$$

$$f_{b-}(x) := I_{(a,b)}(f(x) - f(b-)),$$

provided that $f(a+) := \lim_{x \to a^+} f(x)$ *and* $f(b-) := \lim_{x \to b^-} f(x)$ *exist. Then*

1. *If* $f_{a+} \in I_{a+}^\alpha(L^p(a,b))$, $g_{b-} \in I_{b-}^{1-\alpha}(L^q(a,b))$, $g(a+)$ *exists*, $1/p + 1/q \leq 1$, $0 \leq \alpha \leq 1$, $f \in C^{\alpha - 1/p}$ *if* $\alpha p > 1$, *we define the* fractal integral

$$\int_a^b f\, dg := \int_a^b D_{a+}^\alpha f_{a+}(x) D_{b-}^{1-\alpha} g_{b-}(x)\, dx + f(a+)[g(b-) - g(a+)]. \tag{B.6}$$

2. *If* $f_{a+} \in I_{a+}^\alpha(L^p(a,b))$, $g_{b-} \in I_{b-}^{1-\alpha}(L^q(a,b))$, $1/p + 1/q \leq 1$, $0 \leq \alpha \leq 1$, $g \in C^{1-\alpha-1/q}$, $\alpha p < 1$, *we define the* fractal integral

$$\int_a^b f\, dg := \int_a^b D_{a+}^\alpha f_{a+}(x) D_{b-}^{1-\alpha} g_{b-}(x)\, dx. \tag{B.7}$$

These definitions do not depend on α and agree when f, g satisfy both conditions. For further details, see [238].

For $\alpha = 0$ or $\alpha = 1$, the fractal integrals reduce to the well-known case of Stieltjes integral in the smooth case:

$$\int_a^b f\, dg = f(b-)g(b-) - f(a+)g(a+) - \int_a^b f'(x) g(x)\, dx$$

and

$$\int_a^b f\, dg = \int_a^b f'(x) g(x)\, dx.$$

If g is Hölder continuous on $[a, b]$, then for any $f = \sum_{i=1}^n f_i I_{[x_i, x_{i+1}]}$ with $x_0 = a$, $x_{n+1} = b$, we have that

$$\int_a^b f\, dg = \sum_{i=1}^n f_i (g(x_{i+1}) - g(x_i))$$

if the integral exists in the sense of (B.6). For further properties of fractional integrals, we also refer to [74].

B.2 Fractional calculus on the whole real line

We can also introduce the following left- and right-sided fractional integral and derivative operators on $\mathbb{R}$ for $\alpha \in (0, 1)$ (see [206] and also [54] and [177]).

Definition B.2.1. *Let $\alpha \in (0, 1)$. The fractional integrals I_+^α and I_-^α of a function ϕ on the whole real line are defined, respectively, by*

$$I_+^\alpha f(x) := \frac{1}{\Gamma(\alpha)} \int_{-\infty}^x (x - y)^{\alpha - 1} f(y)\, dy, \quad x \in \mathbb{R},$$

and

$$I_-^\alpha f(x) := \frac{1}{\Gamma(\alpha)} \int_x^\infty (x - y)^{\alpha - 1} f(y)\, dy, \quad x \in \mathbb{R}.$$

The *Marchaud fractional derivatives* D_+^α and D_-^α of a function ϕ on the whole real line are defined by the following limits:

$$D_\pm^\alpha f(x) := \lim_{\epsilon \to 0} D_{\pm, \epsilon}^\alpha f(x), \quad x \in \mathbb{R},$$

where

$$D_{\pm, \epsilon}^\alpha f(x) := \frac{\alpha}{\Gamma(1 - \alpha)} \int_\epsilon^\infty \frac{\phi(t) - \phi(t \mp s)}{s^{1+\alpha}}\, ds, \quad x \in \mathbb{R}.$$

By Theorem 6.1 of [206] we obtain that for $\phi \in L^2(\mathbb{R})$,

$$D_+^\alpha I_+^\alpha f = f, \qquad D_-^\alpha I_-^\alpha f = f.$$

C

Estimation of Hurst parameter

Assume that we have a sequence of data $B_1^{(H)}, B_2^{(H)}, \ldots, B_N^{(H)}$ from observations of a *fBm* B_t^H at time instants $t = 1, 2, \ldots, N$. How can we estimate the Hurst parameter H? There are many approaches that we will summarize in this appendix (see [23], [91], [222], and the references therein for further details).

In what follows we put $Y_1 = B_1^{(H)}, Y_2 = B_2^{(H)}, \ldots, Y_N = B_N^{(H)}$. Since *fBm* has stationary increments,

$$X_1 = Y_1, X_2 = Y_2 - Y_1, \ldots, X_N = Y_N - Y_{N-1}$$

is a stationary Gaussian time sequence. The covariance of this sequence is

$$\rho_H(k) = \mathcal{E}(X_1 X_{k+1}) = \mathcal{E}(X_m X_{m+k})$$
$$= \frac{1}{2}\left[(k+1)^{2H} + (k-1)^{2H} - 2k^{2H}\right] \tag{C.1}$$

We use the following convention to construct a statistic.

> If a statistic $f(m)$ behaves like m^{aH+b} as $m \to \infty$ for some constants a and b, then
> $$\log f(m) \approx (aH + b) \log m + R,$$
> where R is independent of H. We may use linear regression to find the slope $aH + b$ for the log plot of $f(m)$ against the log plot of m. Hence we can find an estimator of H.

C Estimation of Hurst parameter

C.1 Absolute value method

Given positive integers k and m, denote

$$X^m(k) := \frac{1}{m} \sum_{i=(k-1)m}^{km} X_i, \quad k = 1, 2, \ldots, \left[\frac{N}{m}\right],$$

where $\left[\frac{N}{m}\right]$ denotes the integer part of $\frac{N}{m}$, and

$$\bar{X}_N = \frac{1}{N} \sum_{i=1}^{N} X_i.$$

Define

$$AM^{(m)} = \frac{1}{N/m} \sum_{k=1}^{[N/m]} |X^{(m)}(k) - \bar{X}_N|.$$

Then $AM^{(m)}$ behaves like m^{H-1} for large m.

C.2 Variance Method

Use the notation of the above subsection and define

$$\widehat{\mathrm{Var}} X^{(m)} = \frac{1}{N/m} \sum_{k=1}^{[N/m]} \left[X^{(m)}(k) - \bar{X}_N\right]^2.$$

Then $\widehat{\mathrm{Var}} X^{(m)}$ behaves like m^{2H-2} as $m \to \infty$.

C.3 Variance residuals methods

The variance of residuals methods was introduced by [185]. Denote

$$L_{1/2,H}(t) = \int_{-\infty}^{\infty} \left[|t-x|^{H-1/2} - |x|^{H-1/2}\right] dB_x,$$

the 1/2-stable fractional stable motion (see [207]), where B_x is a Brownian motion, and denote

$$A_1 = \int_0^1 L_{1/2,H}(t)\, dt, \qquad A_2 = \int_0^1 t L_{1/2,H}(t)\, dt.$$

Define

$$b = m^{H-1}\left[12 A_2 - 6 A_1\right], \qquad a = m^H \left[4 A_1 - 6 A_2\right].$$

Then $1/m \sum_{i=1}^{m} (Y_i - a - bi)^2$ behaves like m^{2H} in distribution.

C.4 Hurst's rescaled range (R/S) analysis

To use Hurst's R/S analysis we first define the *adjusted range*

$$S(n) = \left[\frac{1}{N}\sum_{i=1}^{k}(Y_i - \bar{Y}_N)^2\right]^{1/2}$$

and

$$RS(N) = \frac{[\max_{1\leq i \leq N} - \min_{1\leq i \leq N}]\{Y_i - i/N Y_N\}}{S(n)}.$$

We have that $RS(N)$ behaves like N^H for large N (in distribution).

C.5 Periodogram method

Let

$$I_N(\nu) = \frac{1}{2\pi N}\left|\sum_{j=1}^{N} X_j e^{ij\nu}\right|^2. \qquad (C.2)$$

When $N \to \infty$, $I_N(\nu)$ is the spectral density of $\rho_H(k)$. Then $I_N(\nu)$ behaves like ν^{1-2H} as $\nu \to 0$.

C.6 Discrete variation method

In this section we present a result from [56].

Let $p \geq 1$ be an integer. A *filter* of length $l + 1$ and order p is an $l + 1$ dimensional vector $a = (a_0, a_1, \ldots, a_l) \in \mathbb{R}^{l+1}$ such that $\sum_{j=0}^{l} a_j j^r = 0$ for all integers r such that $0 \leq r \leq p - 1$, and $\sum_{j=0}^{l} a_j j^p \neq 0$. Given a filter $a = (a_0, a_1, \ldots, a_l)$ define the random variable

$$V_N^a(i) = \sum_{q=0}^{\ell} a_q B_{i-q}^{(H)}$$

and define the kth empirical absolute moment as

$$S_N(k, a) = \frac{1}{(N-\ell)N^{-KH}}\sum_{i=\ell}^{N-1}|V_N^a(i)|^k.$$

Denote

$$\pi_H^a = \mathcal{E}\left[V_N^a(i)\right]^2 = -\frac{1}{2}\sum_{q,r=0}^{\ell} a_q a_r |q-r|^{2H}.$$

Set
$$g_{k,a,N}(t) = \frac{1}{N^{tk}} \pi_t^a E_k,$$
where
$$E_k = \frac{1}{\sqrt{2\pi}} \int_{\mathbb{R}} |x|^k e^{-|x|^2/2} dx = 2^{k/2} \frac{\Gamma(K+1/2)}{\Gamma(1/2)}.$$

Lemma C.6.1. *When N is large, $g_{k,a,N}(t)$ is a monotonic function of t.*

Proof. It is easy to see that
$$\frac{d}{dt} g_{k,a,N}(t) = -k \log N + \frac{k}{2}(\pi_t^a)^{-1} t \sum_{|q-r| \geq 2} a_q a_r |q-r|^{2t} \log |q-r|,$$
which is strictly negative if
$$N > \max_{0 \leq t \leq 1} \exp\left(\frac{\sum_{|q-r| \geq 2} a_q a_r |q-r|^{2t} \log |q-r|}{\sum_{q,r=1}^{\ell} a_q a_r |q-r|^{2t}}\right).$$
□

From this lemma, the inverse $g_{k,a,N}^{-1}(t)$ exists. The k-variation estimator is defined as $\hat{H}_N(k,a) = g_{k,a,N}^{-1}(S_N(k,a))$.

Theorem C.6.2. *As $N \to \infty$,*
$$\hat{H}_N(k,a) \to H \quad a.s.,$$
and if $p \geq H + 1/4$,
$$\sqrt{N} \log(N) \left[\hat{H}_N(k,a) - H\right] \to N\left(0, \frac{A_1(H,k,a)}{k^2}\right)$$
in distribution as $N \to \infty$.

For the proof, we refer to [56].

C.7 Whittle method

For the following three subsections we shall use θ to denote the Hurst parameter H. The spectral density of the time series $X_1 = B_1^\theta$, $X_2 = B_2^\theta - B_1^\theta$, ..., $X_N = B_N^\theta - B_{N-1}^\theta$ is
$$f(x,\theta) := \sum_{k=-\infty}^{\infty} \rho(k) e^{ikx} = (1 - \cos x) \sum_{k=-\infty}^{\infty} |x + 2k\pi|^{-1-2\theta}. \tag{C.3}$$

Set

$$\tilde{f}(x,\theta) = \exp\left(-\frac{1}{2\pi}\int_{-\pi}^{\pi} \log f(x,\theta)\,dx\right) f(x,\theta) \tag{C.4}$$

and

$$L_N^w(\theta) := \int_{-\pi}^{\pi} \frac{I_N(x)}{\tilde{f}(x,\theta)}\,dx + \int_{-\pi}^{\pi} \log\left[\tilde{f}(x,\theta)\right]\,dx,$$

where $I_N(x)$ is defined as in (C.2). Then *the Whittle estimator is the maximizer* $\hat{\theta}$ of $L_N^w(\theta)$, or

$$\hat{\theta} = \arg\max_{0 \leq \theta \leq 1} L_N^w((\theta)).$$

C.8 Maximum likelihood estimator

The maximum likelihood estimator is a commonly used estimator in the theory of statistics. Denote

$$\mathcal{X}_N = (X_1, X_2, \ldots, X_N)^T$$

and the covariance matrix of $\mathcal{X}_N$

$$\Sigma(\theta) = (\sigma_{ij})_{1 \leq i,j \leq n}, \qquad \text{where } \sigma_{ij} = \rho_\theta(|j - i + 1|).$$

The likelihood function $L(\theta)$ is given by

$$L(\theta) = \frac{1}{(2\pi)^{n/2}} \det(\Sigma(\theta))^{-1/2} \exp\left(-\frac{1}{2}\mathcal{X}_N^T \Sigma(\theta)^{-1} \mathcal{X}_N\right).$$

Thus the maximum likelihood estimator $\hat{\theta}_n$ is the maximizer of the following function:

$$\hat{\theta}_n = \arg\max\left\{-\frac{1}{2}\left[\log\det(\Sigma(\theta)) + \mathcal{X}_N^T \Sigma(\theta)^{-1} \mathcal{X}_N\right]\right\},$$

or $\hat{\theta}_n$ is given by the following equation:

$$\frac{d}{d\theta}\log\det(\Sigma(\theta)) + \mathcal{X}_N^T \frac{d}{d\theta}\Sigma(\theta)^{-1}\mathcal{X}_N = 0.$$

The following theorem is proved in [67].

Theorem C.8.1. *As $n \to \infty$, we have*

$$\hat{\theta}_n \to \theta \quad a.s.$$

and

$$n^{1/2}(\hat{\theta}_n - \theta) \to_d \xi$$

where ξ is a mean 0 normal distribution with variance $\sigma^2 = 2D^{-1}$ with

$$D = \frac{1}{2\pi}\int_{-\pi}^{\pi}\left[\frac{\partial}{\partial\theta}\log f(x,\theta)\right]^2 dx\bigg|_{\theta=H}.$$

C.9 Quasi maximum likelihood estimator

In this section we shall give a detailed discussion of the so-called quasi maximum likelihood estimator for the Hurst parameter H. Recall the autocorrelation and the spectral density of the increments of the *fBm* are given by (C.1) and (C.3).

We normalize $f(x,\theta)$ by a constant as in (C.4) and denote the normalized spectral density $\tilde{f}(x,\theta)$ still by $f(x,\theta)$ from now on.

Lemma C.9.1. $f(x,\theta)$ has the following expression:

$$f(x,\theta) = C_\theta(1-\cos x)\sum_{j=1}^{\infty} |2\pi j + x|^{-2\theta-1},$$

and when $x \to 0$, we have

$$f(x,\theta) = c_f |x|^{1-2\theta} + o(|x|^{\min(3-2\theta,2)}).$$

For the proof we refer to [212].

Now we consider

$$L_N(\theta) = \frac{\mathcal{X}_N^T A_N(\theta) \mathcal{X}_N}{N},$$

where

$$A_N(\theta) = (a_{j-k}(\theta))_{1 \le j,k \le n}$$

with

$$a_k(\theta) = \frac{1}{(2\pi)^2} \int_{-\pi}^{\pi} e^{ijx} [f(x,\theta)]^{-1}\,dx.$$

Recalling the definition $I_N(\theta)$ defined by (C.2), we may write

$$L_N(\theta) = \frac{1}{2\pi} \int_{-\pi}^{\pi} [f(x,\theta)]^{-1} I_N(x)\,dx.$$

It is easy to see that as $x \to 0$,

$$f(x,\theta) \approx x^{1-2\theta}, \qquad \frac{d}{d\theta} f(x,\theta) \approx x^{1-2\theta} \ln x.$$

The quasi maximum likelihood estimator $\hat{\theta}_N$ is the minimizer of $L_N(\theta)$:

$$\hat{\theta}_N = \arg\min_{\theta} L_N(\theta).$$

Theorem C.9.2. *1.*

$$\hat{\theta}_N \to \theta_0 = H \qquad \text{as } n \to \infty$$

almost surely.

2.
$$\sqrt{N}(\hat{\theta}_N - \theta_0) \to N(0, \sigma^2)$$

in distribution, where

$$\sigma^2 = 4\pi \left[\int_{-\pi}^{\pi} f(x,\theta) \frac{\partial^2}{\partial \theta^2} f(x,\theta)^{-1} dx \right]^{-1}$$

$$= 4\pi \left[\int_{-\pi}^{\pi} \left[f(x,\theta) \frac{\partial}{\partial \theta} f(x,\theta)^{-1} \right]^2 dx \right]^{-1}.$$

For a proof of this theorem see [99].

D

Stochastic differential equations for fractional Brownian motion

In this appendix we present an overview of some results concerning stochastic differential equations for *fBm*.

Several approaches have been considered in the literature, but a comprehensive theory has not yet been formulated. Here we summarize some of them without aiming for completeness. Since these methods use different techniques, it is in fact quite difficult to formulate them in a systematic way. However, in our opinion this summary can be still of interest for the reader. In particular, we consider here the approaches of [58], [60], [61], [149], [180], [203], [237].

Other possible methods are, for example, the following: In [92] they consider ordinary stochastic differential equations with respect to integrators with finite p-variation when $p \leq 3$. This applies to the case of *fBm* with $H \geq 1/3$. In [39] they investigate for which given drift and volatility the solution of a stochastic differential equation (SDE) in terms of a pathwise integral with respect to *fBm* can be expressed as a monotone transformation of a stationary fractional Ornstein–Uhlenbeck process. In [40] they also obtain a characterization of the maximum domain of attraction for stationary solutions.

A certain number of authors have also studied the problem of defining a stochastic integration for *fBm* in Hilbert space. This goes beyond the aims of this books and we refer the reader to [84], [85], [86], [224], [225] and [226] for a complete survey on the subject.

Stochastic delay differential equations driven by *fBm* with Hurst index $H > 1/2$ are studied in [95].

D.1 Stochastic differential equations with Wiener integrals

Let $K_H(t,s)$ be the reproducing kernel defined in (2.2) for $H > 1/2$ and in (2.3) for $H < 1/2$. We follow here the approach of [58] and study for any Hurst index $H \in (0,1)$ the existence, uniqueness, and regularity of the stochastic equation

$$X(t) = x + \int_0^t K_H(t,s)b(s,X(s))\,ds + \int_0^t K_H(t,s)\sigma(s,X(s))\,dB(s), \quad (D.1)$$

where $B(t)$ is a 1-dimensional standard Brownian motion. Here we note that the $K_H(t,s)$ term appears in the drift just in order to symmetrize the role of b and σ.

The following proposition guarantees that equation (D.1) makes sense:

Proposition D.1.1. *Let $X = (X_t)_{t \in [0,T]}$ and $Y = (Y_t)_{t \in [0,T]}$ be two indistinguishable processes on $(\Omega, \mathcal{F}^{(H)}, \mathbb{P}^H)$, i.e., such that*

$$X = Y \qquad d\mathbb{P}^H \otimes dt \text{ a.s.}$$

For b, σ Lipschitz continuous, then

$$\int_0^t K_H(t,s)b(s,X(s))\,ds = \int_0^t K_H(t,s)b(s,Y(s))\,ds \qquad d\mathbb{P}^H \otimes dt \text{ a.s.}$$

and

$$\int_0^t K_H(t,s)\sigma(s,X(s))\,dB(s) = \int_0^t K_H(t,s)\sigma(s,Y(s))\,dB(s) \qquad d\mathbb{P}^H \otimes dt \text{ a.s.}$$

Proof. The proof follows by applying the Cauchy–Schwartz inequality and the Burkholder–Davis–Gundy inequality. For further details, we refer to Proposition 3.1 of [58]. □

We introduce now the definition of solution. Set $\alpha_p = (1 - p|H - 1/2|)^{-1}$.

Definition D.1.2. *Consider $\alpha_2 = (1 - 2|H - 1/2|)^{-1}$. By a solution of the SDE (D.1) we mean a real-valued adapted stochastic process $X = (X_t)_{t \in [0,T]}$ satisfying the equation (D.1) and such that the function $\psi(t) = E\left[X(t)^2\right]$ belongs to $\cup_{\alpha > \alpha_2} L^\alpha([0,T])$.*

The main result of [58] is then the following theorem.

Theorem D.1.3. *Let $b(t,x), \sigma(t,x)$ be Lipschitz continuous in x, uniformly with respect to t, i.e.,*

$$|b(t,x) - b(t,y)| + |\sigma(t,x) - \sigma(t,y)| \leq L|x-y|,$$

for some $L > 0$. If, in addition, there exists $x_0, y_0 \in \mathbb{R}$ and $\alpha_b > \alpha_1, \alpha_\sigma > 2\alpha_2$ such that

$$b(\cdot, x_0) \in L^{\alpha_b}([0,T]), \qquad \sigma(\cdot, y_0) \in L^{\alpha_\sigma}([0,T]),$$

then the SDE has a unique solution. Moreover, for this solution and for any $p \in \{p \geq 2, p|H - \tfrac{1}{2}| < 1\}$ the function $\psi_p(t) = E[X(t)^p]$ is bounded on $[0,T]$.

Proof. The *uniqueness* of the solution follows by the Jensen inequality, the Burkholder–Davis–Gundy inequality, the Lipschitz continuity of b, σ, and Lemmas D.1.5 and D.1.6.

The *existence* of the solution can be proved by using Picard iteration. For further details, we refer to the proof of Theorem 3.1 in [58]. □

To prove Theorem D.1.3 we need some additional lemmas that we recall here without proof. The Gronwall lemma is here replaced by the following result on the resolvent kernel associated to K_H. Set $H_0 = |H - 1/2|$ and $\Delta_H = \{p \geq 1 : pH_0 < 1\}$, and for any $p \geq 1$ define

$$K_1^p(t,s) := |K_H(t,s)|^p,$$

$$K_{n+1}^p(t,s) := \int_s^t K_1^p(t,u) K_n^p(u,s) \, du.$$

Lemma D.1.4. *For every $p \in \Delta_H$ and for all $t \in [0,T]$, $K_H(t, \cdot)$ belongs to $L^p([0,T])$ and*

$$\sup_{t \in [0,T]} \|K_H(t, \cdot)\|_{L^p([0,T])} < \infty.$$

Lemma D.1.5. *For every $p \in \Delta_H$, the resolvent series*

$$\sum_{n=1}^{+\infty} z^n \int_0^t |K_H(t,s_1)|^p \, ds_1 \int_0^{s_1} |K_H(s_1,s_2)|^p \, ds_2 \cdots \int_0^{s_{n-1}} |K_H(s_{n-1}, s_n)|^p \, ds_n$$

converges for all $z \in \mathbb{C}$.

A detailed proof of this Lemma can be found in [58]. A consequence of Lemma D.1.5 and the Hölder inequality is Lemma D.1.6.

Lemma D.1.6. *For every $p \in \Delta_H$, let $\alpha_p = (1 - pH_0)^{-1}$. If ϕ belongs to $\cup_{\alpha > \alpha_p} L^\alpha([0,T])$, then $K_1^p \phi(t) = \int_0^t K_1^p(t,s) \phi(s) ds$ belongs to $L^\infty([0,T])$.*

Finally, we recall some regularity behavior of the solution, proved in Theorems 3.2 and 3.3 of [58].

Theorem D.1.7. *Let b, σ satisfy the hypothesis of Theorem D.1.3 and assume also that $\alpha_p \geq 2$ and σ is bounded. Then the solution of (D.1) has almost surely continuous sample paths.*

Moreover, we have the continuity with respect to the initial conditions.

Theorem D.1.8. *Let b, σ satisfy the hypothesis of Theorem D.1.3 and denote with X^x, X^y the solution of (D.1), respectively, with, initial condition x and y. There exists a positive constant $c > 0$ such that*

$$\sup_{t \in [0,T]} E\left[|X_t^x - X_t^y|^2\right] \leq c|x - y|^2,$$

and this upper bound is uniform with respect to x, y on any compact set of $\mathbb{R}$.

D.2 Stochastic differential equations with pathwise integrals

Here we summarize the approaches of [149], [180], [203] and [237] for SDEs with respect to pathwise integrals.

We begin with an existence-uniqueness theorem due to [203] for SDEs driven by *fBm* with Hurst index $H > 1/2$, which derives from an analogous result for ordinary differential equations with Hölder continuous forcing. In order to define the Riemann Stieltjes integral $\int f \, dg$ for functions of unbounded variation, we recall the following result of [142] and [232]. For an interesting introduction to Young integrals we refer to [82].

In the sequel for any $0 < \gamma < 1$ we denote by $C^\gamma(\mathbb{R})$ the space of γ-Hölder continuous functions $f : [0,T] \longrightarrow \mathbb{R}$, equipped with the norm

$$\|f\|_\gamma = \sup_{t \in [0,T]} |f(t)| + \sup_{0 \leq s < t \leq T} \frac{|f(t) - f(s)|}{(t-s)^\gamma} < \infty.$$

Theorem D.2.1. *Let $f \in C^\beta(\mathbb{R}), g \in C^\gamma(\mathbb{R})$. If $\beta + \gamma > 1$, then $\int_0^t f \, dg$ exists as a Stieltjes integral for $t > 0$.*

In [203] the previous theorem is proved by using the following result, which turns out also to be the key to prove existence and uniqueness for differential equations as we describe in the sequel.

Theorem D.2.2. *Consider the interval $[s,t]$ divided into 2^n subintervals of equal size and denote s_i^n the ith point of this partition. Let $f \in C^\beta(\mathbb{R}), g \in C^\gamma(\mathbb{R})$ and suppose $\beta + \gamma > 1$. Then*

$$\int_s^t f(u) \, dg(u) = f(s)(g(t) - g(s)) + \sum_{k=1}^\infty \sum_{i=0}^{2^{k-1}-1} \Delta f(s + s_{2i}^k) \Delta g(s + s_{2i+1}^k),$$

where $\Delta f(s + s_n^k) = f(s + s_{n+1}^k) - f(s + s_n^k)$ and the similar expression for Δg.

Proof. The idea behind Theorem D.2.2 is to write a recursion between the Riemann sums on finer partitions of the interval and Riemann sums on coarser partitions of the interval. We refer to [203] for further details. □

Theorem D.2.3. *Let $b, \sigma : \mathbb{R}^+ \times \mathbb{R} \to \mathbb{R}$, $g \in C^\gamma(\mathbb{R})$ with $1/2 < \gamma \leq 1$. Suppose b is globally Lipschitz in t and x, and $\sigma \in C^1(\mathbb{R}^+ \times \mathbb{R})$ with $\sigma, \sigma_t, \sigma_x$ globally Lipschitz in t and x. Then for every $T > 0$ and $\gamma > \beta > 1 - \gamma$, the ordinary differential equation*

$$dx(t) = b(t, x(t)) \, dt + \sigma(t, x(t)) \, dg(t), \qquad x(0) = x_0,$$

has a unique solution in $C^\beta([0,T])$.

Proof. For the proof of this Theorem we refer to the one given in [203, Sections 5.3 and 5.4], by the means of Theorem D.2.2. First local existence and uniqueness are proved and then the global result is shown by applying the local existence and uniqueness repeatedly, taking as initial condition the value of the solution at the end of the previous interval. □

Since $B^{(H)}(t,\omega) \in C^\gamma(\mathbb{R})$ for $\gamma < H$ with probability 1 (see Theorem 1.6.1), Theorem D.2.1 implies that the pathwise stochastic integral $\int_0^t f(s)\, d^-B^{(H)}(s)$ exists for all $f \in C^\beta(\mathbb{R})$ with $\beta > (1-H)$. A special case of this result for functions $f(B^{(H)}(\cdot,\omega))$ is also contained in [149]. Analogously, Theorem D.2.2 holds with probability one also for the pathwise stochastic integral $\int_0^t f(s)\, d^-B^{(H)}(s)$ for all $f \in C^\beta(\mathbb{R})$ with $\beta > (1-H)$. Hence, Theorem D.2.2 and Theorem D.2.3 imply the following existence and uniqueness theorem for SDEs driven by fBm.

Theorem D.2.4. *Let* $b, \sigma : \mathbb{R}^+ \times \mathbb{R} \to \mathbb{R}$, $Z : \Omega \to \mathbb{R}$ *and* $1/2 < H < 1$, $(1-H) < \beta < H$. *Suppose* b *is globally Lipschitz in* t *and* x, *and* $\sigma \in C^1(\mathbb{R}^+ \times \mathbb{R})$ *with* $\sigma, \sigma_t, \sigma_x$ *globally Lipschitz in* t *and* x. *Then for every* $T > 0$ *the SDE*

$$dX(t) = b(t, X(t))\, dt + \sigma(t, X(t))\, d^-B^{(H)}(t), \qquad X(0) = Z,$$

has a unique solution in $C^\beta([0,T])$ *with probability 1.*

If $f(t,\omega) = f(t)$ is deterministic and $f \in L^{\frac{1}{H}}(\mathbb{R}^+)$, then the stochastic integral $\int_0^t f(s)\, d^-B^{(H)}(s)$ exists in $L^2(\mathbb{R}^+ \times \Omega)$ for all $t \geq 0$. If, in addition, $f \in L^{2/(1+\beta)}([0,T])$, where $0 < \beta < 2H-1$, then for almost every ω, the integral $\int_0^t f(s)\, d^-B^{(H)}(s)$ has a t-continuous version for all $t \in [0,T]$. For further details, we refer to [203].

For $H > 1/2$ a solution of the SDE

$$dX(t) = b(t, X(t))\, dt + \sigma(t, X(t))\, d^-B^{(H)}(t), \qquad X(0) = Z \qquad (D.2)$$

for an arbitrary random initial value Z can be obtained under weaker assumptions by using the method developed in [237] for a continuous process X_t with generalized covariation process of the form $[X]_t = \int_0^t q(s)\, ds$, for some continuous random function q. In [237] this approach is actually formulated for m-dimensional driving processes and extends a previous 1-dimensional result by [139].

Remark D.2.5. In [149] a uniqueness and existence result for pathwise SDEs is also provided in the case when the diffusion coefficient is a bounded deterministic function, by using Gronwall's lemma and Picard's iteration.

In [178] they prove the existence and uniqueness of a strong solution for a SDE with constant $\sigma = 1$, where $b(s,x)$ is a bounded Borel function with linear growth in x for $H \leq 1/2$ or a Hölder continuous function of order, respectively, strictly larger than $1-1/2H$ in x and than $H-1/2$ in t for the case $H > 1/2$. Existence of a weak solution with monotonous drift and constant $\sigma = 1$ can be found in [37] for the case $H > 1/2$. The ergodic properties of finite-dimensional systems of SDEs driven by nondegenerate additive *fBm* with arbitrary Hurst index $H \in (0,1)$ are studied in [106]. An SDE with deterministic coefficients with respect to a sequence of *fBms* with Hurst index $H > 1/2$ is solved in [41]. For the problem of identification and estimation of the drift function for linear stochastic systems driven by *fBm* with constant volatility, we refer to [193].

Here we sketch the method presented in [237], but only in the case $m = 1$ for the sake of simplicity. Assume that the following conditions hold:

1. $b(t,x) \in C^1([0,T] \times \mathbb{R}, \mathbb{R})$, $\partial b/\partial x(t,x)$ is locally Lipschitz in $x \in \mathbb{R}$.
2. $\sigma(t,x) \in C^1([0,T] \times \mathbb{R}, \mathbb{R})$, $\sigma(t,x)$ is locally Lipschitz in $x \in \mathbb{R}$.

Consider pathwisely the *auxiliary partial differential equation* on $\mathbb{R}^n \times \mathbb{R} \times [0,T]$ given by

$$\frac{\partial h}{\partial z}(y,z,t) = b(t, h(y,z,t)), \quad h(Y_0, Z_0, t_0) = X_0, \tag{D.3}$$

where $Z_0 = B^{(H)}(0) = 0$ and Y_0 is an arbitrary random vector in $\mathbb{R}^n$. By the classical theory, there exists a (nonunique) local solution $h \in C^1$ in a neighborhood of (Y_0, Z_0, t_0) with $\partial h/\partial y \neq 0$. For any such solution h consider the *ordinary differential equation* in $\mathbb{R}^n$

$$\frac{dY}{dt}(t) = \left[\frac{\partial h}{\partial y}(Y(t), Z(t), t)\right]^{-1} [\sigma(t, h(Y(t), Z(t), t)) - \frac{\partial h}{\partial t}(Y(t), Z(t), t)$$
$$- \frac{1}{2} q(t) b(t, h(Y(t), Z(t), t)) \frac{\partial b}{\partial x}(t, h(Y(t), Z(t), t))] \tag{D.4}$$

with $Y(t_0) = Y_0$, which has a unique solution in a neighborhood of t_0.

Theorem D.2.6. *Under the above conditions any representation*

$$X(t) = h(Y(t), Z(t), t),$$

with h satisfying (D.3) and $Y(t)$ given by the unique solution of (D.4), provides a solution of (D.2). In addition, the solution X of (D.2) is unique in the maximal interval of definition.

Proof. For the proof and further details, we refer to [237]. □

Note that this method reduces the problem of existence of a global solution of (D.2) to a growth condition on the coefficients b and σ for equations (D.3) and (D.4).

By following this approach, in [180] they have established an existence and uniqueness result for equation (D.2) using the a priori estimate (D.6) based

D.2 Stochastic differential equations with pathwise integrals

on the fractional integration by parts formula. For the sake of simplicity, we sketch only the 1-dimensional case. For the multi–dimensional case and further details, we refer to [180]. We introduce now some useful space of functions. Let $0 < \alpha < 1/2$. We denote by $W_0^{\alpha,\infty}(0,T)$ the space of measurable functions $f : [0,T] \to \mathbb{R}$ such that

$$\|f\|_{\alpha,\infty} := \sup_{t \in [0,T]} \left(|f(t)| + \int_0^t \frac{|f(t) - f(s)|}{(t-s)^{\alpha+1}} \, ds \right) < \infty.$$

Note that for all $0 < \epsilon < \alpha$,

$$C^{\alpha+\epsilon}(0,T) \subset W_0^{\alpha,\infty}(0,T) \subset C^{\alpha-\epsilon}(0,T).$$

Moreover, we denote by $W_T^{1-\alpha,\infty}(0,T)$ the space of measurable functions $g : [0,T] \to \mathbb{R}$ such that

$$\|g\|_{1-\alpha,\infty} := \sup_{0 < s < t < T} \left(|\frac{|g(t) - g(s)|}{(t-s)^{1-\alpha}}| + \int_s^t \frac{|g(y) - g(s)|}{(y-s)^{2-\alpha}} \, dy \right) < \infty.$$

Then, for all $0 < \epsilon < \alpha$,

$$C^{1-\alpha+\epsilon}(0,T) \subset W_T^{1-\alpha,\infty}(0,T) \subset C^{1-\alpha}(0,T).$$

For $g \in W_T^{1-\alpha,\infty}(0,T)$ we define

$$\Lambda_\alpha(g) = \frac{1}{\Gamma(1-\alpha)} \sup_{0 < s < t < T} |(D_{t-}^{1-\alpha} g_{t-})(s)|.$$

Finally we denote by $W_0^{\alpha,1}(0,T)$ the space of measurable functions $h : [0,T] \to \mathbb{R}$ such that

$$\|h\|_{\alpha,1} := \int_0^T \frac{|h(t)|}{t^\alpha} \, dt + \int_0^T \int_0^t \frac{|h(y) - h(t)|}{(t-y)^{\alpha+1}} \, dy \, dt < \infty.$$

If $h \in W_0^{\alpha,1}(0,T)$ and $g \in W_T^{1-\alpha,\infty}(0,T)$, then the generalized Stieltjes integral $\int_0^t h \, dg$ exists for all $t \in [0,T]$ and

$$\left| \int_0^t h \, dg \right| \leq \Lambda_\alpha(g) \|f\|_{\alpha,1}.$$

We come back to equation (D.2) and consider the following regularity assumptions for the coefficients:

1. Assumption H_1: $\sigma(t,x)$ is differentiable in x and there exists some constants $M > 0, 0 < \beta, \delta \leq 1$, and for every $N > 0$ there exists $M_N > 0$ such that the following holds:

$$|\sigma(t,x) - \sigma(t,y)| \leq M|x-y|, \quad \forall x \in \mathbb{R}, \forall t \in [0,T],$$
$$|\partial_x \sigma(t,x) - \partial_x \sigma(t,y)| \leq M_N |x-y|^\delta, \quad \forall |x|, |y| \leq N, \forall t \in [0,T],$$
$$|\sigma(t,x) - \sigma(s,x)| + |\partial_x \sigma(t,x) - \partial_x \sigma(s,x)| \leq M|t-s|^\beta,$$
$$\forall x \in \mathbb{R}, \forall t, s \in [0,T].$$

2. Assumption H_2: There exists $b_0 \in L^p([0,T])$ and for every $N > 0$ there exists $L_N > 0$ such that

$$|b(t,x) - b(t,y)| \leq L_N |x - y|, \quad \forall |x|, |y| \leq N, \forall t \in [0,T],$$
$$|b(t,x)| \leq L_0 |x| + b_0(t), \quad \forall |x| \in \mathbb{R}, \forall t \in [0,T].$$

3. Assumption H_3: There exists $\gamma \in [0,1)$ and $M_0 > 0$ such that

$$|\sigma(t,x)| \leq M_0(1 + |x|^\gamma), \quad \forall x \in \mathbb{R}, \forall t \in [0,T].$$

Denote

$$\alpha_0 = \min\left(\frac{1}{2}, \beta, \frac{\delta}{1+\delta}\right).$$

Under these conditions, in [180] the following theorem on the existence and uniqueness of a solution for equation (D.2) is proved. Recently this result has been extend in [39], where they provide a method to find stationary solutions of integral equations driven by $B^{(H)}$ with only Hölder continuous drift and volatility. Their hypotheses on b and σ are weaker than the others usually assumed in literature (see also [163]), but in this case the uniqueness of the solution is in general lost.

Theorem D.2.7. *Suppose X_0 is an $\mathbb{R}$-valued random variable, the coefficients $b(t,x)$ and $\sigma(t,x)$ satisfy the previous hypotheses with $\beta > 1 - H$, $\delta > 1/H - 1$, and assume that the constants $M, M_N,$ and L_N and the function b_0 may depend on ω. Then*

1. *If $\alpha \in (1 - H, \alpha_0)$ and $p \geq 1/\alpha$, then there exists a unique stochastic process $X \in L^0(\Omega; W_0^{\alpha,\infty}(0,T))$ that satisfies the stochastic equation (D.2), and moreover, for $\mathbb{P}^H$-almost all $\omega \in \Omega$*

$$X(\omega, \cdot) \in C^{1-\alpha}([0,T]).$$

2. *If $\alpha \in (1 - H, \alpha_0 \wedge (2 - \gamma)/4)$, $p \geq 1/\alpha$, $X \in L^\infty(\Omega; \mathbb{R})$ and the constants M, M_N, L_N, b_0 are independent of ω, then for all $p \geq 1$,*

$$E\left[\|X\|_{\alpha,\infty}^p\right] < \infty.$$

Theorem D.2.7 is a consequence of the following result for deterministic differential equations.

Theorem D.2.8. *Let $0 < \alpha < 1/2$. Let $g \in W_T^{1-\alpha,\infty}(0,T)$ and consider the deterministic differential equation on $\mathbb{R}$*

$$x(t) = x_0 + \int_0^t b(s, x(s))\, ds + \int_0^t \sigma(s, x(s))\, dg(s), \tag{D.5}$$

where $x_0 \in \mathbb{R}$, and the coefficients b, σ are measurable functions satisfying the assumptions H_1 and H_2, respectively with $p = 1/\alpha$, $\beta > 0$, $\delta \leq 1$, and

$$0 < \alpha < \alpha_0 = \min(\frac{1}{2}, \beta, \frac{\delta}{1+\delta}).$$

Then there exists a unique solution $x \in W_0^{\alpha,\infty}(0,T)$ for equation (D.5). Moreover, the solution is $(1-\alpha)$-Hölder continuous.

Proof. The existence is proved by using a fixed point argument in $W_0^{\alpha,\infty}(0,T)$. The uniqueness is a consequence of the following estimate for functions $g \in W_T^{1-\alpha,\infty}(0,T)$ and $f \in W_0^{\alpha,\infty}(0,T)$. If $h(t) = \int_0^t f(s)\, dg(s)$, then

$$|h(t)| + \int_0^t \frac{|h(t)-h(s)|}{(t-s)^{\alpha+1}}\, ds \leq \Lambda_\alpha(g) c_{\alpha,T} \int_0^t [(t-s)^{-2\alpha} + s^{-\alpha}]$$

$$\left[|f(s)| + \int_0^s \frac{|f(s)-f(r)|}{(s-r)^{\alpha+1}}\, dr\right] ds, \qquad (D.6)$$

where $c_{\alpha,T}$ is a constant depending on α, T. For further details, we refer to the proof of Theorem 5.1 of [180]. □

Hence Theorem D.2.7 is a direct consequence of Theorem D.2.8 because if $H > 1/2$ the random variable

$$G = \frac{1}{\Gamma(1-\alpha)} \sup_{0<s<t<T} |(D_{t-}^{1-\alpha} B^{(H)}(t-))(s)|$$

has moments of all orders (see Lemma 7.4 in [180]). Hence, if $u(t)$ is a stochastic process whose trajectories belongs to $W_T^{\alpha,1}(0,T)$ with $1-H < \alpha < 1/2$, then the pathwise integral $U(t) = \int_0^t u(s)\, d^- B^{(H)}(s)$ exists and we have the estimate

$$\left|\int_0^t u(s)\, d^- B^{(H)}(s)\right| \leq G\|u\|_{\alpha,1}.$$

Moreover, if $u(t)$ is such that its trajectories belongs to $W_0^{\alpha,\infty}(0,T)$, then the pathwise integral $U(t) = \int_0^t u(s)\, d^- B^{(H)}(s)$ is Hölder continuous of order $1-\alpha$ and the estimate (D.6) holds. For further details, we refer to [180].

D.3 Stochastic differential equations via rough path analysis

In [153] an integration theory has been established for a class of nonsmooth paths under the name of *geometric rough paths*. Here we sketch this framework and its application to SDEs for *fBm* according to [61].

D.3.1 Rough path analysis

Let V be an Euclidean space with norm $|\cdot|_V$, and for each integer k let $V^{\otimes k}$ be the kth tensor product endowed with a compatible norm $|\cdot|_{V^{\otimes k}}$. Denote by $T^{(3)}$ the truncated tensor algebra $\mathbb{R} \oplus V \oplus V^{\otimes 2} \oplus V^{\otimes 3}$. A *multiplicative functional* w in $T^{(3)}$ is a map

$$w : \Delta \longrightarrow T^{(3)},$$

where Δ is given by the simplex $\Delta = \{(s,t) \in [0,1] \times [0,1] : 0 \leq s \leq t \leq 1\}$, such that $w_{s,t} = (1, w_{s,t}^1, w_{s,t}^2, w_{s,t}^3)$ satisfies the Chen equality

$$w_{s,u} = w_{s,t} \otimes w_{t,u}, \qquad \forall s \leq t \leq u.$$

Given a continuous path ω in V with finite variation, we can define the multiplicative functional in $T^{(3)}$ by taking

$$w_{s,t}^1 = \omega_t - \omega_s,$$
$$w_{s,t}^2 = \int_{s<t_1<t_2<t} d\omega_{t_1} \otimes d\omega_{t_2},$$
$$w_{s,t}^3 = \int_{s<t_1<t_2<t_3<t} d\omega_{t_1} \otimes d\omega_{t_2} \otimes d\omega_{t_3}.$$

We say that $w_{s,t} = (1, w_{s,t}^1, w_{s,t}^2, w_{s,t}^3)$ is a *smooth rough path* in $T^{(3)}$ over the continuous path ω.

Let $3 < p < 4$. A multiplicative functional w in $T^{(3)}$ is said to have finite *p*-variation if

$$\sup_\pi \sum_k |w_{t_{k-1},t_k}^i|_{V^{\otimes i}}^{p/i} < +\infty, \qquad i = 1,2,3,$$

where the supremum is taken over all the finite partitions π of $[0,1]$. The space of all smooth rough paths is denoted by $\Omega_p^\infty(V)$, and it is endowed with the *p*-variation distance

$$d_p(w,v) = \sup_{i=1,2,3} \left(\sup_\pi \sum_k |w_{t_{k-1},t_k}^i - v_{t_{k-1},t_k}^i|_{V^{\otimes i}}^{p/i} \right)^{i/p},$$

where again the second supremum is taken over all the finite partitions π of $[0,1]$. The space $\Omega_p(V)$ of geometric rough paths is then the closure of $\Omega_p^\infty(V)$ under the *p*-variation distance and an element in $\Omega_p(V)$ is called a *geometric rough path* (of finite *p*-variation).

D.3.2 Stochastic calculus with rough path analysis

In [153] it is proved that solutions of differential equations with smooth vector fields are continuous on $\Omega_p(V)$ under the *p*-variation distance.

Theorem D.3.1. *Let $p \geq 1$ and let $f : \mathbb{R} \times \mathbb{R}^n \to L(\mathbb{R}^d, \mathbb{R}^n)$ be a differentiable function with bounded derivatives up to degree $[p]+1$. Denote by $F(w,x) \in$*

$\Omega_p^\infty(\mathbb{R}^n)$ *the smooth rough path associated with the unique solution y to the differential equation*

$$dy_t = f(t, y_t) d\omega_t, \qquad y_0 = x, \qquad (D.7)$$

where ω is a continuous path in $\mathbb{R}^d$ with finite variation and w is its associated smooth rough path. Then the map, called the Itô map determined by (D.7), $w \to F(w, x)$ is continuous from $\Omega_p^\infty(\mathbb{R}^n) \to \Omega_p^\infty(\mathbb{R}^n)$ with respect to the p-variation distance. Therefore, there is a unique extension of the Itô map $F(w, x)$ to the space $\Omega_p(\mathbb{R}^d)$ of all geometric rough paths.

Theorem D.3.1 is a particular case of the main result of [153]. In [61] they construct a canonical geometric rough path in $T^{(3)}(\mathbb{R}^d)$ associated to a d-dimensional fBm with Hurst index $H > 1/4$ by using the so-called the dyadic approximations. More precisely, the dyadic approximation of $B^{(H)}$ is the piecewise linear path

$$W^m(t) = B^{(H)}(t_{k-1}^m) + [t - (t_{k-1}^m)] \Delta_k^m B^{(H)} \qquad \text{for } t_{k-1}^m \le t < t_k^m, \qquad (D.8)$$

where $t_k^m = k/2^m$, $k = 1, \ldots, 2^m$, and $\Delta_k^m B^{(H)} = B^{(H)}(t_k^m) - B^{(H)}(t_{k-1}^m)$. Since W^m has finite variation, it can be associated to a smooth rough path

$$w_{s,t}^m = (1, w_{s,t}^{m,1}, w_{s,t}^{2,m}, w_{s,t}^{3,m}),$$

where $w_{s,t}^{m,i}$ is the ith iterated path integral of W^m over the interval $[s, t]$.

Theorem D.3.2. *Consider the smooth rough path w^m associated to the dyadic approximation (D.8) of the fBm $B^{(H)}$.*

1. *If $1/4 < H < 1/2$, then for any $p < 4$ such that $Hp > 1$, w^m converges almost surely to a geometric rough path w in p-variation distance and $w_{s,t}^1 = B^{(H)}(t) - B^{(H)}(s)$.*
2. *If $1/4 \ge H$, then even the second level paths $w^{m,2}$ of its dyadic approximations do not converge in $L^1(\Omega, \mathcal{F}^{(H)}, \mathbb{P}^H)$ according to the p-variation distance.*

Proof. The proof relies on the Gaussian properties of $B^{(H)}$. For further details, we refer to Theorem 2 of [61]. □

Theorem D.3.2 is the key in order to use the universal limit Theorem D.3.1 to study pathwise SDEs driven by fBm. For the sake of simplicity, we again consider the case of a 1-dimensional fBm. Let α be a smooth one form and $\alpha^{(k)}$ denote its $(k-1)$th derivative (so that $\alpha^1 = \alpha$). Then the pathwise integral of α along $B^{(H)}$ exists and can be seen as

$$\int_s^t \alpha(B^{(H)}(r)) d^- B^{(H)}(r) = \lim_{|\pi| \to 0} \sum_l \sum_{k=1}^3 \alpha^{(k)}(B^{(H)}(t_{l-1}))(w_{t_{l-1}, t_l}^k),$$

where $0 \leq s \leq t \leq 1$ and $|\pi| = \max_l |t_l - t_{l-1}|$ for a finite partition of $[0,1]$. More precisely, $\int_s^t \alpha(B^{(H)}(r)) d^- B^{(H)}(r)$ is defined as a geometric rough path, whose projection coincides with the above limit.

Let $H > 1/4$ and consider the SDE

$$dX(t) = b(t, X(t)) \, dt + f(t, X(t)) \, d^- B^{(H)}(t), \qquad X(0) = \xi, \qquad (D.9)$$

where all the derivatives of f are bounded. Then if $x^m(t)$ is the unique solution of the following ordinary differential equation

$$dx^m(t) = b(t, x^m(t)) \, dt + f(t, x^m(t)) \, dW^m(t), \qquad x^m(0) = \xi,$$

by Theorem D.3.2 and D.3.1 it follows that $x^m(t)$ converges to a continuous path $x(t)$ uniformly on any finite interval as m goes to ∞ and that $x(0) = \xi$. The limit path $x(t)$ actually gives the strong solution $X(t)$ for (D.9). Moreover, the following stronger result holds.

Theorem D.3.3. *If $H > 1/4$ and $pH > 1$, let $v_t^m = (x_t^{m,1}, \ldots, x_t^{m,k})$ with $v_{s,t}^{m,i}$ denotes the ith iterated path integral of the unique solution x^m over $[s,t]$. Then $v_{s,t}^{m,i}$ converges to the unique limit $v_{s,t}^i$ in p-variation distance, as $m \to \infty$,*

$$\sup_\pi \sum_k |v_{t_{k-1},t_k}^{m,i} - v_{t_{k-1},t_k}^i|^{p/i} \to 0$$

almost surely for any $i = 1, 2, \ldots$, $v_{s,t}^1 = x(t) - x(s)$. Moreover, the solution x is a flow of diffeomorphisms of $\mathbb{R}$ as a function of the initial value ξ.

Proof. For the proof and further details, we refer to [61]. □

References

1. Aase, K., Øksendal, B., Privault, N. and Ubøe, J.: White noise generalizations of the Clark–Haussmann–Ocone theorem with application to mathematical finance. Finance Stoch. 4, 465–496, 2000.
2. Aase, K., Øksendal, B. and Ubøe, J.: Using the Donsker delta function to compute hedging strategies. Potential Anal. 14, 351–374, 2001.
3. Ait Rami, M., Moore, J.B. and Zhou, X.Y.: Indefinite stochastic linear quadratic control and generalized differential Riccati equation. SIAM J. Control Optim. 40, 1296–1311, 2001/02.
4. Albeverio, S., Hu, Y. and Zhou, X.Y.: A remark on non smoothness of self intersection local time of planar Brownian motion. Statist. Probab. Letter 32, 57–65, 1997.
5. Alòs, E., León, A. and Nualart, D.: Stratonovich stochastic calculus with respect to to fractional Brownian motion with Hurst parameter less than 1/2. Taiwanesse J. Math. 5, 609–632, 2001.
6. Alòs, E., Mazet, O. and Nualart, D.: Stochastic calculus with respect to Gaussian processes. Ann. Probab. 29, 766–801, 2000.
7. Alòs, E., Mazet, O. and D. Nualart: Stochastic calculus with respect to fractional Brownian motion with Hurst parameter less than 1/2. Stoch. Proc. Appl. 86, 121–139, 1999.
8. Alòs, E. and Nualart, D.: Stochastic integration with respect to the fractional Brownian motion. Stoch. Stoch. Rep. 75, 129–152, 2003.
9. Arcones, M. A.: On the law of the iterated logarithm for gaussian processes. Journal of Theoretical Probability, 8 (4), 877–904, 1995.
10. Asch, J., and Potthoff, J.: Itô lemma without non-anticipatory conditions, Probab. Theory Rel. Fields 88, 17–46, 1991.
11. Bardina, X., Jolis, M., and Tudor, C.A.: Weak convergence to the fractional Brownian sheet and other two-parameter Gaussian processes. Statist. Probab. Letters 65, 317–329, 2003.
12. Barton, R.J. and Vincent Poor, H.: Signal detection in fractional gaussian noise. IEEE trans. Information Theory 34, 943–959, 1988.
13. Beckenbach, E.F. and Bellman, R.: Inequalities. Springer, 1965.
14. Bender, C.: An Itô formula for generalized functionals of a fractional Brownian motion with arbitrary Hurst parameter. Stoch.Proc. Appl. 104 (1), 81–106, 2003.

15. Bender, C.: The fractional Itô integral, change of measure and absence of arbitrage. Manuscript 2002.
16. Bender, C.: An S-transform approach to integration with respect to fractional Brownian motion. Bernoulli 9, 955–983, 2003.
17. Bender, C.: Construction of a weak arbitrage. Preprint, May 2003.
18. Bender, C.: Explicit solution of a class of linear fractional BSDEs. Syst. Control Letters 54, 671–680, 2005.
19. Bender, C. and Elliott, R.: On the Clark Ocone formula for fractional Brownian motion with Hurst parameter bigger than a half. Stoch. Stoch. Rep. 75, 391–405, 2003.
20. Bender, C. and Elliott, R.: Arbitrage in a discrete version of the Wick-fractional Black-Scholes market. Math. Oper. Res. 29, 935–945, 2004.
21. Bender, C., Sottinen, T., and Valkeila, E.: No-arbitrage pricing beyond semimartingales, WIAS preprint 1110, 2005.
22. Benth, F.E.: Integrals. In the Hida distribution space $(\mathcal{S})^*$. In Lindstrøm, T., Øksendal, B. and Üstünel, A.S. (edi.): Stochastic analysis and related topics, 89–99, Stochastics Monogr., 8, Gordon and Breach, Montreux, 1993.
23. Beran, J.: Statistics for long-memory processes. Monographs on Statistics and Applied Probability, 61. Chapman and Hall, New York, 1994.
24. Berger, M.A. and Mizel, V.J.: An extension of the stochastic integral, Ann. Prob. 10, 235–450, 1982.
25. Berman, S.M.: A central limit theorem for the renormalized self-intersection local time of a stationary process. Probability in Banach spaces, 8 (Brunswick, ME, 1991), 351–363, Progr. Probab., 30, Birkhäuser Boston, Boston, MA, 1992.
26. Berman, S.M.: Self-intersections and local nondeterminism of Gaussian processes. Ann. Probab. 19, 160–191, 1991.
27. Berman, S.M.: Local nondeterminism and local times of general stochastic processes. Ann. Inst. H. Poincaré Sect. B (N.S.) 19, 189–207, 1983.
28. Berman, S.M.: Local nondeterminism and local times of Gaussian processes. Indiana Univ. Math. J. 23, 69–94, 1973/74.
29. Berman, S.M.: Local nondeterminism and local times of Gaussian processes. Bull. Amer. Math. Soc. 79, 475–477, 1973.
30. Biagini, F., Campanino, M., and Fuschini, S.: Discrete approximation of stochastic integrals with respect to fractional Brownian motion with Hurst index $H > 1/2$. Accepted on Stochastics, 2007.
31. Biagini, F., Hu, Y., Øksendal, B., and Sulem, A.: A stochastic maximum principle for processes driven by fractional Brownian motion. Stoch. Proc. Appl. 100, 233–253, 2002.
32. Biagini, F. and Øksendal, B.: Minimal variance hedging for fractional Brownian motion. Methods Appl. Anal. 10, 347–362, 2003.
33. Biagini, F. and Øksendal, B.: Forward integrals and an Ito formula for fractional Brownian motion. Preprint University of Oslo, June 2004.
34. Biagini, F., Øksendal, B, Sulem, A., and Wallner, N.: An introduction to white noise theory and Malliavin calculus for fractional Brownian motion. Proc. Royal Soc., 460, 347–372, 2004.
35. Björk, T. and Hult, H.: A note on Wick products and the fractional Black Scholes model. Finance Stoch. 9, 197–209, 2005.
36. Boufoussi, B. and Lakhel, E.H.: Weak convergence in Besov spaces to fractional Brownian motion. C. R. Acad. Sci. Paris, t. 333, Serie I, 39–44, 2001.

37. Boufoussi, B. and Ouknine, Y.: On a SDE driven by fractional Brownian motion and with monotone drift, Electronic Communications Probab. 8, 122–134, 2003.
38. Brody, D., Syroka J. and Zervos, M.: Dynamical pricing of weather derivatives. Quantitative Finance 2, 189–198, 2002.
39. Buchmann, B. and Klüppelberg, C.: fractional integral equations and state space transforms. Bernoulli 12(3), 431–456, 2004.
40. Buchmann, B. and Klüppelberg, C.: Maxima of stochastic processes driven by fractional Brownian motion. Adv. Appl. Probab. 37, 743–764, 2005.
41. Caithamer, P.: The stochastic wave equation driven by fractional Brownian motion and temporally correlated smooth noise. Stoch. Dyn. 5, 45–64, 2005.
42. Carmona, P., Coutin, L., and G. Montseny, G.: Stochastic integration with respect to fractional Brownian. Ann. I. H. Poincaré-PR 39(1), 27–68, 2003.
43. Carmona, P., Coutin, L., and G. Montseny, : Applications of a representation of long memory Gaussian processes. Rapport LAAS N 97422, November 1997.
44. Carmona, P., Coutin, L., and Montseny, G.: Approximation of some Gaussian processes. Statistical Inference for Stochastic Processes 3, 161–171, 2000.
45. Carmona, P. and Coutin, L.: Integrales stochastiques pour le mouvement brownien fractionnaire. C.R. Acad. ScI.Paris, t. 330, Serie I, 213–236, 2000.
46. Carmona, P. and Coutin, L.: Stochastic integration with respect to fractional Brownian. Ann. Institut Henri Poincaré 39, 27–68, 2003.
47. Carmona, P. and Coutin, L. : The linear Kalman-Bucy filter with respect to Liouville fractional Brownian motion. 14th International Symposium of Mathematical Theory of Networks and Systems (MTNS'2000), Perpignan (France), 19–23, 2000.
48. Chang, Y.C. and Chang, S.: A fast estimation algorithm on the Hurst parameter of discrete-time fractional Brownian motion. IEEE Trans. Signal Process. 50, 554–559, 2002.
49. Carr, P.P. and Jarrow, R.A.: The stop-loss-start-gain paradox and option valuation: A new decomposition into intrinsic and time value. The Review of Financial Studies 3, 469–492, 1990.
50. Cheridito, P.: Regularizing fractional Brownian motion with a view towards stock pricing modeling. Diss. No. 14051, ETH Zürich, 2001.
51. Cheridito, P.: Regularized fractional Brownian motion and option pricing. Preprint ETH Zürich, 2000.
52. Cheridito, P.: Arbitrage in fractional Brownian motion models. Finance Stoch., 7(4), 533–553, 2003.
53. Cheridito, P.: Mixed fractional Brownian motion. Bernoulli 7, 913–934, 2001.
54. Cheridito, P. and Nualart, D.: Stochastic integral of divergence type with respect to the fractional Brownian motion with Hurst parameter 1/2. Ann. Inst. H. Poincaré Probab. Statist. 41, 1049–1081, 2005.
55. Ciesielski, Z., Kerkyacharian, Z.G. and Roynette, B.: Quelques espaces fonctionnels associés à des processus gaussiens. Studia Mathematica 107(2), 171–204, 1993.
56. Coeurjolly, J.F.: Estimating the parameters of a fractional Brownian motion by discrete variations of its sample paths. Stat. Inference Stoch. Process. 4, 199–227, 2001.
57. Comte, F., Coutin, L. and Renault, E.: Affine fractional stochastic volatility models with application to option pricing. Preprint, 2003.

58. Coutin, L. and Decreusefond, L.: Stochastic differential equation driven by a fractional Brownian motion. C.R.I. 331, 75–80, 2000.
59. Coutin, L. and Decreusefond, L.: Abstract non linear filtering theory in the presence of fractional Brownian motion. Ann. Appl. Probab. 9, 1058–1090, 1999.
60. Coutin, L. and Qian, Z.: Stochastic differentials equations for fractional Brownian motions. C.R. Acad. Sci. Paris Sér. I Math. 331, 75–80, 2000.
61. Coutin, L. and Qian, Z.: Stochastic analysis, rough path analysis and fractional Brownian motions. Probab. Theory. Rel. Fields 122, 108–140, 2002.
62. Coutin, L., Nualart, D., and Tudor, C.A.: Tanaka formula for the fractional motion. Stoch. Proc. Appl. 94 (2), 301–315, 2001.
63. Cox, J. and Huang, C.F.: Optimal consumption and portfolio policies when asset prices follow a diffusion process. Journal of Economic Theory 49, 33–83, 1989.
64. Cox, J. and Huang, C.F.: A variational problem arising in financial economics. J. Mathematical Economics 20, 465–487, 1991.
65. Csaki, E., Shi, Z., and Yor, M.: Fractional Brownian motion as "higher order" fractional derivatives of Brownian local times. Limit theorems in probability and statistics, Vol. I (Balatonlelle, 1999), 365–387, Jànos Bolyai Math. Soc., Budapest, 2002.
66. Cutland, N.J., Kopp, P.E., and Willinger, W.: Stock price returns and the Joseph effect: A fractional version of the Black-Scholes model. Seminar on Stochastic Analysis. Random Fields and Applications (Ascona, 1993), 327–351, Progr. Probab. 36, Birkhäuser, Basel, 1995.
67. Dahlhaus, R.: Efficient parameter estimation for self-similar processes. Ann. Statist. 17, 1749–1766, 1989.
68. Dai, W. and Heyde, C.C.: Itô formula with respect to fractional Brownian motion and its application. J. Appl. Math. Stoch. Anal. 9, 439–448, 1996.
69. Da Prato, G. and Jabczyk, J.: Ergodicity for Infinite Dimensional Systems. Cambridge University Press, 1996.
70. Dasgupta, A.: Fractional Brownian motion: Its properties and applications to stochastic integration. Ph.D. Thesis, Dept. of Statistics, University of North Carolina at Chapel Hill, 1997.
71. Decreusefond, L.: A Skohorod Stratonovich integral for the fractional Brownian motion. Proc. of the 7th Workshop on Stochastic Analysis and Related fields, Birkhäuser 2000.
72. Decreusefond, L.: Stochastic integration with respect to fractional Brownian motion. In Theory and Applications of Long-Range Dependence. Birkhäuser Boston, Boston, MA, 203–226, 2003.
73. Decreusefond, L.: Stochastic calculus for Volterra processes. C.R. Acad. Paris Sér.I Math. 334, 903–908, 2002.
74. Decreusefond, L.: Stochastic integration with respect to Volterra processes. Ann. Inst. H. Poincaré Probab. Statist. 41, 123–149, 2005.
75. Decreusefond, L. and Üstünel, A.S.: Application du calcul des variations stochastiques au mouvement brownien fractionnaire. Compte-Rendus de l'Acaémie des Sciences 321, 1605–1608, 1995.
76. Decreusefond, L. and Üstünel, A.S.: Stochastic analysis of the fractional Brownian motion. Potential Analysis 10, 177–214, 1999.
77. Delbaen, F. and Schachermayer, W.: A general verion of the fundamental theorem of asset pricing. Mathematische Annalen 300, 463–520, 1994.

78. Delgado, R. and Jolis, M.: On a Ogawa-type integral with application to the fractional Brownian motion. Stoch. Anal. Appl. 18, 617–634, MR 2001, 2000.
79. Dellacherie, C. and Meyer, P.A.: Probability and Potentials B. North Holland, 1982.
80. Djehiche, B. and Eddahbi, M.: Hedging options in markets models modulated by fractional Brownian motion. Stoch. Anal. Appl. 19, 753–770, 2001.
81. Doukhan, P., Oppenheim, G., and Taqqu, M.S.: Theory and Applications of Long Range-Dependence, Birkhäuser, 2003.
82. Dudley, R.M. and Norvaisa, R.: An introduction to p-variation and Young integrals. Lecture Notes 1, MaPhySto, 1998.
83. Duncan, T.E., Hu, Y., and Pasik–Duncan, B.: Stochastic calculus for fractional Brownian motion. I. Theory. SIAM J. Control Optim. 38, 582–612, 2000.
84. Duncan, T.E., Jakubowski, J., and Pasik-Duncan, B.: An approach to stochastic integration for fractional Brownian motion in a Hilbert Space. Stoch. Dyn., 53–75, 2006.
85. Duncan, T.E., Jakubowski, J., and Pasik-Duncan, B.: Stochastic integration for fractional Brownian motion in a Hilbert space. Proc. Fifteenth International Symposium on Mathematical Theory of Networks and Systems, South Bend, IN, USA, 2002.
86. Duncan, T.E., Maslowski, B., and Pasik-Duncan, B.: Fractional Brownian motion and stochastic equations in a Hilbert Space. Stoch. Dyn. 2, 225–250, 2002.
87. Eddahbi, M., Lacayo, R., Solé, J.L., Tudor, C.A., and Vives, J. Regularity and the asymptotic behaviour of the local time for the d-dimensional fractional Brownian with N parameters. Stoch. Anal. Appl. 23, 383–400, 2005.
88. Elliott, R.C. and Chang, L.: Perpetual American options with fractional Brownian motion. Quant. Finance 4(2), 123–128, 2004.
89. Elliott, R.C. and Van der Hoek, J.: A general fractional white noise theory and applications to finance. Mathematical Finance 13, 301–330, 2003.
90. Elliott, R.J. and Van der Hoek, J.: A basic lemma for stochastic integrals and Itô formulas for processes driven by fractional Brownian motion. Working paper, 2004.
91. Embrechts, P. and Maejima, M.: Selfsimilar processes. Princeton Series in Applied Mathematics. Princeton University Press, Princeton, NJ, 2002.
92. Errami, M. and Russo, F.: N-covariation, generalized Dirichlet processes and calculus with respect to a finite cubic variation processes. Stoch. Proc. Appl. 104 (2), 259–299, 2003.
93. Feyel, D. and de La Pradelle, A.: The FBM Itô's formula through analytic continuation, Electron. J. Probab. 6, MR 2002, 2001.
94. Feyel, D. and de La Pradelle, A.: On fractional Brownian processes. Potential Analysis, 10(3), 273–288, 1999.
95. Ferrante, M. and Rovira, C.: Stochastic delay differential equations driven by fractional Brownian motion with Hurst index 1/2. Bernoulli 12, 85–100, 2006.
96. Fleming, W.H. and Rishel, R.W.: Deterministic and stochastic optimal control. Applications of Mathematics, 1. Springer, Berlin-New York, 1975.
97. Fleming, W.H. and Soner, H.M.: Controlled Markov processes and viscosity solutions. Applications of Mathematics, 25. Springer, New York, 1993.
98. Fouque, J.P., G. Papanicolaou, G., and Sircar, K.R.: Derivatives in financial markets with stochastic volatility. Cambridge University Press, Cambridge, 2000.

99. Fox, R., Taqqu, M., and Murad, S.: Central limit theorems for quadratic forms in random variables having long-range dependence. Probab. Theory Rel. Fields 74, 213–240, 1987.
100. Geman, D. and Horowitz, J.: Occupation densities. Ann. Probab. 1, 1–67, 1980.
101. Gradinaru, M., Nourdin, I, Russo, F., and Vallois, P.: m-Order integrals and generalized Itô's formula; the case of a fractional Brownian motion with any index. Ann. Inst. H. Poincaré Probab. Statist. 41(4), 781–806, 2005.
102. Gradinaru, M., Russo, F., and Vallois, P.: Generalized covariations, local time and Stratonovich Itô's formula for fractional Brownian motion with Hurst index 1/4. Ann. Probab. 31, 1772–1820, 2003.
103. Gripenberg, G. and Norros, I.: On the prediction of fractional Brownian motion. J. Appl. Probab. 33, 400–410, 1996.
104. Guasoni, P.: No Arbitrage with Transaction Costs, with Fractional Brownian Motion and Beyond. Mathematical Finance 16, 569-582, 2003.
105. Guasoni, P., Rasonyi, M. amd Schachermayer, W.: Consistent Price Systems and Face-Lifting Pricing under Transaction Costs. Preprint, 2007.
106. Hairer, M.: Ergodicity of stochastic differential equations driven by fractional Brownian motion. Ann. Probab. 33, 703–758, 2005.
107. Haussman, U.: A Stochastic Maximum Principle for Optimal Control of Diffusions. Longman Scientific and Technical, 1986.
108. Hida, T., Kuo, H.H., Potthoff, J., and Streit, L.: White Noise Analysis. Kluwer, 1993.
109. Holden, H., Øksendal, B., Ubøe, J., and Zhang, T.: Stochastic Partial Differential Equations. Birkhauser, 1996.
110. Hu, Y.: On the self-intersection local time of Brownian motion via chaos expansion. Publicacions Matemàtiques 40, 337–350, 1996.
111. Hu, Y.: Itô–Wiener chaos expansion with exact residual and correlation, variance inequalities. Journal of Theoretical Probability 10, 835–848, 1997.
112. Hu, Y.: Heat equation with fractional white noise potentials, Applied Mathematics and Optimization, 43, 221–243, 2001.
113. Hu, Y.: A class of SPDE driven by fractional white noise. Stochastic processes, physics and geometry: new interplays, II (Leipzig), 317–325, CMS Conf. Proc., 29, Amer. Math. 1999.
114. Hu, Y.: Probability structure preserving and absolute continuity. Annales de l'Institut Henri Poincaré 38, 557–580, 2002.
115. Hu, Y.: Prediction and translation of fractional Brownian motions, Stochastics in Finite and Infinite Dimensions, T. Hida et al. (ed.), Trends Math, Birkhauser, Boston, MA, 153–171, 2001.
116. Hu, Y.: Integral transformation and anticipative calculus for fractional Brownian motions. Mem. Ameri. Math. Soc. 825, 2005.
117. Hu, Y.: Optimal consumption and portfolio in a market where the volatility is driven by fractional Brownian motions. Probability, Finance and Insurance. World Sci. Publishing, River Edge, NJ, 164–173, 2004.
118. Hu, Y.: Option pricing in a market where the volatility is driven by fractional Brownian motions. Recent Developments in Mathematical Finance (Shanghai, 2001), 49–59, World Sci. Publishing, River Edge, NJ, 2002.
119. Hu, Y.: Self-intersection local time of fractional Brownian motions via chaos expansion. J. Math. Kyoto Univ. 41, 233-250, 2001.
120. Hu, Y. and Nualart, D.: renormalized self-intersection local time for fractional Brownian motion. Ann. Probab. 33, 948–983, 2005.

121. Hu, Y. and Øksendal, B.: Fractional white noise calculus and applications to finance. Inf. Dim. Anal. Quant. Probab. 6, 1–32, 2003.
122. Hu, Y. and Øksendal, B.: Chaos expansion of local time of fractional Brownian motions. Stochastic Anal. Appl. 20, 815–837, 2002.
123. Hu, Y., Øksendal, B., and Salopek, D.M.: Weighted local time for fractional Brownian motion and applications to finance. Stoch. Anal. Appl. 23, 15–30, 2005.
124. Hu, Y., Øksendal, B., and Sulem, A.: Optimal portfolio in a fractional Black-Scholes market. In S. Albeverio et al. (ed.): Mathematical Physics and Stochastic Analysis. World Scientific 2000.
125. Hu, Y., Øksendal, B., and Sulem, A.: Optimal consumption and portfolio in a Black-Scholes market driven by fractional Brownian motion. Inf. Dim. Anal. Quant. Probab. 6, 519–536, 2003.
126. Hu, Y., Øksendal, B. and Zhang, T.: Stochastic partial differential equations driven by multiparameter fractional white noise. Stochastic processes, physics and geometry: new interplays, II (Leipzig, 1999), 327–337, CMS Conf. Proc., 29, Amer. Math. Soc., Providence, RI, 2000.
127. Hu, Y., Øksendal, B., and Zhang, T.: General fractional multiparameter white noise theory and stochastic partial differential equations. Comm. Partial Differential Equations 29(1-2), 1–23, 2004.
128. Hu, Y. and Zhou, X.Y.: Linear quadratic control for systems driven by fractional Brownian motions. SIAM J. Control Optim. 43, 2245–2277, 2005.
129. Hurst, H.E.: Long-term storage capacity in reservoirs. Trans. Amer. Soc. Civil Eng. 116, 400–410, 1951.
130. Hurst, H.E.: Methods of using long-term storage in reservoirs. Proc. Inst. Civil Engineers Part I, Chapter 5, 519–590, 1956.
131. K. Itô: Multiple Wiener integrals, J. Math. Soc. Japan 3, 157–164, 1951.
132. Jeganathan, P.: Convergence of functionals of sums of r.v.s to local times of fractional stable motions. Ann. Probab. 32, no. 3A, 1771–1795, 2004.
133. Jost, C.: Transformation formulas for fractional Brownian motion. Stochastic Processes and their Applications 116, 1341–1357, 2006.
134. Jost, C.: On the connection between Molchan-Golosov and Mandelbrot-Van Ness representations of fractional Brownian motion. To appear in the Journal of Integral Equations and Applications.
135. Karatzas, I. and Shreve, S.E.: Methods of Mathematical Finance. Applications of Mathematics, 39. Springer, New York, 1998.
136. Kleptsyna, M.L., Kloeden, P.E., and Anh, V.V.: Existence and uniqueness theorems for fBm stochastic differential equations. Problems Inform. Transmission 34, 332–341, 1999.
137. Kleptsyna, M.L. and Le Breton, A.: Statistical analysis of the fractional Ornstein Uhlenbeck type process. Stat. Inference Stoch. Process. 5, 229–248, 2002.
138. Kleptsyna, M.L., Le Breton, A., and Roubaud, M.C.: General approach to filtering with fractional Brownian noises. Application to linear systems. Stoch. Stoch. Rep. 71, 119–140, 2000.
139. Klingenhöfer, F. and Zähle, M.: Ordinary differential equations with fractal noise. Proc. AMS 127, 1021–1028, 1999.
140. Klüppelberg, C. and Kühn, C.: Fractional Brownian motion as a weak limit of Poisson shot noise processes with applications to finance. Stoch. Proc. Appl. 113(2), 333–351, 2004.

141. Kolmogorov, A.N.: Wienersche Spiralen und einige andere interessante Kurven im Hilbertschen Raum. C.R.(Doklady) Acad. URSS (N.S) 26, 115–118, 1940.
142. Kondurar, V.: Sur íntégrale de Stieltjes. Mat.Sbornik 2, 361–366, 1937.
143. Kubilius, K.: The existence and uniqueness of the solution of the integral equation driven by fractional Brownian motion. Lith. Math. J. 40, 104–110, 2000.
144. Kuo, H.H.: White Noise Distribution Theory. CRC Press, 1996.
145. Kuo, H.H. and Russek, A.: White Noise Approach to stochastic integration. J. Multivariate Anal. 24, 218–236, 1988.
146. Le Breton, A.: Filtering and parameter estimation in a simple linear system driven by a fractional Brownian motion. Statist. Probab. Lett. 38, 263–274, 1998.
147. Léger, S. and Pontier, M.: Drap Brownien fractionnaire. CRAS 329 série I, 893–898, 1999.
148. Lévy Véhel, J. and N'Doye, M.: Le mouvement Brownien multifractionnaire: Variation quadratique et intégrale stochastique. Preprint.
149. Lin, S.J.: Stochastic analysis of fractional Brownian motion. Stoch. Stoch. Rep. 55, 121–140, 1995.
150. Lindstrøm, T.: Fractional Brownian fields as integrals of white noise. Bull. London Math. Soc. 25, 83–88, 1993.
151. Lindstrøm, T.: A weighted random walk approximation to fractional Brownian motion. Preprint, 2007.
152. Lipster, R.Sh. and Shiryaev, A.N.: Theory of Martingales. Kluwer Acad. Publ. Dordrecht, 1989.
153. Lyons, T.: Differential equations driven by rough signals. I. An extension of an inequality of L.C. Young. Math. Res. Lett. 4, 451–464, 1994.
154. Malliavin, P.: Stochastic Analysis. Springer 2003.
155. Mandelbrot, B.B.: Fractals and Scaling in Finance: Discontinuity, Concentration, Risk. Springer Verlag, 1997.
156. Mandelbrot, B. B.and Van Ness, J.W.: Fractional Brownian motions, fractional noises and applications. SIAM Rev. 10, 422–437, 1968.
157. B. Maslowski, B. and Nualart, D.: Evolution equations driven by a fractional Brownian motion. J. Funct. Anal. 202, 277–305, 2003.
158. Mendes, R.V. and Oliveira, M.J.: Fractional volatility and option pricing. Preprint, 2004.
159. Meyer, Y., Sellan, F., and Taqqu, M.S.: Wavelets, generalized white noise and fractional integration. The synthesis of fractional Brownian motion. J. Fourier Anal. Appli. 5, 466–494, 1999.
160. Meyer, P. A.: Quantum probability for probabilist, Lect. Notes in Math. 1538, Springer, 1993.
161. Mémin, J., Mishura, Y., and Valkeila, E.: Inequalities for the moments of Wiener integrals with respect to a fractional Brownian motion. Statist. Probab. Lett. 51, 197–206, 2001.
162. Merton, R.C.: Optimum consumption and portfolio rules in a continuous-time model. J. Econom. Theory 3, 373–413, 1971.
163. Mikosch, T. and Norvaisa, R.: Stochastic integration without probability. Bernoulli 6, 401–434, 2000.
164. Mishura, Y.: Fractional stochastic integration and Black Scholes equation for fractional Brownian model with stochastic volatility. Stoch. Stoch. Rep., 76, 4, 363–381, 2004.

165. Mishura, Y. and Valkeila, E.: An isometric approach to generalized stochastis integrals. Journal of Theoretical Probability 13, 673–693, 2000.
166. Mishura, Y. and Valkeila, E.: Absence of arbitrage in a mixed Brownian-fractional Brownian model. (Russian) Tr. Mat. Inst. Steklova 237, Stokhast. Finans. Mat., 224–233, 2002.
167. Mishura, Y. and Valkeila, E.: Martingale transformation and Girsanov theorem for long memory Gaussian processes. Statist. Probab. Lett. 55, 412–430, 2001.
168. Molchan, G.M.: Gaussian processes with spectra which are asymptotically equivalent to a power of λ. Theory of Probability and its Applications 14, 530–532, 1969.
169. Molchan, G.M.: Maximum of a fractional Brownian motion: Probabilities of small values. Commutations in Mathematical Physics 205, 97–111, 1999.
170. Necula, C.: Option pricing in a fractional Brownian motion environment. Math. Rep. (Bucur.) 6(3), 259–273, 2004.
171. Neveu, J.: Processus aléatoires gaussiens. Montréal, Presses de l'Université de Montréal, 1968.
172. Norros, I., Valkeila, E., and Virtamo, J.: An elementary approach to a Girsanov formula and other analytic results on fractional Brownian motions. Bernoulli 5, 571–587, 1999.
173. Norros, I, Valkeila, E. and Virtamo, J.: A Girsanov-type formula for the fractional Brownian motions. Proc. First Nordic-Symposium on Stochastics, 1996.
174. Nourdin, I.: One-dimensional differential equations driven by a fractional Brownian motion with any Hurst index $H \in (0,1)$. Preprint.
175. Novikov, A. and Valkeila, E.: On some maximal inequalities for fractional Brownian motion. Statist. Probab. Lett. 44(1), 47–54, 1999.
176. Nualart, D.: The Malliavin Calculus and Related Topics. Springer 1995.
177. Nualart, D.: Stochastic integration with respect to fractional Brownian motion and applications. Stochastic Models (Mexico City, 2002), Contemp. Math. 336 Amer. Math. Soc. Providence, RI, 3–39, 2003.
178. Nualart, D. and Ouknine, Y.: Regularization of differential equations by fractional noise. Stoch. Proc. Appl., Vol. 102, 103–116, 2002.
179. Nualart, D. and Pardoux, E.: Stochastic calculus with anticipative integrands. Probab. Theory. Rel. Fields 78, 535–581, 1988.
180. Nualart, D. and Rascanu, A.: Differential Equations driven by fractional Brownian motion, Collectanea Math. 53, 55–81, 2002.
Functional Analysis 149, 200–225, 1997.
181. Nualart, D., Rovira, C., and Tindel, S.: Probabilistic models for vortex filaments based on fractional Brownian motion. Ann. Probab. 31, 1862–1899, 2003.
182. Øksendal, B.: Stochastic Differential Equations. Fifth edition, Springer 2000.
183. Øksendal, B.: Fractional Brownian motion in finance. In B.S. Jensen and T. Palokangas (ed.): Stochastic Economic Dynamics. Cambridge Univ. Press (to appear).
184. Øksendal, B. and Zhang, T.: Multiparameter fractional Brownian motion and quasi-linear stochastic partial differential equations. Stoch. Stoch. Rep. 71, 141–163, 2001.
185. Peng, C.K., Buldyrev, S.V., Simons, M., Stanley, H.E., and Goldberger, A.L.: Mosaic organization of DNA nucleotides. Phys. Rev. E 49, 1685–1689, 1994.
186. Peng, S.: A general stochastic maximum principle for optimal control problems. SIAM J. Control Optim. 28, 966–979, 1990.

318   References

187. Perez, D.G.: The fractional Brownian motion property of the turbulent refraction index and the Fermat's extremal principle. Preprint, 2002.
188. Pipiras, V. and Taqqu, M.S.: Integration questions related to fractional Brownian motion. Prob. Theory Rel. Fields 118, 251–291, 2000.
189. Pipiras, V. and Taqqu, M.S.: Are classes of deterministic integrands for fractional Brownian motion on a interval complete? Bernoulli 7, 873–897, 2001.
190. Pipiras, V. and Taqqu, M.S.: Fractional calculus and its connection to fractional Brownian motion. In P. Doukhan, G. Oppenheim, M. S. Taqqu. (eds.), Theory and Applications of Long-Range Dependence, Birkhäuser, 165–201, 2003.
191. Pipiras, V. and Taqqu, M.S.: Deconvolution of fractional Brownian motion. Journal of Time Series Analysis 23, 487–501, 2002.
192. Potthoff, J. and Timpel, M.: On a dual pair of smooth and generalized random variables. Potential Analysis 4, 637–654, 1995.
193. Prakasa Rao, B.L.S.: Identification for linear stochastic systems driven by fractional Brownian motion. Stoch. Anal. Appl. 22(6), 1487–1509, 2004.
194. Revuz, D. and Yor, M.: Continuous Martingales and Brownian Motions, Third edition, Springer, 1999.
195. Rogers, L.C.G.: Arbitrage with fractional Brownian motion. Math. Finance 7, 95–105, 1997. A.L.
196. Rogers, L.C.G. and Williams, D.: Diffusions, Markov Processes, and Martingales, Volume 1: Foundations. Cambridge University Press, 2000.
197. Rosen, J.: The intersection local time of fractional Brownian motion in the plane. J. Multivariate Anal. 23, 37–46, 1987.
198. Russo, F. and Vallois, P.: Forward, backward and symmetric stochastic integration, Probab. Theory Rel. Fields 97, 403–421, 1993.
199. Russo, F. and Vallois, P.: The generalized covariation process and Itô formula. Stoch. Proc. Appl. 59, 81–104, 1995.
200. Russo, F. and Vallois, P.: Itô formula for $\mathcal{C}^1$-functions of semimartingales. Probab. Theory Rel. Fields 104, 27–41, 1996.
201. Russo, F. and Vallois, P.: Stochastic calculus with respect to continuous finite quadratic variation processes. Stoch. Stoch. Rep. 70, 1–40, 2000.
202. Ruzmaikina, A.A.: Stochastic calculus for fractional Brownian motion. Ph.D. Thesis, Princeton University, 1999.
203. Ruzmaikina, A.A.: Stieltjes integrals of Hölder continuous functions with applications to fractional Brownian motion. J. Statist. Phys. 100, 1049–1069, 2000.
204. Salopek, D.M.: Tolerance to arbitrage. Stoch. Proc. Appl. 76, 217–230, 1998.
205. Salopek, D.M.: A new class of nearly self-financing strategies. Stat. Probab. Lett., 56, 69–75, 2002.
206. Samko, S.G., Kilbas, A.A. and Marichev, O.I.: Fractional Integrals and Derivatives, Theory and Applications. Gordon and Breach Science Publishers, 1993.
207. Samorodnitsky, G. and Taqqu, M. S.: Stable non-Gaussian random processes. Stochastic models with infinite variance. Stochastic Modeling. Chapman & Hall, New York, 1994.
208. Shiryaev, A.: On arbitrage and replication for fractal models: In A. Shiryaev and A. Sulem (editors): Workshop on Mathematical Finance, INRIA, Paris, 1998.
209. Shiryaev, A.: Essentials of Stochastic Finance. World Scientific, 1999.
210. Simonsen, I.: Measuring anti-correlations in the Nordic electricity spot market by wavelets. Physica A 322, 597–606, 2003.

211. Simonsen, I., Hansen, A., and Nes, O.M.: Determination of Hurst exponent by use of wavelet transforms. Phys. Rev. E 58, 2779–2787, 1998.
212. Sinaĭ, J.G.: Self-similar probability distributions. Theor. Probab. Appl. 21, 64–80, 1976.
213. Skorohod, A.V.: On a generalization of a stochastic integral. Theor. Probab. Appl. 20, 219–233, 1975.
214. Sottinen, T.: Fractional Brownian motion, random walks and binary market models. Finance Stoch. 5, 343–355, 2001.
215. Sottinen, T. and Valkeila, E.: On arbitrage and replication in the fractional Black Scholes model. Statist. Decisions 21(2), 93–107, 2003.
216. Sottinen, T. and Valkeila, E.: Fractional Brownian motion as a model in finance, Department of Mathematics, University of Helsinki, Preprint 302, 2001.
217. Stein, E.M: Singular Integrals and Differentiability Properties of Functions, University Press, 1961.
218. Stroock, D.W. and Varadhan, S.R.S.: Multidimensional Diffusion Processes, Springer, 1979.
219. Szego, G.: Orthogonal polynomials. Amer. Math. Soc. Colloq. Pub. 23, Providence, R.I., 1967.
220. Taqqu, M.S.: Weak convergence to fractional Brownian motion and Rosenblatt Process. Z. Wahr. verw. Geb. 31, 287–302, 1975.
221. Taqqu, M.S.: Fractional Brownian motion and long-range dependence. In P. Doukhan, G. Oppenheim, M.S. Taqqu (eds.): Theory and Applications of Long-Range Dependence. Birkäuser, 2003.
222. Taqqu, M.S. and Teverovsky, V.: On estimating the intensity of long-range dependence in finite and infinite variance time series. A practical guide to heavy tails (Santa Barbara, CA, 1995), 177–217. Birkhäuser Boston, Boston, MA, 1998.
223. Thangavelu, S.: Lectures of Hermite and Laguerre Expansions. Princeton University Press, 1993.
224. Tindel, S., Tudor, C.A. and Viens, F.: Stochastic evolution equations with fractional Brownian motion. Probab. Theory Rel. Fields 127, 186–204, 2003.
225. Tudor, C.A.: Stochastic calculus with respect to the infinite dimensional fractional Brownian motion. Preprint, 2002.
226. Tudor, C.A.: Itô formula for the infinite-dimensional fractional Brownian motion. J. Math. Kyoto Univ. 45, no. 3, 531–546, 2005.
227. Valkeila, E.: On some properties of geometric fractional Brownian motion, Preprint, University of Helsinki, May 1999.
228. Venttsel, A.D.: A Course in the Theory of Stochastic Processes. McGraw Hill, New York, 1981.
229. Xiao, Y. and Zhang, T.: Local times of fractional Brownian sheet. Probab. Theory Rel. Fields 124(2), 204–226, 2002.
230. Yaglom, A.M.: Correlation theory of processes with random stationary nth increments. AMS Transl. 2, 8, 87-141, 1958.
231. Yong, J. and Zhou, X.Y.: Stochastic Controls: Hamiltonian Systems and HJB Equations. Springer 1999.
232. Young, L.C.: An inequality of the Hölder type, connected with Stieltjes integration. Acta Math. 67, 251–282, 1936.
233. Wallner, N.: Fractional Brownian Motion and Applications to Finance. Thesis, Philipps-Universität Marburg, March 2001.

234. Walsh, J.B.: An introduction to stochastic partial differential equations. In R.Carmona, H.Kesten, and J.B.Walsh (eds.): École d'Été de Probabilités de Saint-Flour XIV. Springer LNM 1180, pp.265–437, 1984.
235. Wiener, N.: The homogeneous chaos, Amer. J. Math. 60, 897–936, 1941.
236. Willinger, W., Taqqu, M.S., and Teverovsky, V.: Stock market prices and long-range dependence. Fin. and Stoch. 3, 1–13, 1999.
237. Zähle, M.: On the link between fractional and stochastic calculus. In: Stochastic Dynamics, Bremen 1997, H.Crauel and M.Gundlach (eds.), Springer, 305–325, 1999.
238. Zähle, M.: Integration with respect to fractal functions and stochastic calculus I. Probab. Theory Rel. Fields 111, 333–374, 1998.
239. Zähle, M.: Integration with respect to fractal functions and stochastic calculus II. Math. Nachr. 225, 145–183, 2001.
240. Zähle, M.: Forward integrals and stochastic differential equations. Progr. Probab. 52, 293–302, 2002.
241. Zähle, M.: Stochastic differential equations with fractal noise. Math. Nachr. 278(9), 1097–1106, 2005.

Index of symbols and notation

$I_{[0,t]}$, 274
$K_H(t,s)$, 7
$(\mathcal{S})^{(-q)}$, 187
$(\mathcal{S})^{(k)}$, 186
$(\mathcal{S})$, 187, 276
$(\mathcal{S})^*$, 187
$(\mathcal{S})_H$, 55
$(\mathcal{S})_H^*$, 55
$(\hat{u}, \hat{X})$, 213
$(\mathcal{S})^*$, 277
$<\omega, f>$, 99, 273
$A(t)$, 222
B, 5
$B(x)$, 181
$B^{(H)}(x)$, 184
$B_p^{\alpha,\beta}$, 14
$C^\alpha([0,T])$, 37
$C^\alpha(\mathbb{R})$, 11
$C^\gamma(\mathbb{R})$, 300
$C_n(f)$, 14
$D(x,\varepsilon)$, 243
D^-f, 253
$D^{(H),k}$, 38
D_{a+}^α, 285
D_{b-}^α, 286
D_γ, 281
D_t, 282
D_t^ϕ, 65
D_+^α, 288
D_-^α, 288
$D_{\Phi g}$, 65
$D_\gamma^{(H)}$, 110
$D_{n,t}^{(H)}$, 121

$F \diamond_P G$, 61
$F \diamond_Q G$, 61
$G(x,y)$, 189
$G_{t-s}(x,y)$, 194
H, 5
$H(t, \hat{X}(t), \hat{u}(t), \hat{p}(t), \hat{q}(\cdot))$, 213
H_0, 249, 299
$I(X, [0,T])$, 13
$I_-^{H-1/2}$, 48
I_+^α, 288
I_-^α, 288
I_{a+}^α, 285
I_{b-}^α, 285
I_{a+}^n, 285
I_{b-}^n, 285
$I_{[0,t]}$, 49
$I_{[0,x]}(y)$, 182
$I_{[0,x_i]}(y_i)$, 182
$J(u)$, 220
$J(z,u)$, 221
$J_{\lambda^*}(F,c)$, 236
K_H^*, 30
$K_1^p \phi(t)$, 299
K_H^*, 35
$K_H^{*,a}$, 43
$K_\varepsilon(x)$, 178
$L(\theta)$, 293
$L^1(a,b)$, 285
$L^2([0,T])$, 26
$L^2(\mathbb{P}^H)$, 26
$L^p(\Omega, \mathcal{H}^{\otimes k})$, 38
$L_H^2(\mathbb{R}^d)$, 184
$L_\phi^2([0,T])$, 48

Index of symbols and notation

$L^2_\phi(\mathbb{R})$, 48
$L^2_\phi(\mathbb{R}_+)$, 48
$L_{1/2,H}(t)$, 290
$Lip_p(\alpha,\beta)$, 14
M, 100
$M[0,t](x)$, 101
$M_H f(x)$, 183
M_H, 100
$M_{H_j} f_j(x_j)$, 183
$M_{t+} D_{t+} \psi(t)$, 139
$P_{t,T}(x)$, 87
$R_H(t,s)$, 23, 47
$S(t)$, 222
$S_0(t)$, 170, 219
$S_1(t)$, 170, 219
$S_2(t)$, 219
S_H, 37
$S_\mathcal{K}$, 44
$U(t,x)$, 194
$V^\theta(t)$, 170, 175
$W^{(H)}(t)$, 104
$W_0^{\alpha,\infty}(0,T)$, 303
$X^{\diamond n}$, 58
$X^m(k)$, 290
$Z(t)$, 222
$[X,Y]_t$, 125
$[X]_t^{(\alpha)}$, 125
$B^{(H)}$, 5
$B^{(H)}(\psi)$, 30
Δ, 306
$\Delta U(x)$, 189
$\Delta V(t_k)$, 173
Δ_H, 299
$\mathcal{F}$, 273
$\mathcal{F}_t^{(H)}$, 23, 49, 170
$\mathcal{F}g(\xi)$, 183
Γ, 6, 24
$\mathfrak{K}_H^*$, 28
Ω, 273
$\mathbb{P}$, 99, 118, 273
$\mathbb{P}^H$, 48, 87
$\mathbb{P}^{H,\gamma}$, 61
$\mathbb{Q}$, 234
$\mathcal{S}'(\mathbb{R}^d)$, 181
$\mathcal{S}(\mathbb{R}^d)$, 181
$\mathcal{S}_p(X,\pi)$, 12
$\operatorname{Tr} D^{(H)} u$, 128
$\mathcal{V}_p(X,[0,T])$, 13
$W^{(H)}(x)$, 187

$\mathcal{X}_N$, 293
α_p, 299
$\bar{D}$, 189
$\bar{S}_1(t)$, 172
$\bar{V}_1(t)$, 173
$\bar{\theta}_1(t)$, 173
$\beta(\alpha,\gamma)$, 24
$\mathcal{D}$, 173
$\mathcal{F}^{(H)}$, 207
$\mathcal{F}_t$, 232
$\mathcal{G}_t$, 232
$\mathcal{L}_\phi^{(m)}(0,T)$, 88
$\mathcal{L}_\phi(0,T)$, 70
$\mathcal{L}_t^{(B^{(H)})}$, 250
$\mathcal{L}_t^{(X)}$, 255
$\mathcal{L}_\phi^{(2)}(0,T)$, 219
$\delta(x)$, 240
δ_{jk}, 239
δ, 40
$\delta_{1/2}$, 40
$\delta_{B^{(H)}(t)}(x)$, 240
$D_t^{(H)}$, 38, 84, 111, 112
$\diamond$, 57, 119, 278
$\operatorname{dom} \delta$, 40
$\operatorname{dom} \delta_H$, 38
$\operatorname{dom}^* \delta_H$, 44
$\mathcal{I}_2^H$, 28
$\ell_T(x)$, 239
$\ell_T^{(H)}(x)$, 239, 245
$\ell_t^{(X)}$, 255
$\tilde{E}\left[G|\mathcal{F}_t^{(H)}\right]$, 84
$\tilde{E}_{\hat{\mathbb{P}}^H}\left[\cdot \mid \mathcal{F}_t^{(H)}\right]$, 209
$E_\mathbb{Q}[G]$, 235
$\eta_\alpha(x)$, 185
$\eta_n(x)$, 185
$\exp^\diamond(<\omega, Mf>)$, 106
$\exp^\diamond(X)$, 58
$\frac{d\hat{\mathbb{P}}^H}{d\hat{\mathbb{P}}}$, 208, 223
$\frac{d\hat{\mathbb{P}}}{d\mathbb{P}}$, 176
$\hat{B}^{(H)}(t)$, 61, 223
$\hat{H}_N(k,a)$, 292
$\hat{L}_H^2(\mathbb{R}_+^n)$, 80
$\hat{\mathbb{P}}$, 109
$\hat{\mathbb{P}}^H$, 208
$\hat{\theta}_n$, 293
$\hat{L}^2(\mathbb{R}^n)$, 274

Index of symbols and notation

$\hat{p}(t)$, 213
$\hat{q}(t)$, 213
$\int_0^T Y(t)\,d^- B^{(H)}(t)$, 137
$\int_0^T u(s)\,d^\circ B^{(H)}(s)$, 124
$\int_0^T u(s)\,d^+ B^{(H)}(s)$, 124
$\int_0^T u(s)\,d^- B^{(H)}(s)$, 124
$\int_0^t h(s)\,d^- B^{(H)}(s)$, 125
$\int_{\mathbb{R}} Y(t)\,dB_t^{(H)}$, 64
$\int_{\mathbb{R}} Y(t)\,\delta B(t)$, 280
$\int_{\mathbb{R}} Y(t) \diamond W^{(H)}(t)\,dt$, 64
$\int_{\mathbb{R}^d} f(x)\,dB^{(H)}(x)$, 184
$\int_{\mathbb{R}} Z(t)\,dt$, 277
$\int_{\mathbb{R}} Y(t) \diamond W^{(H)}(t)\,dt$, 104
$\int_{\mathbb{R}} Y(t,\omega)\,dB^{(H)}(t)$, 104
λ, 121
$\lambda(\tau)$, 260
λ_1, 243
λ_d, 243
$\langle S_1(t), \psi \rangle$, 172
$\langle\langle G, \psi \rangle\rangle$, 55
$\langle f, g \rangle_H$, 27, 47
$\{e_i\}_{i=1}^\infty$, 51
$\mathbb{L}_{1,2}^{(H)}$, 138
$\mathcal{M}$, 111
$\mathbb{D}^{k,p}(\mathcal{H})$, 38
$\mathbb{D}_H^{1,2}$, 38
$\mathbb{D}_H^{k,p}$, 38
$\mathbb{L}^{1,2}$, 39
$\mathcal{B}(\mathcal{S}'(\mathbb{R}))$, 99, 273
$\mathcal{H}_\alpha$, 112, 185, 275
$\mathcal{S}'(\mathbb{R})$, 99, 273
$\mathcal{S}(\mathbb{R})$, 99, 273
$\mathcal{T}_1$, 262
$\mathcal{T}_2$, 262
$\mathcal{T}_3$, 262
$\mu(\tau)$, 260
∇_t^ϕ, 86
∂D, 189
$\phi(s,t)$, 47
$\phi_H(s,t)$, 47
$\mathcal{E}$, 27, 50
$\mathcal{H}$, 25
$\mathcal{H}_2$, 29
$\mathcal{J}$, 52, 275

$\mathcal{J}^{(N)}$, 53
$\mathcal{K}$, 44
$\rho(\tau)$, 260
ρ_H, 9
$\mathbb{D}_H^{1,2}$, 112
$L_H^{2,(m)}(\mathbb{R})$, 88
$\theta_0(t)$, 173
$\theta_1(t)$, 173
$\tilde{B}(t)$, 274
$\varepsilon(f)$, 50
$\varepsilon^{(j)}$, 183
$\mathcal{J}^H$, 26
$\widetilde{\operatorname{Var}} X^{(m)}$, 290
$\tilde{\mathcal{H}}_\alpha$, 54
$\tilde{B}(x)$, 182
$\tilde{B}^{(H)}(x)$, 183
$\mathcal{J}_1^H$, 28
$\xi^{\otimes \alpha}(x)$, 276
$\xi^{\otimes \alpha}(x)$, 276
ξ_n, 52
$\xi_n(x)$, 275
$b_H(t)$, 179
$d_p(w,v)$, 306
$e_\alpha(x)$, 185
e_k, 103
$e_n(x)$, 185
$f^\diamond(X)$, 282
$f_{a+}(x)$, 287
$f_{b-}(x)$, 287
$h_n(x)$, 51, 275
$h_n^{H,f}(t)$, 76
i, 273
i_1, 27
i_2, 27
i_3, 29
m_t^w, 250
$m_j g(\cdot)$, 182
m_t, 250
$p(t)$, 208
$q(t)$, 208
v_t^m, 308
$v_{s,t}^{m,i}$, 308
w, 306
$w_{s,t}^1$, 306
$w_{s,t}^2$, 306
$w_{s,t}^3$, 306
$w_{s,t}$, 306

Index

1/2-Stable fractional Stable motion, 290
L^p Estimate, 75
M operator, 99
R/S analysis, 291
ϕ-Derivative, 65
ϕ-Differentiable, 66
dt-Integrable, 153, 277
k-th empirical absolute moment, 291
p-Variation, 12
p-Variation distance, 306, 307

Absolute value method, 290
Adjoint equation, 212
Adjoint operator, 28, 31, 38, 43
Adjusted range, 291
Admissible pair, 211
Anti-persistence, 9
Arbitrage, 170, 171
Autocorrelation, 294
Autocovariance functions, 9
Auxiliary partial differential equation, 302

Backward integral, 124
Banach space, 32
Bank account, 266
Besov space, 14
Black Scholes market, 170
Bochner Minlos theorem, 273
Budget constraint, 223

Chain rule, 282
Clark Hausmann Ocone formula, 83
Complete, 219

Constrained optimization problem, 225
Consumption process, 222
Continuous trajectories, 6
Corrected functions, 287
Correlation, 8
Covariance, 9, 23
Covariance matrix, 293
Covariation, 124

Derivative, 26
  directional, 110
  fractional, 31
  fractional stochastic, 112
  Hida Malliavin, 112
  Malliavin, 281
  operator, 37
  stochastic, 37
Differentiable, 282
Differentiation, 115, 121
Dirac delta function, 258
Dirac measure, 29
Directional derivative, 62, 110, 281
Dirichlet Laplacian, 189
Discrete variation method, 291
Divergence operator, 38
Divergence-type integral, 41
Domain, 38
Domain of attraction, 297
Donsker delta function, 240
Doob Meyer decomposition, 75
Dual operator, 37
Dyadic approximations, 307

Estimation of Hurst parameter, 289

Estimator, 289
European call option, 267
Existence-uniqueness theorem, 300
Exponential function, 50
Extended forward integral, 125
Extended Wiener integral, 34

Feedback gain, 216
Filter, 291
Finite length, 275
Forward integral, 124, 137
Fourier transform, 183
Fractal integral, 287
Fractional
   backward stochastic differential
      equation, 208, 212
   Black Scholes formula, 267
   Black Scholes market, 266
   Brownian fields, 181
   Brownian motion, 5, 49
   Clark derivative, 86
   conditional expectation, 84
   Girsanov formula, 60
   Hermite transform, 241
   Hida distribution function space, 55
   Hida test function space, 55
   integration by parts, 95, 122
   Itô formula, 107
   Liouville derivatives, 285
   representation, 30
   stochastic derivative, 112
   stochastic differential equation, 64
   stochastic gradient, 113, 121
   stochastic integral of Itô type, 64
   stochastic maximum principle, 213
   stochastic Sobolev spaces, 112
   white noise, 56, 104, 187
   Wick calculus, 64
   Wick Itô Skorohod integral, 64
   Wiener Itô chaos expansion, 53
Fractional heat equation, 194
   quasi-linear, 198
Fractional integrals
   left-sided version, 285
   of order $\alpha > 0$, 285
   right-sided version, 285
Fubini theorem, 45
Function
   Hölder continuous, 11, 37
   monotonic, 292
   slowly varying, 10
Fundamental theorem of stochastic
     calculus, 283
fWIS integral, 64

Gamma function, 6, 24
Gaussian process, 6
Gaussianity, 23
Generalized
   SLSG portfolio , 267
   total wealth process, 173
   value process, 219
   wealth process, 222
Generalized expectation, 244
Generalized forward integral, 137
Generalized Stieltjes integral, 303
Geometric fractional Brownian motion,
     64, 255, 266
Geometric rough path, 305, 306
Girsanov Theorem, 109
Global solution, 302
Green function, 189, 194
Growth condition, 302

Hölder continuity, 11
Hölder continuous, 300, 305
Hamiltonian, 212
Heat kernel, 258
Heat operator, 194
Hermite
   functions, 52, 103, 185, 275
   polynomials, 51, 185, 275
Hermite polynomial, 252
Hida distributions, 187
Hida Malliavin derivative, 63, 111, 282
Hida space, 276, 277
Hida test functions, 187
Homogeneous chaos, 75
Homogeneous increments, 5
Hurst index, 5, 6
Hurst parameter, 289
Hurst's rescaled range analysis, 291

Index of p-variation, 13
Inductive topology, 187, 277
Inner product space
   complete, 36
   incomplete, 36

Integral representation, 6
Integration, 115, 121
Integration by parts, 116, 283
Intermittency, 9
Invariance principle, 14
Itô Formula, 71, 161
  for the divergence integral, 161, 162
  for the fWIS integral, 162
  for the WIS integral, 163
Itô map, 307
Itô Skorohod isometry, 284
Iterated integrals, 78
Iterated Itô integral, 274
Iterated path integral, 308

Kolmogorov criterion, 11

Lagrange multiplier method, 235
Likelihood function, 293
Linear operator, 150
Linear quadratic control, 216
Linear regression, 289
Lipschitz continuous, 298
Local nondeterminism property, 262
Local time
  approximated self-intersection, 258
  fractional, 243
  renormalized self-intersection, 259
  self-intersection, 258
  weighted, 250
Locally Lipschitz, 302
Long range dependence, 9

Malliavin derivative, 138, 281
Marchaud fractional derivatives, 288
Markov linear feedback control, 216
Maximum likelihood estimator, 293
Memory, 9
Method
  absolute value, 290
  Discrete variation, 291
  periodogram, 291
  variance, 290
  variance residual, 290
  Whittle, 292
Meyer inequality, 77
Meyer Tanaka formula, 254
Minimal variance hedging problem, 218
Multi-indices, 275

Multidimensional
  fractional Wick Itô Skorohod integral, 88
  fractional Wick Itô Skorohod Isometry, 88
  WIS integral, 120
  WIS isometry, 122
Multiparameter fractional
  Brownian motion, 181
  white noise calculus, 185
Multiple integral, 82
Multiplicative functional, 306

Normal distribution, 293

Observed wealth, 173
Optimal
  consumption, 229
  consumption rate, 238
  control, 212
  pair, 212, 213
  terminal wealth, 229
Ordinary differential equation, 302

Path differentiability, 11
Pathwise integration model, 170, 180
Performance functional, 211
Periodogram method, 291
Persistence, 9
Poisson equation, 189
Polarization technique, 68
Polynomial fractional Brownian
  functionals, 50
Portfolio, 170, 232
  admissible, 170, 233
  generalized, 173
  pathwise self-financing, 170
  self-financing, 233
  Wick Itô Skorohod admissible, 176
  Wick Itô Skorohod self-financing, 176
  WIS admissible, 219
  WIS self-financing, 176
Process
  adjoint, 212
  finite quadratic variation, 125
  fractional forward, 144
  mixing, 43
  zero quadratic variation, 125
Projective topology, 187, 277

Quasi maximum likelihood estimator, 294
Quasi-conditional expectation, 84, 177

Regularity behavior, 299
Relation between symmetric and fWIS integrals, 159
Relation between the (generalized) forward and WIS integral, 160
Relation between the symmetric integral and the divergence I, 158
Relation between the symmetric integral and the divergence II, 159
Representation, 27, 148
reproducing kernel Hilbert space, 25, 147
Resolvent
    kernel, 299
    series, 299
Riccati equation, 217
Riemann Stieltjes integral, 300
Riemann sum, 68, 300
Risk free investment, 170
Risky investment, 170
RKHS, 25, 148
Rough path analysis, 305

Schauder basis, 14
Schwartz space, 47
Second level paths, 307
Self-similarity, 10
Semimartingale, 12
Shadow price, 212
Skorohod integrable, 280
Skorohod integral, 278–280
    extended, 280
Smooth cylindrical random variables, 37, 44, 150
Smooth rough path, 306
Solution, 194, 298, 301, 308
Space of tempered distributions, 48, 99
Spectral
    density, 10
    representation, 7, 19
Spectral density, 291, 294
Stationary Gaussian time sequence, 289
Statistic, 289
Statistical fractal dimension, 11
Stochastic
    derivative, 37, 130
    derivative in the direction n, 120
    distributions, 277
    gradient, 63, 84, 111, 282
    integral in L^2, 68
Stochastic calculus with rough path analysis, 306
Stochastic delay differential equations, 297
Stochastic differential equation, 301
Stochastic differential equations, 297
    via rough path analysis, 305
    with pathwise integrals, 300
    with Wiener integrals, 297
Stochastic maximum principle, 211
Stochastic volatility, 232
Stock, 266
Stock price
    generalized, 172
    observed, 172
Stop-loss–start-gain, 266
Stratonovich integral, 143
Strong α-variation, 125
Strong arbitrage, 176
Strong arbitrage free, 219
Strong solution, 301, 308
Symmetric Fredholm integral equation of the first kind, 221
Symmetric integral, 123, 128
Symmetrization, 279
Symmetrized tensor product, 276

Tempered distributions, 273
Terminal condition, 211
Test functions, 276
Theorem
    Wiener Itô chaos expansion I, 274
    Wiener Itô chaos expansion II, 276
Total wealth process, 175
Translation operator, 59
Twisted scalar product, 27, 148

Ucp, 125, 126
Unbounded variation, 300
Unconstrained optimization problem, 225
Uniform convergence in probability, 125

Value function, 229
Variance, 293

Variance Method, 290
Variance residuals methods, 290

Weak arbitrage, 178
Weak semimartingale, 13
Weak solution, 302
Wealth process, 170
Weyl representation, 286
White noise, 277
   probability measure, 99, 273
Whittle estimator, 293
Whittle method, 292
Wick chain rule, 111, 283
Wick Itô Skorohod
   complete, 177
   fractional Black Scholes market, 176
   integrable, 104
   integral, 104
   integration model, 172, 180
   isometry, 117
   self-financing, 222
   self-financing property, 174
Wick product, 50, 96, 156, 278
Wick version, 282
Wiener Helix, 5
Wiener integral, 23, 26, 148
   of first type, 28, 29
   of second type, 28, 29
WIS exponential, 105
WIS integrable, 104
WIS integral, 104, 120

Young integrals, 300